Mathematics: Key Concepts and Applications

Mathematics: Key Concepts and Applications

Editor: Victor Nason

New York

Published by NY Research Press
118-35 Queens Blvd., Suite 400,
Forest Hills, NY 11375, USA
www.nyresearchpress.com

Mathematics: Key Concepts and Applications
Edited by Victor Nason

International Standard Book Number: 978-1-63238-573-4 (Hardback)

Cataloging-in-Publication Data

Mathematics : key concepts and applications / edited by Victor Nason.
    p. cm.
Includes bibliographical references and index.
ISBN 978-1-63238-573-4
1. Mathematics. I. Nason, Victor.
QA7 .M38 2018
510--dc23

Contents

Preface VII

Chapter 1   **Approximation properties of modified Szász–Mirakyan operators in polynomial weighted space** 1
Prashantkumar Patel, Vishnu Narayan Mishra and Mediha Örkcü

Chapter 2   **Skew polynomial rings over σ-skew Armendariz rings** 11
V. K. Bhat and Meeru Abrol

Chapter 3   **Generalised colouring sums of graphs** 18
Johan Kok, N. K. Sudev and K. P. Chithra

Chapter 4   **On I-convergent sequence spaces defined by a compact operator and a modulus function** 29
Vakeel A. Khan, Mohd Shafiq and Rami Kamel Ahmad Rababah

Chapter 5   **On the convergence of extended Newton-type method for solving variational inclusions** 41
M. H. Rashid

Chapter 6   **Boundedness on Orlicz space of Toeplitz type operators related to multiplier operator and mean oscillation** 60
Ouyang Difei

Chapter 7   **Some new results on inner product quasilinear spaces** 69
Hacer Bozkurt and Yılmaz Yılmaz

Chapter 8   **On some properties of Toeplitz matrices** 79
Dan Kucerovsky, Kaveh Mousavand and Aydin Sarraf

Chapter 9   **Taylor expansions for the generating function of Catalan-like numbers** 91
Lily Li Liu and Xiaoli Li

Chapter 10   **Graded fuzzy topological spaces** 103
Ismail Ibedou

Chapter 11   **Attractor and self-similar group of generalized fuzzy contraction mapping in fuzzy metric space** 116
R. Uthayakumar and A. Gowrisankar

Chapter 12   **Direct product of general intuitionistic fuzzy sets of subtraction algebras** 128
Muhammad Gulistan, Shah Nawaz and Syed Zaheer Abbas

Chapter 13   **Injective module based on rough set theory** 138
Arvind Kumar Sinha and Anand Prakash

Chapter 14   **Bounds on Hankel determinant for starlike and convex functions with respect to symmetric points**..**145**
Ambuj K. Mishra, Jugal K. Prajapat and Sudhananda Maharana

Chapter 15   **Reflection principle for classical solutions of the homogeneous real Monge–Ampère equation**...**153**
Mika Koskenoja

Chapter 16   **Exploring Riemann's functional equation: Revised Version 2.0**..**168**
Michael Milgram

Chapter 17   **Sharp bounds for the Neuman-Sándor mean in terms of the power and contraharmonic means**..**199**
Wei-Dong Jiang and Feng Qi

Chapter 18   **Minimum-energy wavelet frames generated by the Walsh polynomials**.........................**205**
Sunita Goyal and Firdous A. Shah

   Permissions

   List of Contributors

   Index

Preface

Mathematics is concerned with the study of abstract numbers as well as real quantities. It studies various sets of numbers such as natural numbers, rational numbers, real numbers, integers, etc. Pure mathematics studies the inherent nature of abstract entities while applied mathematics uses methods that are derived from mathematics in diverse fields such as engineering, computer science, business, etc. This book will also provide interesting topics for research which readers can take up. It attempts to assist those with a goal of delving into the field of mathematics.

All of the data presented henceforth, was collaborated in the wake of recent advancements in the field. The aim of this book is to present the diversified developments from across the globe in a comprehensible manner. The opinions expressed in each chapter belong solely to the contributing authors. Their interpretations of the topics are the integral part of this book, which I have carefully compiled for a better understanding of the readers.

At the end, I would like to thank all those who dedicated their time and efforts for the successful completion of this book. I also wish to convey my gratitude towards my friends and family who supported me at every step.

Editor

Approximation properties of modified Szász–Mirakyan operators in polynomial weighted space

Prashantkumar Patel[1,2]*, Vishnu Narayan Mishra[1,3] and Mediha Örkcü[4]

*Corresponding author: Prashantkumar Patel, Department of Applied Mathematics & Humanities, S. V. National Institute of Technology, Surat 395 007, Gujarat, India; Department of Mathematics, St. Xavier's College, Ahmedabad 380 009, Gujarat, India

E-mail: prashant225@gmail.com

Reviewing editor: Lishan Liu, Qufu Normal University, China

Abstract: We introduce certain modified Szász–Mirakyan operators in polynomial weighted spaces of functions of one variable. We studied approximation properties of these operators.

Subjects: Advanced Mathematics; Analysis - Mathematics; Mathematics & Statistics; Pure Mathematics; Science

Keywords: Szász–Mirakyan operators; rate of convergence; weighted approximation; polynomial weight

2000 Mathematics subject classifications: primary 41A25; 41A30 ; 41A36

1. Introduction

Becker (1978) studied approximation problems for functions $f \in C_p$ and Szász–Mirakyan operators

$$S_n(f, x) = e^{-nx} \sum_{k=0}^{\infty} \frac{(nx)^k}{k!} f\left(\frac{k}{n}\right),$$

(1.1)

$x \in \mathbb{R}_0 = [0, \infty), n \in \mathbb{N}$, where C_p with fixed $p \in \mathbb{N}_0 = \mathbb{N} \cup \{0\}$ is space generated by the weighted function

ABOUT THE AUTHORS

Prashantkumar Patel is an assistant professor at SXCA and doing the PhD in Mathematics from NIT, Surat under VNM. His area of scientific interest includes approximation theory with positive linear operators and q-calculus which is proved by his research articles.

Vishnu Narayan Mishra received the PhD in Mathematics from IIT, Roorkee. His research interests are in the areas of pure and applied mathematics. He has published more than 90 research articles in reputed international journals of mathematical and engineering sciences. He is a referee and an editor of several international journals in frame of Mathematics. He guided many postgraduate and PhD students. Citations of his research contributions can be found in many books and monographs, PhD thesis and scientific journal articles.

Mediha Örkcü received the PhD in Mathematics from Gazi University Institute of Science Department of Mathematics during 2007–2011. Her research interest is Approximation theory. She has published many research articles in reputed international journals of Mathematics.

PUBLIC INTEREST STATEMENT

In this work, we define new sequence of operators depending on a parameter. We prove that these newly defined sequence of operators are positive and linear. Using the moments of these operators, we estimate continuous signals (functions). The admissible value of the involved parameter allows us to make appropriate choice of it, in order to have better approximation. We approximate these sequence of operators in terms of the modulus of continuity and the modulus of smoothness in polynomial weighted space.

$$\omega_0(x) = 1, \quad \omega_p(x) = \left(1 + x^p\right)^{-1} \quad \text{if } p \geq 1,$$

for $x \in \mathbb{R}_0$, and B_p be the set of all functions $f: \mathbb{R}_0 \to \mathbb{R}$ for which $f\omega_p$ is bounded on $\mathbb{R}_0$ and the norm is given by the following formula:

$$\|f\|_p = \sup_{x \in \mathbb{R}_0} \omega_p(x)|f(x)|.$$

Moreover, C_p be the set of all $f \in B_p$ for which $f\omega_p$ is a uniformly continuous function on $\mathbb{R}_0$. The spaces B_p and C_p are called polynomial weighted spaces.

Becker (1978) theorems on degree of approximation of $f \in C_p$ by the operators S_n were proved. From these theorems, it was deduced that

$$\lim_{n \to \infty} S_n(f, x) = f(x),$$

(1.2)

for every $f \in C_p$, $p \in \mathbb{N}_0$ and $x \in \mathbb{R}_0$. Moreover, the convergence (1.2) is uniform on every interval $[x_1, x_2]$, $x_1, x_2 \geq 0$.

Jain (1972) introduced generalization of Szász–Mirakyan operators (1.1) with help of a Poisson type distribution, as follows:

$$J_n^{[\beta]}(f, x) = \sum_{k=0}^{\infty} \omega_\beta(k, nx) f\left(\frac{k}{n}\right),$$

(1.3)

where $x \in \mathbb{R}_0 := [0, \infty)$, $n \in \mathbb{N}$, $0 \leq \beta < 1$ and

$$\omega_\beta(k, \alpha) = \frac{\alpha}{k!}(\alpha + k\beta)^{k-1} e^{-(\alpha + k\beta)} \text{ for } \alpha \in \mathbb{R}_0, k \in \mathbb{N}_0 = \mathbb{N} \cup \{0\}.$$

(1.4)

The convergence properties and degree of approximation properties of $J_n^{[\beta]}$ were examined by Jain (1972) for $f \in C(\mathbb{R}_0)$, the set of all real valued continuous functions f on $\mathbb{R}_0$. In the particular case $\beta = 0$, $J_n^{[\beta]}$ turn out to well known the Szász–Mirakyan operators (Szász, 1950) which defined by (1.1). Kantorovich type extension of the operators (1.3) was discussed in Umar and Razi (1985). Various other generalization and its approximation properties of similar type of operators are studied in Agratini (2013, 2014), Mishra and Patel (2013), Mishra, Khatri, Mishra, and Deepmala (2013), Örkcü (2013), Patel and Mishra (2014, 2015), Rempulska and Tomczak (2009), Tarabie (2012), Bardaro and Mantellini (2006, 2009). In this paper, we modify operators $J_n^{[\beta]}$ given by (1.3), i.e. we consider operators

$$J_n^{[\beta]}(f; a_n, b_n; x) = \sum_{k=0}^{\infty} \omega_\beta(k, a_n x) f\left(\frac{k}{b_n}\right), \quad x \in \mathbb{R}_0, \quad n \in \mathbb{N}$$

(1.5)

for $f \in C\big([0, \infty)\big)$, where $(a_n)_{n=1}^{\infty}$ and $(b_n)_{n=1}^{\infty}$ are given increasing and unbounded numerical sequence such that $a_n \geq 1$, $b_n \geq 1$ and $\left(\dfrac{a_n}{b_n}\right)_1^{\infty}$ is non decreasing and

$$\frac{a_n}{b_n} = 1 + o\left(\frac{1}{b_n}\right).$$

(1.6)

If $a_n = b_n = n$ for all $n \in \mathbb{N}$, then the operators (1.5) reduce to the operators (1.3).

The paper is organized as follows. In our manuscript, we shall study approximation properties of operators (1.5). In Section 2, we shall examine moments of the operators $J_n^{[\beta]}(f; a_n, b_n; x)$. We discuss approximation properties of the operators (1.5) in Section 3.

2. Moments of $J_n^{[\beta]}(f; a_n, b_n; x)$

In order to obtain moments of $J_n^{[\beta]}(f; a_n, b_n; x)$, we need some background results, which are as follows:

LEMMA 1　(Jain, 1972) Let $0 < \alpha < \infty$, $0 \le \beta < 1$ and let the generalized Poisson distribution given by (1.4). Then

$$\sum_{k=0}^{\infty} \omega_\beta(\alpha, k) = 1. \tag{2.1}$$

LEMMA 2　(Jain, 1972) Let $0 < \alpha < \infty$, $0 \le \beta < 1$. Suppose that

$$S(r, \alpha, \beta) := \sum_{k=0}^{\infty} (\alpha + \beta k)^{k+r-1} \frac{e^{-(\alpha+\beta k)}}{k!}, \quad r = 0, 1, 2, \ldots$$

and

$$\alpha S(0, \alpha, \beta) := 1.$$

Then

$$S(r, \alpha, \beta) = \alpha S(r - 1, \alpha, \beta) + \beta S(r, \alpha + \beta, \beta). \tag{2.2}$$

Also,

$$S(r, \alpha, \beta) = \sum_{k=0}^{\infty} \beta^k (\alpha + k\beta) S(r - 1, \alpha + k\beta, \beta). \tag{2.3}$$

From (2.2) and (2.3), when $0 \le \beta < 1$, we get

$$S(1, \alpha, \beta) = \frac{1}{1 - \beta};$$

$$S(2, \alpha, \beta) = \frac{\alpha}{(1 - \beta)^2} + \frac{\beta^2}{(1 - \beta)^3};$$

$$S(3, \alpha, \beta) = \frac{\alpha^2}{(1 - \beta)^3} + \frac{\alpha\beta^2}{(1 - \beta)^4} + \frac{\beta^3 + 2\beta^4}{(1 - \beta)^5}; \tag{2.4}$$

$$S(4, \alpha, \beta) = \frac{\alpha^3}{(1 - \beta)^4} + \frac{6\alpha^2\beta^2}{(1 - \beta)^5} + \frac{4\alpha\beta^3 + 11\alpha\beta^4}{(1 - \beta)^6} + \frac{\beta^4 + 8\beta^5 + 6\beta^6}{(1 - \beta)^7}.$$

In the following lemma, we have computed moments up to fourth order.

LEMMA 3　Let $0 \le \beta < 1$, then the following equalities hold:

(1) $J_n^{[\beta]}(1; a_n, b_n; x) = 1$;

(2) $J_n^{[\beta]}(t; a_n, b_n; x) = \dfrac{a_n x}{b_n(1 - \beta)}$;

(3) $J_n^{[\beta]}(t^2; a_n, b_n; x) = \dfrac{x^2 a_n^2}{(1 - \beta)^2 b_n^2} + \dfrac{x a_n}{(1 - \beta)^3 b_n^2}$;

(4) $J_n^{[\beta]}(t^3; a_n, b_n; x) = \dfrac{x^3 a_n^3}{(1 - \beta)^3 b_n^3} + \dfrac{3x^2 a_n^2}{(1 - \beta)^4 b_n^3} + \dfrac{x(1 + 2\beta) a_n}{(1 - \beta)^5 b_n^3}$;

(5) $J_n^{[\beta]}(t^4; a_n, b_n; x) = \dfrac{x^4 a_n^4}{(1 - \beta)^4 b_n^4} + \dfrac{6x^3 a_n^3}{(1 - \beta)^5 b_n^4} + \dfrac{x^2(7 + 8\beta) a_n^2}{(1 - \beta)^6 b_n^4} + \dfrac{x\left(1 + 8\beta + 6\beta^2\right) a_n}{(1 - \beta)^7 b_n^4}$.

Proof　Using equalities (2.1), (2.4–2.7) and by simple commutation, we obtain

$$J_n^{[\beta]}(1; a_n, b_n; x) = \sum_{k=0}^{\infty} \omega_\beta(k, a_n x) = 1;$$

$$J_n^{[\beta]}(t; a_n, b_n; x) = \frac{a_n x}{b_n} \sum_{k=0}^{\infty} \frac{1}{k!}(a_n x + k\beta + \beta)^k e^{-(a_n x + k\beta + \beta)}$$

$$= \frac{a_n x}{b_n} S(1, a_n x + \beta, \beta)$$

$$= \frac{a_n x}{b_n(1 - \beta)};$$

$$J_n^{[\beta]}(t^2; a_n, b_n; x) = \sum_{k=0}^{\infty} \frac{a_n x}{k!}(a_n x + k\beta)^{k-1} e^{-(a_n x + k\beta)} \frac{k^2}{b_n^2}$$

$$= \frac{a_n x}{b_n^2} \left[S(1, a_n x + \beta, \beta) + S(2, a_n x + 2\beta, \beta) \right]$$

$$= \frac{a_n x}{b_n^2} \left[\frac{1}{1 - \beta} + \frac{a_n x + 2\beta}{(1 - \beta)^2} + \frac{\beta^2}{(1 - \beta)^3} \right]$$

$$= \frac{x^2 a_n^2}{(1 - \beta)^2 b_n^2} + \frac{x a_n}{(1 - \beta)^3 b_n^2};$$

$$J_n^{[\beta]}(t^3; a_n, b_n; x) = \sum_{k=0}^{\infty} \frac{a_n x}{k!}(a_n x + k\beta)^{k-1} e^{-(a_n x + k\beta)} \frac{k^3}{b_n^3}$$

$$= \frac{a_n x}{b_n^3} \left[S(1, a_n x + \beta, \beta) + 3S(2, a_n x + 2\beta, \beta) + S(3, a_n x + 3\beta, \beta) \right]$$

$$= \frac{x^3 a_n^3}{(1 - \beta)^3 b_n^3} + \frac{3x^2 a_n^2}{(1 - \beta)^4 b_n^3} + \frac{x(1 + 2\beta)a_n}{(1 - \beta)^5 b_n^3};$$

$$J_n^{[\beta]}(t^4; a_n, b_n; x) = \sum_{k=0}^{\infty} \frac{a_n x}{k!}(a_n x + k\beta)^{k-1} e^{-(a_n x + k\beta)} \frac{k^4}{b_n^4}$$

$$= \frac{a_n x}{b_n^4} \left[S(1, a_n x + \beta, \beta) + 7S(2, a_n x + 2\beta, \beta) \right.$$

$$\left. + 6S(3, a_n x + 3\beta, \beta) + S(4, a_n x + 4\beta, \beta) \right]$$

$$= \frac{x^4 a_n^4}{(1 - \beta)^4 b_n^4} + \frac{6x^3 a_n^3}{(1 - \beta)^5 b_n^4} + \frac{x^2(7 + 8\beta)a_n^2}{(1 - \beta)^6 b_n^4} + \frac{x\left(1 + 8\beta + 6\beta^2\right)a_n}{(1 - \beta)^7 b_n^4}.$$

LEMMA 4   Let $0 \le \beta < 1$, then the following equalities hold:

(1) $J_n^{[\beta]}(t - x; a_n, b_n; x) = \left(\dfrac{a_n}{b_n(1 - \beta)} - 1 \right)x;$

(2) $J_n^{[\beta]}((t - x)^2; a_n, b_n; x) = x^2 \left(\dfrac{a_n}{(1 - \beta)b_n} - 1 \right)^2 + \dfrac{x a_n}{(1 - \beta)^3 b_n^2};$

(3)
$$J_n^{[\beta]}((t - x)^3; a_n, b_n; x) = x^3 \left(\frac{a_n}{(1 - \beta)b_n} - 1 \right)^3 + \frac{3x^2 a_n}{b_n^2(1 - \beta)^3} \left(\frac{a_n}{(1 - \beta)b_n} - 1 \right)$$

$$+ \frac{x a_n(1 + 2\beta)}{(1 - \beta)^5 b_n^3};$$

(4) $J_n^{[\beta]}((t - x)^4; a_n, b_n; x) = x^4 \left(\dfrac{a_n}{(1 - \beta)b_n} - 1 \right)^4 + \dfrac{6 a_n x^3}{(1 - \beta)^3 b_n^2} \left(\dfrac{a_n}{(1 - \beta)b_n} - 1 \right)^2$

$$+ \frac{a_n x^2}{(1 - \beta)^5 b_n^3} \left(\frac{a_n(7 + 8\beta)}{(1 - \beta)b_n} - 4 - 8\beta \right) + x\left(\frac{a_n(1 + 8\beta + 6\beta^2)}{(1 - \beta)^7 b_n^4} \right).$$

Proof of the above lemma, follows from the linearity of the operators $J_n^{[\beta]}(f; a_n, b_n; x)$.

By equality (1.6) and $\lim_{n \to \infty} \beta_n = 0$, we obtain

$$\lim_{n\to\infty} b_n J_n^{[\beta_n]}(t - x; a_n, b_n; x) = 0;$$

$$\lim_{n\to\infty} b_n J_n^{[\beta_n]}((t - x)^2; a_n, b_n; x) = x;$$

$$\lim_{n\to\infty} b_n J_n^{[\beta_n]}((t - x)^3; a_n, b_n; x) = 0;$$

$$\lim_{n\to\infty} b_n^2 J_n^{[\beta_n]}((t - x)^4; a_n, b_n; x) = 3x^2,$$

for every $x \in \mathbb{R}_0$.

3. Approximation properties

LEMMA 5　*Let $r \in \mathbb{N}$ be fixed number. Then there exist positive numerical coefficients $\lambda_{r,j,\beta}$, $1 \leq j \leq r$, depending only on r and j such that*

$$J_n^{[\beta]}(t^r; a_n, b_n; x) = \frac{1}{b_n^r(1 - \beta)^r} \sum_{j=1}^{r} \frac{\lambda_{r,j,\beta}}{(1 - \beta)^{j-1}} (a_n x)^j,$$

for all $x \in \mathbb{R}_0$ and $n \in \mathbb{N}$. Moreover, we have $\lambda_{r,1,\beta} = 1 = \lambda_{r,r,\beta}$.

The proof follows by a mathematical induction argument.

LEMMA 6　*For given $p \in \mathbb{N}_0$ and $(a_n)_{n=1}^{\infty}$ and $(b_n)_{n=1}^{\infty}$ there exists a positive constant $M_1(b_1, p, \beta)$ such that*

$$\left\| J_n^{[\beta]}\left(\frac{1}{\omega_p(t)}; a_n, b_n; \cdot\right) \right\|_p \leq M_1(b_1, p, \beta), \quad n \in \mathbb{N}. \tag{3.1}$$

Moreover, for every $f \in C_p$, we have

$$\left\| J_n^{[\beta]}(f; a_n, b_n; \cdot) \right\|_p \leq M_1(b_1, p, \beta)\|f\|_p, \quad n \in \mathbb{N}. \tag{3.2}$$

The formula (1.4), (1.5) and the inequality (3.2), show that $J_n^{[\beta]}$, $n \in \mathbb{N}$ is a positive linear operator from the space C_p into C_p, $p \in \mathbb{N}_0$.

Proof　If $p = 0$, then $\left\| J_n^{[\beta]}\left(\frac{1}{\omega_0(t)}; a_n, b_n; \cdot\right) \right\|_0 = \sup_{x \in \mathbb{R}_0} |J_n^{[\beta]}(1; a_n, b_n; x)| = 1.$

If $p \geq 1$, then by (1.5), (1.6), Lemma 3 and Lemma 5, we get

$$\omega_p(x) J_n^{[\beta]}\left(\frac{1}{\omega_p(t)}; a_n, b_n; x\right) = \omega_p(x)\{1 + J_n^{[\beta]}(t^p; a_n, b_n; x)\}$$

$$= \frac{1}{1 + x^p}\left\{1 + \frac{1}{b_n^p(1 - \beta)^p} \sum_{j=1}^{p} \frac{\lambda_{r,j,\beta}}{(1 - \beta)^{j-1}} (a_n x)^j\right\}$$

$$= \frac{1}{1 + x^p} + \frac{1}{(1 - \beta)^p} \sum_{j=1}^{p} \frac{\lambda_{r,j,\beta}}{(1 - \beta)^{j-1}} \frac{1}{b_n^{p-j}} \left(\frac{a_n}{b_n}\right)^j \frac{x^j}{1 + x^p}$$

$$\leq 1 + \frac{1}{(1 - \beta)^p} \sum_{j=1}^{p} \frac{\lambda_{r,j,\beta}}{(1 - \beta)^{j-1}} \frac{1}{b_1^{p-j}} = M_1(b_1, p, \beta),$$

for all $x \in \mathbb{R}_0$ and $n \in \mathbb{N}$. From this, (3.1) follows.

By (1.5) and definition of norm, we have

$$\|J_n^{[\beta]}(f; a_n, b_n; \cdot)\|_p \leq \|J_n^{[\beta]}(\frac{1}{\omega_p(t)}; a_n, b_n; \cdot)\|_p \|f\|_p,$$

for every $f \in C_p$, $p \in \mathbb{N}$ and $n \in \mathbb{N}$. From (3.1), the inequalities (3.2) is achieved.

Theorem 1 *For every $p \in \mathbb{N}_0$ there exists a positive constant $M_2(b_1, p, \beta)$ such that*

$$\omega_p(x) J_n^{[\beta]}\left(\frac{(t-x)^2}{\omega_p(t)}; a_n, b_n; x\right) \leq M_2(b_1, p, \beta)\left[x^2\left(\frac{a_n}{(1-\beta)b_n} - 1\right)^2 + \frac{x}{(1-\beta)^3 b_n}\right], \tag{3.3}$$

for all $x \in \mathbb{R}_0$ and $n \in \mathbb{N}$.

Proof If $p = 0$, then (3.3) follows from values of $J_n^{[\beta]}\left((t-x)^2; a_n, b_n; x\right)$.

Let $J_n^{[\beta]}(f; x) = J_n^{[\beta]}(f; a_n, b_n; x)$. Notice that

$$J_n^{[\beta]}\left(\frac{(t-x)^2}{\omega_p(t)}; x\right) = J_n^{[\beta]}\left((t-x)^2; x\right) + J_n^{[\beta]}\left(t^p(t-x)^2; x\right). \tag{3.4}$$

For $p = 1$, we get

$$J_n^{[\beta]}\left(\frac{(t-x)^2}{\omega_1(t)}; x\right) = J_n^{[\beta]}\left((t-x)^2; x\right) + J_n^{[\beta]}\left(t(t-x)^2; x\right)$$

$$= J_n^{[\beta]}\left((t-x)^2; x\right) + J_n^{[\beta]}\left((t-x)^3; x\right) + x J_n^{[\beta]}\left((t-x)^2; x\right)$$

$$= (1+x) J_n^{[\beta]}\left((t-x)^2; x\right) + J_n^{[\beta]}\left((t-x)^3; x\right).$$

Therefore,

$$(1+x) J_n^{[\beta]}\left(\frac{(t-x)^2}{\omega_1(t)}; x\right) = x^2\left(\frac{a_n}{(1-\beta)b_n} - 1\right)^2 + \frac{xa_n}{(1-\beta)^3 b_n^2} + \frac{x^3}{1+x}\left(\frac{a_n}{(1-\beta)b_n} - 1\right)^3$$

$$+ \frac{3x^2 a_n}{(1+x)b_n^2(1-\beta)^3}\left(\frac{a_n}{(1-\beta)b_n} - 1\right) + \frac{xa_n(1+2\beta)}{(1+x)(1-\beta)^5 b_n^3}$$

$$\leq M_2(b_1, p, \beta)\left[x^2\left(\frac{a_n}{(1-\beta)b_n} - 1\right)^2 + \frac{x}{(1-\beta)^3 b_n}\right].$$

If $p \geq 2$, then by Lemma 5, we get

$$\omega_p(x) J_n^{[\beta]}\left(t^p(t-x)^2; x\right) = \omega_p(x)\left\{J_n^{[\beta]}\left(t^{p+2}; x\right) - 2x J_n^{[\beta]}\left(t^{p+1}; x\right) + x^2 J_n^{[\beta]}\left(t^p; x\right)\right\}$$

$$= \frac{x}{b_n(1-\beta)}\left\{\frac{1}{b_n^{p+1}(1-\beta)^{p+1}}\sum_{j=1}^{p+1}\frac{\lambda_{p+2,j,\beta}}{(1-\beta)^{j-1}}a_n^j\frac{x^{j-1}}{1+x^p}\right.$$

$$- \frac{2}{b_n^p(1-\beta)^p}\sum_{j=1}^{p}\frac{\lambda_{p+1,j,\beta}}{(1-\beta)^{j-1}}a_n^j\frac{x^j}{1+x^p}$$

$$\left.+ \frac{1}{b_n^{p-1}(1-\beta)^{p-1}}\sum_{j=1}^{p-1}\frac{\lambda_{p,j,\beta}}{(1-\beta)^{j-1}}a_n^j\frac{x^{j+1}}{1+x^p}\right\} + \frac{1}{(1-\beta)^{2p+3}}\left(\frac{a_n}{b_n}\right)^{p+2}\frac{x^{p+2}}{1+x^p}$$

$$- \frac{2}{(1-\beta)^{2p+1}}\left(\frac{a_n}{b_n}\right)^{p+1}\frac{x^{p+2}}{1+x^p} + \frac{1}{(1-\beta)^{2p-1}}\left(\frac{a_n}{b_n}\right)^{p}\frac{x^{p+2}}{1+x^p}$$

$$= \frac{x}{b_n(1-\beta)}\left\{\frac{1}{b_n^{p+1}(1-\beta)^{p+1}}\sum_{j=1}^{p+1}\frac{\lambda_{p+2,j,\beta}}{(1-\beta)^{j-1}}a_n^j\frac{x^{j-1}}{1+x^p}\right.$$

$$- \frac{2}{b_n^p(1-\beta)^p}\sum_{j=1}^{p}\frac{\lambda_{p+1,j,\beta}}{(1-\beta)^{j-1}}a_n^j\frac{x^j}{1+x^p}$$

$$\left.+ \frac{1}{b_n^{p-1}(1-\beta)^{p-1}}\sum_{j=1}^{p-1}\frac{\lambda_{p,j,\beta}}{(1-\beta)^{j-1}}a_n^j\frac{x^{j+1}}{1+x^p}\right\}$$

$$+ \frac{x^{p+2}}{1+x^p}\left(\frac{a_n}{b_n}\right)^{p}\frac{1}{(1-\beta)^{2p-1}}\left(\frac{a_n}{b_n(1-\beta)} - 1\right)^2.$$

Since $0 \leq \dfrac{a_n}{b_n} \leq 1$ for $n \in \mathbb{N}, (1-\beta)^{-1} \leq (1-\beta)^{-3}$, we have

$$
\begin{aligned}
\omega_p(x)J_n^{[\beta]}\left(t^p(t-x)^2;x\right) \leq &\ \frac{x}{b_n(1-\beta)^3}\left\{\sum_{j=1}^{p+1}\frac{\lambda_{p+2,j,\beta}}{b_1^{p-j+1}(1-\beta)^{p+j}} + 2\sum_{j=1}^{p}\frac{\lambda_{p+1,j,\beta}}{b_1^{p-j}(1-\beta)^{p+j-1}}\right. \\
&\left. + \sum_{j=1}^{p-1}\frac{\lambda_{p,j,\beta}}{b_1^{p-j-1}(1-\beta)^{p+j-2}}\right\} + \frac{x^2}{(1-\beta)^{2p-1}}\left(\frac{a_n}{b_n(1-\beta)}-1\right)^2. \\
\leq &\ M_2(b_1,p,\beta)\left\{x^2\left(\frac{a_n}{b_n(1-\beta)}-1\right)^2 + \frac{x}{b_n(1-\beta)^3}\right\}.
\end{aligned}
$$

(3.5)

for $x \in \mathbb{R}_0, n \in \mathbb{N}$. Using (3.5) in (3.4), we obtain (3.3) for $p \geq 2$.

Thus, the proof is completed.

Now, we approximate $J_n^{[\beta]}\left(f;a_n,b_n;x\right)$ using the modulus of continuity $\omega_1(f,C_p)$ and the modulus of smoothness $\omega_2(f,C_p)$ of function $f \in C_p$, $p \in \mathbb{N}_0$

$$\omega_1(f,C_p,t):=\sup_{0\leq h\leq t}\|\triangle_h f(\cdot)\|_p, \quad \omega_2(f,C_p,t):=\sup_{0\leq h\leq t}\|\triangle_h^2 f(\cdot)\|_p,$$

for $t \geq 0$, where

$$\triangle_h f(x) = f(x+h)-f(x), \quad \triangle_h^2 f(x) = f(x)-2f(x+h)+f(x+2h).$$

Let

$$\xi_{n,\beta}(x) = x^2\left(\frac{a_n}{b_n(1-\beta)}-1\right)^2 + \frac{x}{b_n(1-\beta)^3}, \quad x \in \mathbb{R}_0, x \in \mathbb{N}. \tag{3.6}$$

THEOREM 2   *Suppose that $f \in C_p^2$ with a fixed $p \in \mathbb{N}_0$. Then there exists a positive constant $M_3(b_1,p,\beta)$ such that*

$$\omega_p(x)|J_n^{[\beta]}\left(f;a_n,b_n;x\right)-f(x)| \leq \|f'\|_p\left|\frac{a_n}{b_n(1-\beta)}-1\right|x + \|f''\|_p M_3(b_1,p,\beta)\xi_{n,\beta}(x), \tag{3.7}$$

for all $x \in \mathbb{R}_0, n \in \mathbb{N}$.

Proof   Notice that $J_n^{[\beta]}\left(0;a_n,b_n;x\right) = f(0), n \in \mathbb{N}$, which implies (3.7) for $x = 0$.

Let $x > 0$ and let $J_n^{[\beta]}(f;x) = J_n^{[\beta]}\left(f;a_n,b_n;x\right)$. For $f \in C_p^2$ and $t \in \mathbb{R}_0$,

$$f(t) = f(x) + f'(x)(t-x) + \int_x^t (t-u)f''(u)du. \tag{3.8}$$

Applying $J_n^{[\beta]}(f;x)$ on both sides, we obtain

$$J_n^{[\beta]}(f(t);x) = f(x) + f'(x)J_n^{[\beta]}((t-x);x) + J_n^{[\beta]}\left(\int_x^t(t-u)f''(u)du;x\right).$$

Notice that

$$\left|\int_x^t(t-u)f''(u)du\right| \leq \|f''\|_p\left(\frac{1}{\omega_p(t)}+\frac{1}{\omega_p(x)}\right)(t-x)^2.$$

Now, using above inequality, we have

$\omega_p(x)|J_n^{[\beta]}(f(t); x) - f(x)| \le \|f'\|_p J_n^{[\beta]}((t-x); x)$

$$+ \|f''\|_p \omega_p(x) J_n^{[\beta]}\left(\left(\frac{1}{\omega_p(t)} + \frac{1}{\omega_p(x)}\right)(t-x)^2; x\right)$$

$$\le \|f'\|_p J_n^{[\beta]}((t-x); x)$$

$$+ \|f''\|_p\left(\omega_p(x) J_n^{[\beta]}\left(\frac{(t-x)^2}{\omega_p(t)}; x\right) + J_n^{[\beta]}\left((t-x)^2; x\right)\right).$$

Now, using (3.3) and (3.6), we get

$$\omega_p(x)|J_n^{[\beta]}(f(t); x) - f(x)| \le \|f'\|_p\left|\frac{a_n}{b_n(1-\beta)} - 1\right|x + \|f''\|_p \xi_{n,\beta}(x) M_3(b_1, n, \beta).$$

Thus, the proof is completed.

COROLLARY 1   Let $\rho(x) = (1+x^2)^{-1}, x \in \mathbb{R}_0$. Suppose that $f \in C_p^2$ with a fixed $p = 2$. Then there exists a positive constant $M_4(b_1, p, \beta)$ such that

$$\|[J_n^{[\beta]}(f; a_n, b_n; x) - f(x)]\rho\|_2 \le \left(1 - \frac{a_n}{b_n(1-\beta)}\right)\|f'\|_2$$

$$+ M_4(b_1, p, \beta)\|f''\|_2 b_n^{-1}(1-\beta)^{-3}, n \in \mathbb{N} \tag{3.9}$$

THEOREM 3   Suppose that $f \in C_p$ with a fixed $p \in \mathbb{N}_0$. Then there exists a positive constant $M_5(b_1, p, \beta)$ such that

$$\omega_p|J_n^{[\beta]}(f; a_n, b_n; x) - f(x)| \le \left|\frac{a_n}{b_n(1-\beta)} - 1\right|x\left(\xi_{n,\beta}(x)\right)^{-1/2}\omega_1\left(f; C_p; \sqrt{\xi_{n,\beta}(x)}\right)$$

$$+ M_5(b_1, p, \beta)\omega_2\left(f; C_p; \sqrt{\xi_{n,\beta}(x)}\right),$$

for all $x > 0$ and $n \in \mathbb{N}$, where $\xi_{n,\beta}(\cdot)$ is defined in (3.6). For $x = 0$, it follows that $J_n^{[\beta]}(f; a_n, b_n; 0) = f(0)$.

Proof   We shall apply the Steklov function f_h for $f \in C_p$:

$$f_h(x) = \frac{4}{h^2}\int_0^{h/2}\int_0^{h/2}\left[f(x+s+t) - f(x+2(s+t))\right]dsdt,$$

$x \in \mathbb{R}_0, h > 0$, for which we have

$$f_h'(x) = \frac{1}{h^2}\int_0^{h/2}\left[8\triangle_{h/2} f(x+s) - 2\triangle_h f(x+2s)\right]ds,$$

$$f_h''(x) = \frac{1}{h^2}\left[8\triangle_{h/2}^2 f(x) - \triangle_h^2 f(x)\right].$$

Hence, for $h > 0$, we have

$$\|f_h - f\|_p \le \omega_2(f, C_p; h), \tag{3.10}$$

$$\|f_h'\|_p \le 5h^{-1}\omega_1(f, C_p; h)\frac{\omega_p(x)}{\omega_p(x+h)}, \tag{3.11}$$

$$\|f_h''\|_p \le 9h^{-2}\omega_2(f, C_p; h), \tag{3.12}$$

which show that $f_h \in C_p^2$ if $f \in C_p$. By denoting $J_n^{[\beta]}(f; a_n, b_n; x)$ by $J_n^{[\beta]}(f; x)$ we can write

$$\omega_p(x)|J_n^{[\beta]}(f; x) - f(x)| \le \omega_p(x)\{|J_n^{[\beta]}(f - f_h; x)| + |J_n^{[\beta]}(f_h; x) - f_h(x)|$$

$$+ |f_h(x) - f(x)|\} := A_1 + A_2 + A_3,$$

for $x > 0, h > 0$ and $n \in \mathbb{N}$. By (3.2) and (3.9), we have

$$A_1 \leq M_1(b_1, p, \beta)\|f - f_h\|_p \leq M_1(b_1, p, \beta)\omega_2(f, C_p; h),$$
$$A_3 \leq \omega_2(f, C_p; h).$$

Applying Theorem 2, inequalities (3.10) and (3.11), we get

$$A_2 \leq \|f'\|_p \left| \frac{a_n}{b_n(1 - \beta)} - 1 \right| x + \|f''\|_p M_3(b_1, p, \beta)\xi_{n,\beta}(x)$$

$$\leq \omega_1(f, C_p; h) \frac{\omega_p(x)}{\omega_p(x + h)} \frac{5x}{h} \left| \frac{a_n}{b_n(1 - \beta)} - 1 \right| + \frac{9}{h^2}\omega_2(f, C_p; h)M_3(b_1, p, \beta)\xi_{n,\beta}(x)$$

Combining these and setting $h = \sqrt{\xi_{n,\beta}(x)}$, for fixed $x > 0$ and $n \in \mathbb{N}$, we obtain the desired result.

THEOREM 4   Let $f \in C_p, p \in \mathbb{N}_0$, and let $\rho(x) = (1 + x^2)^{-1}$ for $x \in \mathbb{R}_0$. Then there exists a positive constant $M_6(b_1, p, \beta)$ such that

$$\|[J_n^{[\beta]}(f; a_n, b_n; x) - f]\rho\|_p \leq \left(1 - \frac{a_n}{b_n(1 - \beta)} \right) \sqrt{b_n}\, \omega_1\left(f, C_p; 1/\sqrt{b_n(1 - \beta)^3} \right)$$

$$+ M_6(b_1, p, \beta)\omega_2\left(f, C_p; 1/\sqrt{b_n(1 - \beta)^3} \right), \quad n \in \mathbb{N}.$$

From Theorems 3 and 4, we derive the following corollary:

COROLLARY 2   Let $f \in C_p, p \in \mathbb{N}_0, \beta_n \to 0$ as $n \to \infty$. Then for $J_n^{[\beta_n]}$ defined by (1.5), we have

$$\lim_{n\to\infty} J_n^{[\beta_n]}(f; a_n, b_n; x) = f(x), \quad x \in \mathbb{R}_0. \tag{3.13}$$

Furthermore, the convergence of (3.12) is uniformly on every interval $[x_1, x_2]$, where $x_2 > x_1 \geq 0$.

Remark 1   The error of approximation of a function $f \in C_p, p \in \mathbb{N}_0$ by $J_n^{[\beta]}(f; a_n, b_n; .)$ where $a_n = n^r + \frac{1}{n}$ and $b_n = n^r, r > 1$ is smaller than by the operators (1.3).

Funding
The authors received no direct funding for this research.

Author details
Prashantkumar Patel[1,2]
E-mail: prashant225@gmail.com
ORCID ID: http://orcid.org/0000-0002-8184-1199
Vishnu Narayan Mishra[1,3]
E-mails: vishnu_narayanmishra@yahoo.co.in;
vishnunarayanmishra@gmail.com
ORCID ID: http://orcid.org/0000-0002-2159-7710
Mediha Örkcü[4]
E-mail: medihaakcay@gazi.edu.tr

[1] Department of Applied Mathematics & Humanities, S. V. National Institute of Technology, Surat, 395 007, Gujarat, India.
[2] Department of Mathematics, St. Xavier's College, Ahmedabad, 380 009, Gujarat, India.
[3] L. 1627 Awadh Puri Colony Beniganj, Phase-III, Opposite - Industrial Training Institute (ITI), Ayodhya Main Road, Faizabad , Uttar Pradesh, 224 001, India.
[4] Faculty of Sciences, Department of Mathematics, Gazi University, 06500, Teknikokullar, Ankara, Turkey.

References
Agratini, O. (2013). Approximation properties of a class of linear operators. *Mathematical Methods in the Applied Science, 36*, 2353–2358.
Agratini, O. (2014). On an approximation process of integral type. *Applied Mathematics and Computation, 236*, 195 201.
Bardaro, C., & Mantellini, I. (2006). Approximation properties in abstract modular spaces for a class of general sampling-type operators. *Applicable Analysis, 85*, 383–413.
Bardaro, C., & Mantellini, I. (2009). A Voronovskaya-type theorem for a general class of discrete operators. *Journal of Mathematics, 39*, 1411–1442.
Becker, M. (1978). Global approximation theorems for Szász–Mirakjan and Baskakov operators in polynomial weight spaces. *Indiana University Mathematics Journal, 27*, 127–142.
Jain, G. C. (1972). Approximation of functions by a new class of linear operators. *Journal of the Australian Mathematical Society, 13*, 271–276.
Mishra, V. N., Khatri, K., Mishra, L. N., & Deepmala (2013). Inverse result in simultaneous approximation by Baskakov–Durrmeyer–Stancu operators. *Journal of Inequalities and Applications, 2013*, 586.
Mishra, V. N., & Patel, P. (2013). Some approximation properties of modified Jain–Beta operators. *Journal of Calculus of Variations, 2013*, 8 p.
Örkcü, M. (2013). q-Szász–Mirakyan–Kantorovich operators of functions of two variables in polynomial weighted spaces. In *Abstract and applied analysis* (Vol. 2013, 9 p.). Hindawi.

Retrieved from http://www.hindawi.com/journals/
aaa/2013/823803/

Patel, P., & Mishra, V. N. (2014). Jain–Baskakov operators
and its different generalization. *Acta Mathematica
Vietnamica.* doi:10.1007/s40306-014-0077-9

Patel, P., & Mishra, V. N. (2015). On new class of linear and
positive operators. *Bollettino dell'Unione Matematica
Italiana, 8,* 81–96. doi:10.1007/s40574-015-0026-0

Rempulska, L., & Tomczak, K. (2009). Approximation by
certain linear operators preserving x^2. *Turkish Journal of
Mathematics, 33,* 273–281.

Szász, O. (1950). Generalization of S. Bernstein's polynomials
to the infinite interval. *Journal of Research of the National
Bureau of Standards, 45,* 239–245.

Tarabie, S. (2012). On Jain–Beta linear operators. *Applied
Mathematics & Information Sciences, 6,* 213–216.

Umar, S., & Razi, Q. (1985). Approximation of function
by a generalized Szasz operators. *Communications
de la Faculté Des Sciences de L'Université D'Ankara:
Mathématique, 34,* 45–52.

2

Skew polynomial rings over σ-skew Armendariz rings

V.K. Bhat[1]* and Meeru Abrol[1]

*Corresponding author: V.K. Bhat, School of Mathematics, SMVD University, P/o SMVD University, Katra 182320, Jammu and Kashmir, India

E-mail: vijaykumarbhat2000@yahoo. com

Reviewing editor: Igor Yakov Subbotin, National University, USA

Abstract: This article concerns skew polynomial rings over Armendariz rings and σ-skew Armendariz ring. Let R be a Noetherian, Armendariz, prime ring. In this paper we prove that R and the polynomial ring $R[x]$ are 2-primal. Further we prove that if σ is an endomorphism of a ring R, then (1) R is a σ-skew Armendariz ring implies that $R[x;\sigma]$ is a $\bar{\sigma}$-skew Armendariz ring, where $\bar{\sigma}$ is an extension of σ to $R[x;\sigma]$. (2) R is a σ-rigid implies that $R[x;\sigma]$ is a 2-primal.

Subjects: Advanced Mathematics; Algebra; Fields & Rings; Mathematics & Statistics; Science

Keywords: Noetherian ring; Armendariz rings; endomorphism; 2-primal ring

2010 Mathematics subject classifications: Primary 16-XX; Secondary 16S36; 16N40; 16P40; 16W20

1. Introduction and preliminaries

A ring R always means an associative ring with identity $1 \neq 0$, unless otherwise stated. Let σ be an endomorphism of ring R. The skew polynomial ring or Ore extension of endomorphism type is denoted by $R[x;\sigma]$. The prime radical and the set of nilpotent elements of R are denoted by $P(R)$ and $N(R)$ respectively. The ring of integers is denoted by $\mathbb{Z}$, unless otherwise stated.

We begin with the following:

A ring R is called 2-primal if the prime radical of R coincides with the set of nilpotent elements of R i.e. $P(R) = N(R)$ or if the prime radical is completely semi-prime (see Birkenmeier, Heatherly, & Lee,

ABOUT THE AUTHORS

V.K. Bhat received the PhD degree in Mathematics from University of Jammu, India. Currently, he is a professor at Shri Mata Vaishno Devi University, Katra, India. Bhat is a researcher in the area of Algebra (ring theory) and theoretical computer science. He has published about 80 research papers in refereed journals. He has guided eight research students for their PhD degree. He has completed two research projects funded by Department of Atomic Energy, Govt. of India and University Grants Commission, Govt. of India. He has attended a number of conferences and research schools in various countries.

Meeru Abrol received the MPhil degree in Mathematics from University of Jammu, India. Currently, she is an associate professor at Govt. Women College, Ghandhi Nagar, Jammu, India. Her research area is Algebra (ring theory). Her work is supported under QIP of UGC, Govt. of India.

PUBLIC INTEREST STATEMENT

One of the earliest examples in non-commutative algebra was skew polynomial rings also known as Ore extensions. Skew polynomial rings have invited attention from Mathematicians and in this area considerable work has been done and investigations are on.

The characterization of ideals and prime ideals (in particular associated prime ideals, completely prime ideals and minimal prime ideals), and 2-primal property of Ore extensions has lead to the extension of certain notions from commutative setup to non-commutative setup. Ore extensions constitute an important class of rings, appearing in extensions of differential calculus, in non-commutative geometry, in quantum groups and algebras and as a uniting framework for many algebras appearing in physics and engineering models.

1993 for more details). An ideal I of a ring R is called completely semi-prime if $a^2 \in I$ implies $a \in I$ for $a \in R$. We also note that any reduced ring is 2-primal and a commutative ring is also 2-primal. The class of 2-primal rings is closed under subrings by Birkenmeier et al. (1993, Proposition 2.2).

Example 1.1   Let $R = \begin{pmatrix} \mathbb{Z} & \mathbb{Z} \\ 0 & \mathbb{Z} \end{pmatrix}$. Here $P(R) = \begin{pmatrix} 0 & \mathbb{Z} \\ 0 & 0 \end{pmatrix}$ and any nilpotent element of R is of the form $\begin{pmatrix} 0 & a \\ 0 & 0 \end{pmatrix}$. Hence R is 2-primal.

Example 1.2   Let $R = \left\{ \begin{pmatrix} a & b \\ 0 & a \end{pmatrix} \mid a, b \in \mathbb{Z}_4 \right\}$. R is not a reduced ring as $\begin{pmatrix} 2 & 1 \\ 0 & 2 \end{pmatrix}$ is a non-zero nilpotent element and hence R is not 2-primal.

Krempa (1996) has investigated the relation between minimal prime ideals and completely prime ideals of a ring R. With this he proved the following:

THEOREM 1.3   *For a ring R the following conditions are equivalent:*

  (1) *R is reduced.*
  (2) *R is semiprime and all minimal prime ideals of R are completely prime.*
  (3) *R is a subdirect product of domains.*

According to Krempa (1996) an endomorphism of a ring R is called rigid if $a\sigma(a) = 0$ implies that $a = 0$ for $a \in R$. We call a ring σ-rigid if there exists a rigid endomorphism σ of R.

Example 1.4   Let $R = \mathbb{Z}[\sqrt{2}]$. Then $\sigma : R \to R$ defined as

$$\sigma(a + b\sqrt{2}) = a - b\sqrt{2} \text{ for } a + b\sqrt{2} \in R$$

is an endomorphism of R. Further $(a + b\sqrt{2})\sigma(a + b\sqrt{2}) = 0$ implies that $(a + b\sqrt{2})(a - b\sqrt{2}) = 0$ i.e. $a^2 - 2b^2 = 0$ which gives $a = 0, b = 0$. Hence $a + b\sqrt{2} = 0$. Thus R is a σ-rigid ring.

Example 1.5   (Bhat, 2011, Example 1)   Let $R = \begin{pmatrix} F & F \\ 0 & F \end{pmatrix}$, where F is a field. Let $\sigma : R \to R$ an automorphism be defined by

$$\sigma\left(\begin{pmatrix} a & b \\ 0 & c \end{pmatrix} \right) = \begin{pmatrix} a & 0 \\ 0 & 0 \end{pmatrix} \text{ for } a, b, c \in F.$$

Let $0 \neq a \in F$. Then $\begin{pmatrix} 0 & a \\ 0 & 0 \end{pmatrix} \sigma\left(\begin{pmatrix} 0 & a \\ 0 & 0 \end{pmatrix} \right) = \begin{pmatrix} 0 & 0 \\ 0 & 0 \end{pmatrix}$.

But $\begin{pmatrix} 0 & a \\ 0 & 0 \end{pmatrix} \neq \begin{pmatrix} 0 & 0 \\ 0 & 0 \end{pmatrix}$.

Hence R is not a σ-rigid ring.

Also R is said to be σ-compatible if for each $a, b \in R$, $ab = 0$ implies and is implied by $a\sigma(b) = 0$. Also a ring R is σ-rigid if and only if R is σ-compatible and reduced. Moreover, R is σ-rigid if and only if $R[x;\sigma]$ is reduced (Hong, Kim, & Kwak, 2003, Proposition 3).

Example 1.6   Let D be an integral domain. Consider the commutative ring

$$R = \left\{ \begin{pmatrix} a & d \\ 0 & a \end{pmatrix} \mid a, d \in D \right\}.$$

Let σ be an automorphism of R defined by

$$\sigma\left(\begin{pmatrix} a & d \\ 0 & a \end{pmatrix}\right) = \begin{pmatrix} a & ud \\ 0 & a \end{pmatrix},$$

where u is a fixed element of D. Then R is σ-compatible.

Example 1.7 Let $R = \mathbb{Z}_2 \oplus \mathbb{Z}_2$ be a commutative ring, where $\mathbb{Z}_2$ is the ring of integers modulo 2. Let $\sigma : R \to R$ be defined by

$\sigma((a,b)) = (b,a)$ for $a,b \in \mathbb{Z}_2$.

Then σ is an automorphism of R. Now $(0,1)(1,0) = (0,0)$. But $(0,1)\sigma((1,0)) = (0,1)(0,1) \neq (0,0)$. Hence R is not σ-compatible.

2. Armendariz rings

The notion of Armendariz rings was introduced by Rege and Chhawchharia (1997). They defined a ring R to be an Armendariz ring if whenever polynomial

$$f(x) = a_0 + a_1 x + \ldots + a_m x^m \in R[x],$$
$$g(x) = b_0 + b_1 x + \ldots + b_n x^n \in R[x]$$

satisfy $f(x)g(x) = 0$, then $a_i b_j = 0$ for each i,j. (The converse is always true.) This ring was named so because as Armendariz (1974, Lemma 2) had noted that a reduced ring satisfies this condition. In addition to reduced rings, quotient rings over a commutative P.I.D. are Armendariz (Rege & Chhawchharia, 1997, Theorem 2.2). But every $n \times n$ full matrix ring over any ring is not Armendariz, where $n \geq 2$ (Rege & Chhawchharia, 1997). Note that Armendariz rings are defined through polynomial rings over them. Also subrings of Armendariz rings are Armendariz. Anderson and Camillo (1998) has found a relation between an Armendariz ring and reduced ring as:

THEOREM 2.1 (Anderson & Camillo, 1998, Theorem 7) *If R is a prime ring which is left and right Noetherian, then R is Armendariz if and only if R is reduced.*

With this we prove the following:

THEOREM 2.2 *Let R be a Noetherian Armendariz prime ring. Then R is 2-primal.*

Proof By Theorem (2.1), R is a reduced ring. We know that a reduced ring is 2-primal. Hence R is 2-primal. □

The converse is not true.

Example 2.3 Let $R = (\mathbb{Z}/8\mathbb{Z} \oplus \mathbb{Z}/8\mathbb{Z})$. Then R is a commutative ring and hence 2-primal. Let $f(x) = (\overline{4}, \overline{0}) + (\overline{4}, \overline{1})x$. Now

$$f(x).f(x) = [(\overline{4}, \overline{0}) + (\overline{4}, \overline{1})x][(\overline{4}, \overline{0}) + (\overline{4}, \overline{1})x] = 0. \text{ But } (\overline{4}, \overline{0})(\overline{4}, \overline{1}) \neq 0.$$

Hence R is not an Armendariz ring.

THEOREM 2.4 *Let R be a Noetherian prime ring. If R is an Armendariz ring, then $P(R)$ is completely semi-prime.*

Proof As proved in Theorem (2.1), R is a reduced ring and by using Theorem (1.3), the result follows. □

The converse of the above is not true.

Example 2.5   Let F be a field, $R = F \times F$. Here $P(R)$ is completely semi-prime, as R is a reduced ring. Let $f(x) = (1, 0) + (0, 1)x, g(x) = (0, 2) + (2, 0)x$. Then

$$f(x).g(x) = [(1, 0) + (0, 1)x].[(0, 2) + (2, 0)x] = 0.$$

But $(1, 0)(2, 0) \neq 0$. Hence R is not an Armendariz ring.

Concerning polynomial rings over some kinds of rings, we have the following results:

(1) A ring R is reduced if and only if $R[x]$ is reduced.
(2) A ring R is 2-primal if and only if $R[x]$ is 2-primal (Birkenmeier et al., 1993, Proposition 2.6).
(3) A ring R is abelian if and only if $R[x]$ is abelian (Kim & Lee, 2000, Theorem 8). Note that a ring R is said to be abelian if every idempotent of it is central.

Recall from Anderson and Camillo (1998) that:

THEOREM 2.6   *(Anderson & Camillo, 1998, Theorem 2)*   *A ring R is Armendariz if and only if $R[x]$ is Armendariz.*

Hilbert's Basis Theorem (1890) states that:

THEOREM 2.7   *(Goodearl & Warfield, 2004, Theorem 1.9)*   *If R is a Noetherian ring, then $R[x]$ is a Noetherian ring.*

These help us to prove a relation between Armendariz rings and 2-primal rings as:

THEOREM 2.8   *Let R be a Noetherian Armendariz prime ring. Then $R[x]$ is 2-primal.*

Proof   By Theorem (2.6), $R[x]$ is Armendariz. Also using Hilbert's Basis Theorem, it follows that $R[x]$ is a Noetherian ring. Therefore, by Theorem (2.2), $R[x]$ is 2-primal.   □

The converse is not true.

Example 2.9   Let F be a field. Let $R = \begin{pmatrix} F & F \\ 0 & F \end{pmatrix}$. Then $P(R) = \begin{pmatrix} 0 & F \\ 0 & 0 \end{pmatrix}$ and any nilpotent element of R is of the form $\begin{pmatrix} 0 & a \\ 0 & 0 \end{pmatrix}$. Hence R is 2-primal. By Proposition (2.6) of Birkenmeier et al. (1993), $R[x]$ is 2-primal. Let

$$f(x) = \begin{pmatrix} 1 & 0 \\ 0 & 0 \end{pmatrix} + \begin{pmatrix} 1 & -1 \\ 0 & 0 \end{pmatrix}x \in R[x]$$

and

$$g(x) = \begin{pmatrix} 0 & 0 \\ 0 & 1 \end{pmatrix} + \begin{pmatrix} 0 & 1 \\ 0 & 1 \end{pmatrix}x \in R[x].$$

Now $f(x).g(x) = 0$, but $\begin{pmatrix} 1 & 0 \\ 0 & 0 \end{pmatrix}\begin{pmatrix} 0 & 1 \\ 0 & 1 \end{pmatrix} \neq 0$. Hence R is not an Armendariz ring.

3. σ-skew Armendariz rings

Recall that $R[x;\sigma]$ is the usual polynomial ring with coefficients in R, in which multiplication is subject to the relation $xa = \sigma(a)x$ for all $a \in R$. We take any $f(x) \in R[x;\sigma]$ to be of the form $f(x) = \sum_{i=0}^{n} a_i x^i$. By Anderson and Camillo (1998, Theorem 2), polynomial rings over Armendariz rings are also Armendariz. There is a natural motivation to investigate the nature of skew polynomial ring over a Armendariz ring, but the fact is that a skew polynomial ring over an Armendariz ring need not be Armendariz as follows:

Example 3.1   (Kim & Lee, 2000, Example 6)   Let $\mathbb{Z}_2$ be the ring of integers modulo 2 and consider the ring $R = \mathbb{Z}_2 \oplus \mathbb{Z}_2$ with the usual addition and multiplication. Then R is a commutative reduced ring; hence R is Armendariz by (1974, Lemma 1). Now let $\sigma : R \rightarrow R$ be defined by

$$\sigma((a, b)) = (b, a) \text{ for } a, b \in \mathbb{Z}_2.$$

Then σ is an automorphism of R. We claim that $R[x; \sigma]$ is not Armendariz. Let

$$f(y) = (1, 0) + [(1, 0)x]y \in R[x; \sigma][y] \text{ and } g(y) = (0, 1) + [(1, 0)x]y \in R[x; \sigma][y].$$

Then $f(y)g(y) = 0$, but $(1, 0)[(1, 0)x] \neq 0$. Therefore, $R[x; \sigma]$ is not an Armendariz ring.

We now discuss σ-skew Armendariz rings (σ an endomorphism of a ring R) and their extensions. Recall (Hong et al, 2003) that a ring R with an endomorphism σ is called a σ-skew Armendariz ring if for

$$p = \sum_{i=0}^{m} a_i x^i \text{ and } q = \sum_{j=0}^{n} b_j x^j \text{ in } R[x; \sigma],$$

$pq = 0$ implies that $a_i \sigma^i(b_j) = 0$, for all $0 \leq i \leq m$ and $0 \leq j \leq n$. It is also known as skew Armendariz ring with endomorphism σ. Every subring of a σ-skew Armendariz ring is σ-skew Armendariz.

Example 3.2   (Hong et al., 2003, Example 1)   Consider the commutative ring

$$R = \left\{ \begin{pmatrix} a & t \\ 0 & a \end{pmatrix} | a \in \mathbb{Z}, t \in \mathbb{Q} \right\}.$$

Let σ be an automorphism of R defined by

$$\sigma\left(\begin{pmatrix} a & t \\ 0 & a \end{pmatrix} \right) = \begin{pmatrix} a & t/2 \\ 0 & a \end{pmatrix}.$$

Then R is a σ-skew Armendraiz ring.

Example 3.3   Let F be a field and $R = \begin{pmatrix} F & F \\ 0 & F \end{pmatrix}$ a ring. Define an endomorphism $\sigma : R \rightarrow R$ by

$$\sigma\left(\begin{pmatrix} a & b \\ 0 & c \end{pmatrix} \right) = \begin{pmatrix} a & -b \\ 0 & c \end{pmatrix} \text{ for } a, b, c \in F.$$

For $p = \begin{pmatrix} 1 & 0 \\ 0 & 0 \end{pmatrix} + \begin{pmatrix} 1 & 1 \\ 0 & 0 \end{pmatrix} x$, $q = \begin{pmatrix} 0 & 0 \\ 0 & -1 \end{pmatrix} + \begin{pmatrix} 0 & 1 \\ 0 & 1 \end{pmatrix} x \in R[x; \sigma]$, we have $pq = 0$. But $\begin{pmatrix} 1 & 1 \\ 0 & 0 \end{pmatrix} \sigma\left(\begin{pmatrix} 0 & 0 \\ 0 & -1 \end{pmatrix} \right) \neq 0$. Hence R is not a σ-skew Armendariz ring.

Let σ be an endomorphism of ring R. Then σ can be extended to an endomorphism (say $\overline{\sigma}$) of $R[x; \sigma]$ by

$$\overline{\sigma}\left(\sum_{i=0}^{m} x^i a_i \right) = \sum_{i=0}^{m} x^i \sigma(a_i).$$

We now prove the following Theorem:

THEOREM 3.4   *Let R be a ring, σ an endomorphism of R such that R is a σ-skew Armendariz ring. Then $R[x;\sigma]$ is a $\bar{\sigma}$-skew Armendariz ring.*

Proof   Let

$$f(x) = a_0 + a_1 x + \dots + a_m x^m,$$
$$g(x) = b_0 + b_1 x + \dots + b_n x^n \in R[x;\sigma]$$

be such that $f(x)g(x) = 0$. Then

$$a_i \sigma^i(b_j) = 0, \quad 0 \le i \le m, \quad 0 \le j \le n. \tag{3.1}$$

Let $f(y), g(y) \in R[x;\sigma][y]$ be   such   that $f(y)g(y) = 0$.   Let $f(y) = f_0 + f_1 y + \dots + f_n y^n$ and $g(T) = g_0 + g_1 y + \dots + g_m y^m$ where $f_i, g_i \in R[x;\sigma]$. Let

$$k = deg(f_0) + \dots + deg(f_n) + deg(g_0) + \dots + deg(g_m).$$

Then

$$f(x^k) = f_0 + f_1 x^k + \dots + f_n x^{kn} \in R[x;\sigma]$$

and

$$g(x^k) = g_0 + g_1 x^k + \dots + g_m x^{km} \in R[x;\sigma]$$

and the set of coefficients of the $f_i's$(res., $g_i's$) equals the set of coefficients of the $f(x^k)$(res., $g(x^k)$). Now $f(y)g(y) = 0$ and $xr = \sigma(r)x$, $f(x^k)g(x^k) = 0$. Since R is σ-skew Armendariz using Equation (3.1), the result follows.   □

Also from Hong et al. (2003):

THEOREM 3.5   *(Hong et al., 2003, Proposition 3)*   *Let σ be an endomorphism of R. Then $R[x;\sigma]$ is reduced if and only if R is σ-rigid.*

With this we prove the following:

THEOREM 3.6   *Let σ be an endomorphism of ring R such that R is a σ-rigid. Then $R[x;\sigma]$ is a 2-primal.*

Proof   By Theorem (3.5), $R[x;\sigma]$ is reduced and hence it is 2-primal.   □

The converse is not true.

Example 3.7   Consider a commutative polynomial ring over $\mathbb{Z}_2$. Let $R = \mathbb{Z}_2[x]$ and $\sigma{:}R \to R$ be an endomorphism defined by $\sigma(f(x)) = f(0)$.

Then

(1) R is 2-primal as $P(R) = \{0\}$.

(2) R is σ-skew Armendariz.

Consider $R[y;\sigma] = \mathbb{Z}_2[x][y;\sigma]$. Let

$$p = f_0 + f_1 y + \dots + f_m y^m \in R[y;\sigma]$$

and

$$q = g_0 + g_1 y + \dots + g_n y^n \in R[y;\sigma].$$

Assume that $pq = 0$. Suppose there is $f_s \neq 0$ and $f_0 = \dots = f_{s-1} = 0$, where $0 \leq s \leq m$. Since

$$f_0 g_s + f_1 \sigma(g_{s-1}) + \dots + f_s \sigma^s(g_0) = 0$$

where $f_k = 0$ if $m < k$, we have $f_s \sigma^s(g_0) = 0$ and $f_s g_0(0) = 0$. Then $g_0(0) = 0$. Again

$$f_0 g_{s+1} + f_1 \sigma(g_s) + \dots + f_s \sigma^s(g_1) + f_{s+1} \sigma^{s+1}(g_0) = 0,$$

gives $f_s \sigma^s(g_1) = f_s g_1(0) = 0$ and so $g_1(0) = 0$, by the same method as above. Continuing this process, we have

$$g_0(0) = g_1(0) = \dots = g_n(0) = 0.$$

Thus, for each $0 \leq j \leq n, f_i \sigma^i(g_j) = 0$, for all $0 \leq i \leq m$, since $f_i = 0$ for $0 \leq i \leq s-1$ and $\sigma^i(g_j) = 0$, for all $s \leq i \leq m$.

(3) R is not σ-rigid.

$(xy)^2 = x\sigma(x)y^2 = x.0.y^2 = 0$, but $xy \neq 0$. Thus $R[y;\sigma]$ is not reduced and hence R is not σ-rigid.

We now give an example of a 2-primal ring which is not a σ-skew Armendariz ring.

Example 3.8   Let $R = R_1 \oplus R_2$ where R_1 and R_2 are reduced rings. Then $R[x;\sigma]$ is 2-primal, because $P(R) = \{(0,0)\}$. Let $\sigma:R \to R$ be an endomorphism defined by

$$\sigma((a,b)) = (b,a).$$

Take $f(x) = (0,1) - (0,1)x$ and $g(x) = (1,0) + (0,1)x$.   Then $f(x).g(x) = 0$.   But $(0,1)\sigma((1,0)) = (0,1)(0,1) = (0,1) \neq (0,0)$. Hence $R[x;\sigma]$ is not a σ-skew Armendariz ring. Note that σ here is not rigid.

Funding
The authors received no direct funding for this research.

Author details
V.K. Bhat[1]
E-mail: vijaykumarbhat2000@yahoo.com
ORCID ID: http://orcid.org/0000-0001-8423-9067
Meeru Abrol[1]
E-mail: meeru.abrol@yahoo.in
[1] School of Mathematics, SMVD University, P/o SMVD
  University, Katra 182320, Jammu and Kashmir, India.

References
Anderson, D. D., & Camillo, V. (1998). Armendariz rings and Gaussian rings. *Communications in Algebra, 26*, 2265–2272.
Armendariz, E. P. (1974). A note on extensions of Baer and P. P. - rings. *Journal of the Australian Mathematical Society, 18*, 470–473.
Bhat, V. K. (2011). On 2-primal ore extensions over Noetherian σ (*)-rings. *Buletinul Academiei de Stiinte a Republicii Moldova. Matematica, 65*, 42–49.
Birkenmeier, G. F., Heatherly, H. E., & Lee, E. K. (1993). Completely prime ideals and associated radicals. In S. K. Jain & S. T. Rizvi (Eds.), *Proceedingd of Biennal Ohio State - Denison Conference 1992* (pp. 102–129). Singapore: World Scientific.
Goodearl, K. R., & Warfield, R.,B. (2004). *An Introduction to non-commutative Noetherian rings*. Cambridge: Cambridge University Press.
Hong, C. Y., Kim, N. K., & Kwak, T. K. (2003). On skew Armendariz rings. *Communications in Algebra, 31*, 103–122.
Kim, N. K., & Lee, Y. (2000). Armendariz rings and reduced rings. *Journal of Algebra, 223*, 477–488.
Krempa, J. (1996). Some examples of reduced rings. *Algebra Colloquium, 3*, 289–300.
Rege, M. R., & Chhawchharia, S. (1997). Armendariz rings. *Proceedings of the Japan Academy, Ser. A, Mathematical Sciences, 73*, 14–17.

3

Generalised colouring sums of graphs

Johan Kok[1], N.K. Sudev[2]* and K.P. Chithra[3]

Corresponding author: N.K. Sudev, Department of Mathematics, Vidya Academy of Science & Technology, Thalakkottukara, Thrissur 680501, India
E-mail: sudevnk@gmail.com
Reviewing editor: Gerald Williams, University of Essex, UK

Abstract: The notion of the b-chromatic number of a graph attracted much research interests and recently a new concept, namely the b-chromatic sum of a graph, denoted by $\varphi'(G)$, has also been introduced. Motivated by the studies on b-chromatic sum of graphs, in this paper we introduce certain new parameters such as χ-chromatic sum, χ^+-chromatic sum, b^+-chromatic sum, π-chromatic sum and π^+-chromatic sum of graphs. We also discuss certain results on these parameters for a selection of standard graphs.

Subjects: Advanced Mathematics; Combinatorics; Discrete Mathematics; Mathematics Statistics; Science

Keywords: chromatic number; χ-chromatic sum; χ^+-chromatic sum; b-chromatic number; b-chromatic sum; b^+-chromatic sum; Thue chromatic number; π-chromatic sum; π^+-chromatic sum

AMS subject classifications: 05C62; 05C05; 05C20; 05C38

1. Introduction

For general notations and concepts in graph theory and digraph theory, we refer to Bondy and Murty (1976), Chartrand and Lesniak (2000), Chartrand and Zhang (2009), Gross and Yellen (2006), Harary (1969), West (2001). Unless mentioned otherwise, all graphs mentioned in this paper are non-trivial, simple, connected, finite and undirected graphs.

Graph colouring has become a fertile research area since its introduction in the second half of nineteenth century. It has numerous theoretical and practical applications. Let us first recall the fact that in a *proper colouring* of a graph G, no two adjacent vertices in G can have the same colour. The

ABOUT THE AUTHORS

Johan Kok is registered with the South African Council for Natural Scientific Professions (South Africa) as a professional scientist in both Physical Science and Mathematical Science. His main research areas are in Graph Theory and the reconstruction of motor vehicle collisions.

N.K. Sudev has been working as a Professor (Associate) in the Department of Mathematics, Vidya Academy of Science and Technology, Thrissur, India, for the last fifteen years. His primary research areas are Graph Theory and Combinatorics.

K.P. Chithra is an independent researcher in Mathematics. Her primary research area is also Graph Theory.

PUBLIC INTEREST STATEMENT

Graph colouring attracted wide interest among researchers since its introduction in the second half of the nineteenth century. A number of interesting extremal graph theoretic problems were researched well. Colouring sums are more recent and allows for applications where colours may be technologies of kind with some relation between the distinct technologies. It is envisaged that colour products and other mathematical relations between colours will naturally follow as enhanced research fields. It is foreseen that the modelling of metabolic or artificial intelligent structures as "colours" within larger real or virtual living structures of which certain components are modelled as graphs will reveal interesting applications. Colouring sums are extremely useful in many practical problems in project management, communication, routing and transportation, assignments, distributions etc.

minimum number of colours in a proper colouring of a graph G is called the *chromatic number of G*, denoted by $\chi(G)$.

Consider a *proper k-colouring* of a graph G and denote the set of k colours by $C = \{c_1, c_2, c_3, \ldots, c_k\}$. Also, consider the disjoint subsets of $V(G)$, defined by $V_{c_i} = \{v_j : v_j \mapsto c_i, v_j \in V(G), c_i \in C\}, 1 \le i \le k$. Clearly, we can see that $V(G) = \bigcup\limits_{i=1}^{k} V_{c_i}$.

The notion of the *b-colouring* of a graph and the parameter *b-chromatic number*, $\varphi(G)$, of a graph $G(V, E)$, has been introduced in Irving and Manlove (1999) as follows. Let G be a graph on n vertices, say $v_1, v_2, v_3, \ldots, v_n$. The *b-chromatic number* of G is defined as the maximum number k of colours that can be used to colour the vertices of G, such that we obtain a proper colouring and each colour i, with $1 \le i \le k$, has at least one element x_i which is adjacent to a vertex of every colour j, $1 \le j \ne i \le k$. Such a colouring is called a *b-colouring* of G (see Effatin & Kheddouci, 2003; Irving & Manlove, 1999).

The concept of b-chromatic number has attracted much attention and many studies have been made on this parameter (see Effatin & Kheddouci, 2003; Irving & Manlove, 1999; Kok & Sudev, in press; Kouider & Mahéo, 2002; Vaidya & Isaac, 2014, 2015; Vivin & Vekatachalam, 2015).

2. General colouring sum of graphs

The notion of the *b-chromatic sum* of a given graph G, denoted by $\varphi'(G)$, has been introduced in Lisna and Sunitha (2015) as the minimum of sum of colours $c(v)$ of v for all $v \in V$ in a b-colouring of G using $\varphi(G)$ colours. Some results on b-chromatic sums proved in Lisna and Sunitha (2015), which are relevant and useful results in our present study, are listed below.

THEOREM 2.1   *(Lisna & Sunitha, 2015) The b-chromatic sum of a path P_n, $n \ge 2$ is*

$$\varphi'(P_n) = \begin{cases} \frac{3}{2}(n-1) + 3, & \text{if } n \ge 5, n \text{ is odd,} \\ \frac{3n}{2} + 1, & \text{if } n \ge 6, n \text{ is even,} \\ 4, & \text{if } n = 3, \\ \frac{3n}{2}, & \text{if } n \in \{2, 4\}. \end{cases}$$

THEOREM 2.2   *Lisna & Sunitha, 2015 The b-chromatic sum of a cycle C_n is given by*

$$\varphi'(C_n) = \begin{cases} \frac{3n}{2} + 3, & \text{if } n \text{ is even}, n \ne 4, \\ \frac{3}{2}(n-1) + 3, & \text{if } n \text{ is odd,} \\ 6, & \text{if } n = 4. \end{cases}$$

THEOREM 2.3   *Lisna & Sunitha, 2015 The b-chromatic sum of a wheel graph W_{n+1} is*

$$\varphi'(W_{n+1}) = \begin{cases} \frac{3(n-1)}{2} + 7, & \text{if } n \text{ is odd,} \\ \frac{3n}{2} + 7, & \text{if } n \text{ is even}, n \ne 4, \\ 9, & \text{if } n = 4. \end{cases}$$

THEOREM 2.4   *(Lisna & Sunitha, 2015) For a complete bipartite graph $K_{m,n}$ assume without loss of generality that $m \ge n$, then $\varphi'(K_{m,n}) = m + 2n$.*

This interesting new invariant motivates us for studying similar concepts in graph colouring. This leads us to define the concept of the general colouring sum of graphs as follows.

Definition 2.5   Let $C = \{c_1, c_2, c_3, \ldots, c_k\}$ allows a b-colouring S of a given graph G. Clearly, there are k! ways of allocating the colours to the vertices of G. The *colour weight* of colour, denoted by $\theta(c_i)$, is the number of times a particular colour c_i is allocated to vertices. Then, the *colouring sum* of a colouring S of a given graph G, denoted by $\omega(S)$, is defined to be $\omega(S) = \sum\limits_{i=1}^{k} i\,\theta(c_i)$.

In view of the above definition, the b-chromatic sum of a graph G can be viewed as

$$\varphi'(G) = \min \left\{ \sum_{i=1}^{k} i\,\theta(c_i) \right\}, \text{where this sum varies over all b-colourings of G.}$$

In view of Definition 2.5, in this paper we introduce certain other colouring sums of graphs similar to the b-chromatic sum of graphs.

3. χ-Chromatic sum of certain graphs

The notion of the χ-chromatic sum of a graph G with respect to a proper k-colouring of G is introduced as follows.

Definition 3.1   Let $C = \{c_1, c_2, \dots, c_k\}$ be a proper colouring of a graph G. Then, the χ-chromatic sum of G, denoted by $\chi'(G)$, is defined as $\chi'(G) = \min \left\{ \sum_{i=1}^{k} i\,\theta(c_i) \right\}$ where the sum varies over all minimum proper colourings of G.

In the following discussion, we investigate the χ-chromatic sum of certain fundamental graph classes. First, we determine the χ-chromatic sum of path graphs in the following theorem.

THEOREM 3.2   *The χ-chromatic sum of a path P_n is given by*

$$\chi'(P_n) = \begin{cases} 1, & \text{if } n = 1, \\ \frac{3n}{2}, & \text{if } n \text{ is even,} \\ \frac{3n-1}{2}, & \text{if } n \text{ is odd.} \end{cases}$$

Proof   Being a bipartite graph, the vertices of a path graph P_n can be coloured using two colours, say c_1 and c_2. Then, we need to consider the following cases.

(1) Assume that $n = 1$. Then, $P_n \cong K_1$ with a single vertex say v_1. Colour this vertex by the colour c_1. Hence, $\theta(c_1) = 1$. Therefore, $\chi'(P_n) = 1$.

(2) Let n be an even integer. Then, the vertices of path P_n can be coloured alternatively by the colours c_1 and c_2 and hence $\theta(c_1) = \theta(c_2) = \frac{n}{2}$. Therefore, $\chi'(P_n) = 1 \cdot \frac{n}{2} + 2 \cdot \frac{n}{2} = \frac{3n}{2}$.

(3) Let $n > 1$ be an odd integer. Without loss of generality, label the vertices of P_n with odd subscripts by the colour c_1 and the vertices with even subscripts by the colour c_2. Then, $\theta(c_1) = \frac{n+1}{2}$ and $\theta(c_2) = \frac{n-1}{2}$. Therefore, $\chi'(P_n) = 1 \cdot \frac{n+1}{2} + 2 \cdot \frac{n-1}{2} = \frac{3n-1}{2}$. $\square$

In a similar way, the χ-chromatic sum of a cycle graph C_n can be determined as follows.

THEOREM 3.3   *The χ-chromatic sum of a cycle C_n is $\chi'(C_n) = 3 \lceil \frac{n}{2} \rceil$.*

Proof   Let C be a proper colouring of the cycle C_n. If n is even, C must contain at least two colours, say c_1 and c_2 and if n is odd, then C must contain at least three colours, say c_1, c_2 and c_3. Then, we consider the following cases.

(1) Let n be an odd integer. Now, we can assign the colour c_1 to the vertices having odd subscripts other than n, the colour c_2 to the vertices having even subscripts and the colour c_3 to the vertex v_n. Hence $\theta(c_1) = \theta(c_2) = \frac{n-1}{2}$ and $\theta(c_3) = 1$. Therefore, $\chi'(G) = 1 \cdot \frac{n-1}{2} + 2 \cdot \frac{n-1}{2} + 3 \cdot 1 = 3 \cdot \frac{n+1}{2}$.

(2) Let n be an even integer. Then, as explained in the previous result, we can assign the colour c_1 to the vertices having odd subscripts and the colour c_2 to the vertices having even subscripts. Hence $\theta(c_1) = \theta(c_2) = \frac{n}{2}$. Therefore, $\chi'(C_n) = 1 \cdot \frac{n}{2} + 2 \cdot \frac{n}{2} = 3 \cdot \frac{n}{2}$. Combining the above two cases, we have $\chi'(C_n) = 3 \cdot \lceil \frac{n}{2} \rceil$. $\square$

A wheel graph, denoted by W_{n+1}, is defined to be the join of a cycle C_n and a trivial graph K_1. That is, $W_{n+1} = C_n + K_1$. The χ-chromatic sum of a wheel graph is determined in the following theorem.

THEOREM 3.4   *The χ-chromatic sum of a wheel graph W_{n+1} is given by*

$$\chi'(W_{n+1}) = \begin{cases} \frac{3n+11}{2}, & \text{if } n \text{ is odd,} \\ \frac{3n+6}{2}, & \text{if } n \text{ is even.} \end{cases}$$

Proof   Let us denote the central vertex of the wheel W_{n+1} by v and the vertices of the outer cycle of W_{n+1} by $v_1, v_2, v_3, \ldots, v_n$. Let C be a minimal proper colouring of W_{n+1}. Then, C must contain three colours, say c_1, c_2, c_3, if n is even and it must contain four colours, say c_1, c_2, c_3, c_4, if n is odd. Hence, we have the following two cases.

(1) Let n be an even integer. Then, in the outer cycle, $\frac{n}{2}$ vertices have colour c_1 and the other $\frac{n}{2}$ vertices have the colour c_2. But the central vertex being adjacent to all vertices of the outer cycle must be coloured using a new colour say c_3. Therefore, $\theta(c_1) = \theta(c_2) = \frac{n}{2}$ and $\theta(c_3) = 1$. Hence, $\chi'(G) = 1 \cdot \frac{n}{2} + 2 \cdot \frac{n}{2} + 3 = \frac{3n+6}{2}$.

(2) Let n be an odd integer. Then, in the outer cycle C_n, $\frac{n-1}{2}$ vertices have colour c_1 and $\frac{n-1}{2}$ vertices have the colour c_2 and the remaining one vertex has the colour c_3. As mentioned in the above case, the central vertex v must be coloured using a new colour say c_4. Therefore, $\theta(c_1) = \theta(c_2) = \frac{n-1}{2}$ and $\theta(c_3) = \theta(c_4) = 1$ and hence $\chi'(G) = 1 \cdot \frac{n-1}{2} + 2 \cdot \frac{n-1}{2} + 3 + 4 = \frac{3n+11}{2}$. □

The following result describes the χ-chromatic sum of a complete graph K_n.

Proposition 3.5   The χ-chromatic sum of a complete graph K_n is $\chi'(K_n) = \frac{n(n+1)}{2}$.

Proof   We know that in a proper colouring of K_n, every vertex has distinct colours. That is, $\chi(K_n) = n$. Therefore, $\theta(c_i) = 1$, for all $1 \le i \le n$. Hence, we have $\chi'(K_n) = \sum_{i=1}^{n} i = \frac{n(n+1)}{2}$. □

The χ-chromatic sum of a complete bipartite graph is determined in the following result.

Proposition 3.6   The χ-chromatic sum of a complete bipartite graph $K_{m,n}, m \ge n$ is $\chi'(K_{m,n}) = m + 2n$.

Proof   Assume that G be the complete bipartite graph with a bipartition (X, Y) such that $|X| \ge |Y|$. As a bipartite graph, G is 2-colourable. Since $|X| \ge |Y|$, label every vertex in X by the colour c_1 and every vertex of Y by the colour c_2. Hence, $\theta(c_1) = |X|$ and $\theta(c_2) = |Y|$. Therefore, $\chi'(G) = |X| + 2|Y|$. □

Let us now recall the definition of a Rasta graph defined in Kok, Sudev, and Sudev (in press) as follows.

Definition 3.7 (Kok et al., in press)   For a l-term sum set $\{t_1, t_2, t_3, \ldots, t_l\}$ with $t_1 > t_2 > t_3 > \ldots > t_l > 1$, define the directed graph $G^{(l)}$ with vertices $V(G^{(l)}) = \{v_{i,j} : 1 \le j \le t_i, 1 \le i \le l\}$ and the arcs, $A(G^{(l)}) = \{(v_{i,j}, v_{(i+1),m}) : 1 \le i \le (l-1), 1 \le j \le t_i \text{ and } 1 \le m \le t_{(i+1)}\}$.

In Kok and Sudev (in press), it is shown that for a Rasta graph R corresponding to the underlying graph of $G^{(l)}$ the chromatic number $\varphi(R) = 2$. Assume, without loss of generality, that $\sum_{i=i}^{\lceil \frac{l}{2} \rceil} t_{(2i-1)} \ge \sum_{i=i}^{\lceil \frac{l}{2} \rceil} t_{2i}$ if l is even and $\sum_{i=i}^{\lceil \frac{l}{2} \rceil} t_{(2i-1)} \ge \sum_{i=i}^{\lfloor \frac{l}{2} \rfloor} t_{2i}$ if l is odd. Then, the χ-chromatic sum of R is determined in the following theorem.

THEOREM 3.8   *The χ-chromatic sum of a Rasta graph R is given by*

$$\chi'(R) = \begin{cases} \sum\limits_{i=1}^{\frac{l}{2}} t_{(2i-1)} + 2 \sum\limits_{i=1}^{\frac{l}{2}} t_{2i}, & \text{if } l \text{ is even,} \\ \sum\limits_{i=1}^{\lceil \frac{l}{2} \rceil} t_{(2i-1)} + 2 \sum\limits_{i=1}^{\lfloor \frac{l}{2} \rfloor} t_{2i}, & \text{if } l \text{ is odd.} \end{cases}$$

Proof

(1) Let l be an even integer. Since all vertices corresponding to $t_{(2i-1)}$, $1 \leq i \leq \frac{l}{2}$ are non-adjacent and hence we can colour these vertices by c_1. Also, the remaining vertices, corresponding to t_{2i}, $1 \leq i \leq \frac{l}{2}$ are also non-adjacent among themselves and these vertices can be coloured using the colour c_2. That is, $\theta(c_1) = \sum\limits_{i=1}^{\frac{l}{2}} t_{(2i-1)}$ and $\theta(c_2) = \sum\limits_{i=1}^{\frac{l}{2}} t_{2i}$. Therefore, in this case $\chi'(G) = \sum\limits_{i=1}^{\frac{l}{2}} t_{(2i-1)} + 2 \sum\limits_{i=1}^{\frac{l}{2}} t_{2i}$.

(2) Let l be an odd integer. Then, as explained in the above case, the $\lceil \frac{l}{2} \rceil$ vertices corresponding to $t_{(2i-1)}$; $1 \leq i \leq \lceil \frac{l}{2} \rceil$ are non-adjacent among themselves and hence we can colour these vertices by c_1. The remaining $\lfloor \frac{l}{2} \rfloor$ vertices corresponding to t_{2i}; $1 \leq i \leq \lfloor \frac{l}{2} \rfloor$ are also non-adjacent among themselves and hence we can colour these vertices by c_2. Therefore, $\theta(c_1) = \sum\limits_{i=1}^{\lceil \frac{l}{2} \rceil} t_{(2i-1)}$ and $\theta(c_2) = \sum\limits_{i=1}^{\lfloor \frac{l}{2} \rfloor} t_{2i}$ and hence $\chi'(G) = \sum\limits_{i=1}^{\lceil \frac{l}{2} \rceil} t_{(2i-1)} + 2 \sum\limits_{i=1}^{\lfloor \frac{l}{2} \rfloor} t_{2i}$.   □

4. The χ^+-chromatic sum of certain graphs

We now define a new colouring sum, namely χ^+-chromatic sum of a given graph G as follows.

Definition 4.1 Let $C = \{c_1, c_2, \ldots, c_k\}$ be a proper colouring of a graph G. Then, the χ^+-chromatic sum of a graph G, denoted by $\chi^+(G)$, is defined as $\chi^+(G) = \max \left\{ \sum\limits_{i=1}^{k} i\, \theta(c_i) \right\}$, where the sum varies over all minimum proper colourings of G.

Analogous to the studies on χ-chromatic sum of certain graphs, here we study the χ^+-chromatic sum of the corresponding graphs.

THEOREM 4.2   *For $n \geq 1$, the χ^+-chromatic sum of a path P_n is given by*

$$\chi^+(P_n) = \begin{cases} 1, & \text{if } n = 1, \\ \frac{3n}{2}, & \text{if } n \text{ is even,} \\ \frac{3n+1}{2}, & \text{if } n \text{ is odd.} \end{cases}$$

Proof If $n = 1$, we can assign c_1 to its unique vertex, which shows that $\chi^+(P_n) = 1$. Hence, let $n > 1$. As stated earlier, every path P_n, $n \geq 2$ is 2-colourable. Then, we have to consider the following cases.

(1) If n is even, as mentioned in Theorem 3.2, the vertices can be coloured alternatively by the colours c_1 and c_2 and hence in this case, $\chi^+(P_n) = \frac{3n}{2}$.

(2) If n is odd, then the mutually non-adjacent $\frac{n-1}{2}$ vertices are coloured by c_1 and the remaining mutually non-adjacent $\frac{n+1}{2}$ vertices can be coloured by the colour c_2. Therefore, $\chi^+(P_n) = 1 \cdot \frac{n-1}{2} + 2 \cdot \frac{n+1}{2} = \frac{3n+1}{2}$. This completes the proof.   □

The following is an immediate consequence of Theorem 3.2 and Theorem 4.2.

COROLLARY 4.3 For a path P_n, $n \geq 1$ it follows that, $\chi^+(P_n) = \chi'(P_n)$ if $n = 1$ or even, else $\chi^+(P_n) = \chi'(P_n) + 1$.

In the following result, let us determine the χ^+-chromatic sum of cycles.

THEOREM 4.4   *The χ^+-chromatic sum of a cycle C_n is given by*

$$\chi^+(C_n) = \begin{cases} \frac{3n}{2}, & \text{if } n \text{ is even,} \\ \frac{5n-3}{2}, & \text{if } n \text{ is odd.} \end{cases}$$

Proof As stated earlier, if n is even, then C_n is 2-colourable and if n is odd, C_n is 3-colourable. Then, we have to consider the following cases.

(1) Let n be an even integer. Then, the vertices of C_n can be alternatively coloured by two colours c_1 and c_2. We can see that exactly $\frac{n}{2}$ vertices in C_n have the colours c_1 and c_2 each. Therefore, $\theta(c_1) = \theta(c_2) = \frac{n}{2}$. Therefore, $\chi^+(C_n) = \frac{3n}{2}$.

(2) Let n be an odd integer. Then, we can assign colour c_3 to $\frac{n-1}{2}$ vertices, colour c_2 to $\frac{n-1}{2}$ vertices and colour c_1 to one vertex, which provides a 3-colouring such that $\theta(c_1) = 1, \theta(c_2) = \theta(c_3) = \frac{n-1}{2}$. Therefore, $\chi^+(C_n) = 5 \cdot \frac{n-1}{2} + 1 = \frac{5n-3}{2}$. □

The following theorem describes the χ^+-chromatic sum of a wheel graph W_{n+1}.

THEOREM 4.5 *The χ^+-chromatic sum of a wheel graph W_{n+1} is given by*

$$\chi^+(W_{n+1}) = \begin{cases} \frac{5n+2}{2}, & \text{if } n \text{ is even,} \\ \frac{7n-1}{2}, & \text{if } n \text{ is odd.} \end{cases}$$

Proof Let $v_1, v_2, v_3, \ldots, v_n$ be the vertices of the outer cycle the wheel graph and v be its central vertex. We have already mentioned in Theorem 3.4 that if n is even, then W_{n+1} is 3-colourable and if n is odd, then W_{n+1} is 4-colourable. Then, we have the following cases.

(1) Let n be an even integer. Then, we can assign the colour c_3 to $\frac{n}{2}$ vertices of the outer cycle, the colour c_2 to the remaining $\frac{n}{2}$ vertices of the outer cycle and the colour c_1 to the central vertex. Hence, $\theta(c_3) = \theta(c_2) = \frac{n}{2}$ and $\theta(c_1) = 1$. Therefore, $\chi^+(W_{n+1}) = 3 \cdot \frac{n}{2} + 2 \cdot \frac{n}{2} + 1 = \frac{5n+2}{2}$.

(2) Let n be an odd integer. Then, we can assign colour c_3 to the $\frac{n-1}{2}$ non-adjacent vertices, assign colour c_3 to the $\frac{n-1}{2}$ non-adjacent vertices, colour c_3 for the remaining single vertex and colour c_4 to the central vertex, so that we get $\theta(c_3) = \theta(c_4) = \frac{n-1}{2}, \theta(c_2) = 1$ and $\theta(c_1) = 1$. Therefore, we have $\chi'(W_{n+1}) = 4 \cdot \frac{n-1}{2} + 3 \cdot \frac{n-1}{2} + 2 \cdot 1 + 1 \cdot 1 = \frac{7n-1}{2}$. □

The following result is an obvious and straightforward result on the χ^+-chromatic sum of complete graphs.

Proposition 4.6 The χ^+-chromatic sum of a complete graph K_n is given by $\chi^+(K_n) = \chi'(G) = \frac{n(n+1)}{2}$.

Proof Note that $\chi(K_n) = n$ and hence as mentioned in Theorem 3.8, all vertices have distinct colours. That is, we have $\theta(c_i) = 1$; for all $1 \le i \le n$. Hence, $\chi^+(K_n) = \sum_{i=1}^{n} i = \frac{n(n+1)}{2}$. □

An obvious and straightforward result on the χ^+-chromatic sum of complete bipartite graphs is given below.

THEOREM 4.7 *Consider the χ^+-chromatic sum of a complete bipartite graph $K_{m,n}$, $m \ge n \ge 1$, $\chi^+(K_{m,n}) = 2m + n$.*

Proof Since $n \ge m$ the maximum sum is obtained by allocating colour c_2 to the n non-adjacent vertices and c_1 to the m non-adjacent vertices. So $\theta(c_1) = n$ and $\theta(c_2) = m$. Therefore, $\chi^+(K_{m,n}) = 2m + n$. □

The χ^+-chromatic sum of Rasta graph can be determined as in the following theorem.

THEOREM 4.8 *The χ^+-chromatic sum of Rasta graph R is given by*

$$\chi^+(R) = \begin{cases} 2\sum_{i=1}^{\frac{l}{2}} t_{(2i-1)} + \sum_{i=1}^{\frac{l}{2}} t_{2i}, & \text{if } l \text{ is even,} \\ 2\sum_{i=1}^{\lceil\frac{l}{2}\rceil} t_{(2i-1)} + \sum_{i=1}^{\lfloor\frac{l}{2}\rfloor} t_{2i}, & \text{if } l \text{ is odd.} \end{cases}$$

Proof

(1) Let l be an even integer. Since all $\frac{n}{2}$ vertices, corresponding to $t_{(2i-1)}$, for all $1 \le i \le \frac{l}{2}$ are non-adjacent, these vertices can be coloured using the colour c_2. By the same reason, the colour c_1 is allocated to the vertices corresponding to $t_{2i}, 1 \le i \le \frac{l}{2}$. Hence, $\theta(c_1) = \sum_{i=1}^{\frac{l}{2}} t_{2i}$ and $\theta(c_2) = \sum_{i=1}^{\lceil \frac{l}{2} \rceil} t_{(2i-1)}$. Hence, $\chi^+(R) = 2\sum_{i=1}^{\frac{l}{2}} t_{(2i-1)} + \sum_{i=1}^{\frac{l}{2}} t_{2i}$ for the even values of n.

(2) If l is an odd integer, then the $\frac{n+1}{2}$ mutually non-adjacent vertices can be coloured using c_2 and the remaining $\frac{n-1}{2}$ mutually non-adjacent vertices can be coloured using c_1. Hence, $\chi^+(R) = 2\sum_{i=1}^{\lceil \frac{l}{2} \rceil} t_{(2i-1)} + \sum_{i=1}^{\lfloor \frac{l}{2} \rfloor} t_{2i}$, for the odd values of n. □

5. b^+-Chromatic Sum of Certain Graphs

Analogous to the χ-chromatic sum and χ^+-chromatic sum of graphs, we can also define the b^+ chromatic sum as follows.

Definition 5.1 The b^+-chromatic sum of a graph G, denoted by $\phi^+(G)$, is defined as $\phi^+(G) = \max\{\sum_{i=1}^{k} i\,\theta(c_i)\}$, where the sum varies over a minimal b-colouring using $\phi(G)$ colours.

Now, for determining the respective values of φ^+ for different graph classes, we use the proof techniques followed in Lisna and Sunitha (2015). Reversing the colouring pattern explained in Lisna and Sunitha (2015), we work out the b^+-chromatic sum of given graph classes. Hence, we have the following results.

THEOREM 5.2 *The b^+-chromatic sum of a path $P_n, n \ge 2$ is given by*

$$\varphi^+(P_n) = \begin{cases} \frac{5n-3}{2}, & \text{if } n \ge 5, n \text{ is odd,} \\ \frac{5n-2}{2}, & \text{if } n \ge 6, n \text{ is even,} \\ 5, & \text{if } n = 3, \\ \frac{3n}{2}, & \text{if } n \in \{2,4\}. \end{cases}$$

Proof We know that a b-colouring of a path P_n requires at most three colours. If $1 < n \le 4$, the b-chromatic number of P_n is 2. In this context, the following cases are to be considered.

(1) Let n be even. That is, $n = 2, 4$. If $n = 2$, then, one of its two vertices has colour c_1 and the other vertex has colour c_2. Hence, the b^+-chromatic sum of P_2 is $2 \cdot 1 + 1 \cdot 1 = 3$. If $n = 4$, Let $C_1 = \{v_2, v_4\}$ and $C_2 = \{v_1, v_3\}$ be the colour classes of the colours c_1 and c_2, respectively, so that $C = \{c_1, c_2\}$ is a b-colouring of P_n. Then, the b^+-chromatic sum of P_4 is given by $2 \cdot 2 + 1 \cdot 2 = 6$. Combining these two cases, it follows that $\varphi'(P_n) = \varphi^+(P_n) = \frac{3n}{2}$, for $n = 2, 4$.

(2) Let $n = 3$. Then, let $C_1 = \{v_2\}$ and $C_2 = \{v_1, v_3\}$, so that $C = \{c_1, c_2\}$ is a b^+-colouring of P_n. Then, the b^+-chromatic sum of P_4 is given by $2 \cdot 2 + 1 \cdot 1 = 5$. If $n \ge 5$, the b-chromatic number of a path P_n is 3. Hence, we have to consider the following cases.

(3) Let $n \ge 5$ and n be odd. Now, let $C = \{c_1, c_2, c_3\}$ be a colouring on P_n such that $C_1 = \{v_3\}$ be the colour class of the colour c_1, $C_2 = \{v_2, v_5, v_7, \ldots, v_n\}$ be the colour class of the colour c_2 and $C_3 = \{v_1, v_4, v_6, \ldots, v_{n-1}\}$ be the colour class of colour c_3. Clearly, this colouring is a b^+-colouring of P_n. Then, we have $\theta(c_1) = 1, \theta(c_2) = \frac{n-1}{2}$ and $\theta(c_3) = \frac{n-1}{2}$. Hence, for $n \ge 5$ and n is odd, $\varphi^+(P_n) = \frac{3}{2}(n-1) + \frac{2}{2}(n-1) + 1 = \frac{5n-3}{2}$.

(4) Let $n \ge 5$ and n be even. Here, assume that $C = \{c_1, c_2, c_3\}$ be a colouring on P_n such that the colour classes C_1, C_2 and C_3 are exactly as defined in the previous case. This colouring is obviously a b^+ colouring of P_n. Then, it follows that $\theta(c_1) = 1, \theta(c_2) = \frac{n-2}{2}$ and $\theta(c_3) = \frac{n}{2}$. Hence, for $n \ge 6$, k is even, $\varphi^+(P_n) = 3 \cdot \frac{n}{2} + 2 \cdot \frac{n-2}{2} + 1 = \frac{5n-2}{2}$. □

Similarly, the b^+-chromatic sum of a cycle C_n is determined in the following theorem.

THEOREM 5.3   The b^+-chromatic sum of a cycle C_n is given by

$$
\varphi^+(C_n) = \begin{cases} 6, & \text{if } n = 4, \\ \frac{5n-3}{2}, & \text{if } n \text{ is odd}, \\ \frac{5n-6}{2}, & \text{if } n \text{ is even}, n \neq 4. \end{cases}
$$

Proof   First, let $n = 4$. It is to be noted that the b-chromatic number of the cycle C_4 is 2, where the vertices v_1 and v_3 have colour c_1 and the vertices v_2 and v_4 have the colour c_2. Therefore, the b^+-chromatic sum of C_4 is $2 \cdot 2 + 2 \cdot 1 = 6$.

Next, assume that $n \neq 4$. We know that the b-chromatic number of a cycle $C_n, n \neq 4$ is 3. Let $C = \{c_1, c_2, c_3\}$ be a b-colouring of a given cycle C_n. Here, we have to consider the following cases.

  (1) Let n be odd. Now a b-colouring which forms the colour classes, $C_1 = \{v_3\}, C_2 = \{v_1, v_4, v_6, \ldots, v_{n-1}\}$ and $C_3 = \{v_2, v_5, v_7, \ldots, v_n\}$, yield the desirable b-colouring such that $\theta(c_1) = 1$, $\theta(c_2) = \frac{n-1}{2}$ and $\theta(c_3) = \frac{n-1}{2}$. Therefore, here the b^+-chromatic sum is given by $3 \cdot \frac{n-1}{2} + 2 \cdot \frac{n-1}{2} + 1 \cdot 1 = \frac{5n-3}{2}$.

  (2) Let n be even. Now, a b-colouring which forms the colour classes, $C_1 = \{v_3, v_n\}$, $C_2 = \{v_1, v_4, v_6, \ldots, v_{n-2}\}$ and $C_3 = \{v_2, v_5, v_7, \ldots, v_{n-1}\}$, yield the desirable b-colouring such that $\theta(c_1) = 2, \theta(c_2) = \frac{n-2}{2}$ and $\theta(c_3) = \frac{n-2}{2}$. Therefore, we have $\varphi^+(C_n) = 3 \cdot \frac{n-2}{2} + 2 \cdot \frac{n-2}{2} + 2 = \frac{5n-6}{2}$. This completes the proof. $\square$

Now, the b^+-chromatic sum of a wheel graph W_{n+1} is determined in the following result.

THEOREM 5.4   The b^+-chromatic sum of a wheel graph W_{n+1} is given by

$$
\varphi^+(W_{n+1}) = \begin{cases} 11, & \text{if } n = 4, \\ \frac{7n-1}{2}, & \text{if } n \text{ is odd}, \\ \frac{7n-4}{2}, & \text{if } n \text{ is even}, n \neq 4. \end{cases}
$$

Proof   We have already stated that the b-chromatic number of the cycle C_4 is 3. Therefore, a b-colouring of $W_5 = C_4 + K_1$ must contain 3 colours, say c_1, c_2 and c_3. Let the corresponding colour classes be $C_1 = \{v\}, C_2 = \{v_1, v_3\}$ and $C_3 = \{v_2, v_4\}$, where v is the central vertex of the wheel graph. Then, $\theta(c_1) = 1$, $\theta(c_2) = 2$ and $\theta(c_3) = 2$. Hence, $\varphi^+(W_5) = 1 \cdot 1 + 2 \cdot 2 + 3 \cdot 2 = 11$. Next, assume that $n \neq 4$. Then, every b-colouring of W_{n+1} must contain 4 colours. Let $C = \{c_1, c_2, c_3, c_4\}$ be the required colouring of G. Then, we have to consider the following cases.

  (1) Assume that n is odd. Then, colour the vertices of W_{n+1} using the colours in C in such a way that the corresponding colour classes are $C_1 = \{v\}$, $C_2 = \{v_3\} C_3 = \{v_1, v_4, v_6, \ldots, v_{n-1}\}$ and $C_4 = \{v_2, v_5, v_7, \ldots, v_n\}$. Therefore, we have $\theta(c_1) = \theta(c_2) = 1$ and $\theta(c_3) = \frac{n-1}{2}$ and $\theta(c_4) = \frac{n-1}{2}$. Then, we have $\varphi^+(W_{n+1}) = 4 \cdot \frac{n-1}{2} + 3 \cdot \frac{n-1}{2} + 2 + 1 = \frac{7n-1}{2}$.

  (2) Assume that n is even. Colour the vertices of W_{n+1} in such a way that the corresponding colour classes are $C_1 = \{v\}, C_2 = \{v_3, v_n\} C_3 = \{v_1, v_4, v_6, \ldots, v_{n-2}\}$ and $C_4 = \{v_2, v_5, v_7, \ldots, v_{n-1}\}$. Then, we have $\theta(c_1) = 1$, $\theta(c_2) = 2$ and $\theta(c_3) = \theta(c_4) = \frac{n-2}{2}$. Hence, $\varphi^+(W_{n+1}) = 4 \cdot \frac{n-2}{2} + 3 \cdot \frac{n-2}{2} + 2 \cdot 2 + 1 = \frac{7n-4}{2}$. $\square$

The following theorem describes the φ^+-chromatic number of a complete bipartite graph.

THEOREM 5.5   The b^+-chromatic sum of a complete bipartite graph $K_{m,n}$, $m \geq n$ is $\varphi^+(K_{m,n}) = 2m + n$.

Proof   The result follows directly from the proof of Theorem 4.7. $\square$

The b-chromatic sum and the b^+-chromatic sum of Rasta Graph R is determined in the theorem given below.

THEOREM 5.6   *The b-chromatic sum of a Rasta graph R is given by*

$$\varphi'(R) = \begin{cases} \sum\limits_{i=1}^{\lceil\frac{l}{2}\rceil} t_{(2i-1)} + 2\sum\limits_{i=1}^{\lceil\frac{l}{2}\rceil} t_{2i}, & \text{if } l \text{ is even,} \\ \sum\limits_{i=1}^{\lceil\frac{l}{2}\rceil} t_{(2i-1)} + 2\sum\limits_{i=1}^{\lfloor\frac{l}{2}\rfloor} t_{2i}, & \text{if } l \text{ is odd,} \end{cases}$$

and the b^+-chromatic sum of R is given by

$$\varphi^+(R) = \begin{cases} 2\sum\limits_{i=1}^{\lceil\frac{l}{2}\rceil} t_{(2i-1)} + \sum\limits_{i=1}^{\lceil\frac{l}{2}\rceil} t_{2i}, & \text{if } l \text{ is even,} \\ 2\sum\limits_{i=1}^{\lceil\frac{l}{2}\rceil} t_{(2i-1)} + \sum\limits_{i=1}^{\lfloor\frac{l}{2}\rfloor} t_{2i}, & \text{if } l \text{ is odd.} \end{cases}$$

Proof   The proof follows directly from the proofs of Theorem 3.8 and 4.8.   □

6. Two Thue chromatic sums of a path
A finite sequence $S = (q_1, q_2, q_3, \ldots, q_t)$ of symbols of any alphabet is known to be *non-repetitive* if for all its subsequences $(r_1, r_2, r_3, \ldots, r_{2m})$; $1 \le m \le \frac{t}{2}$, the condition $r_i \ne r_{2i}, \forall 1 \le i \le m$, holds.

Let G be a simple undirected graph on n vertices and let a minimum set of colours C allow a proper vertex colouring of G. If the sequence of vertex colours of any path of even and finite length in G is non-repetitive, then this proper colouring is said to be a *Thue colouring* of G (see Alon, Grytczuk, Hauszczak & Riordan, 2002).

The *Thue chromatic number* of G, denoted $\pi(G)$, is defined as the minimum number of colours required for a Thue colouring of G.

It is known that $\pi(P_1) = 1$, $\pi(P_2) = \pi(P_3) = 2$ and for $n \ge 4$, $\pi(P_n) = 3$. Determining $\pi'(P_n)$ is a hard problem, hence the problem is very hard for graphs in general.

The following lemma is the motivation for our further discussions in this paper.

LEMMA 6.1   Škrabul'áková, in press *Up to equivalence, there is exactly one non-repetitive 3-colouring of the cycle C_{11}.*

In view of this lemma, we restrict our further discussion to the path P_{11}. Let the vertices of P_n be labelled from left to right to be $v_1, v_2, v_3, \ldots v_{11}$. Recall that the *colouring sum* of a colouring S is defined by $\omega(S) = \sum\limits_{i=1}^{k} i\,\theta(c_i)$. The possible minimum Thue colourings of P_{11} are listed below.

(1) $S_1 = (c_1, c_2, c_1, c_3, c_1, c_2, c_3, c_1, c_3, c_2, c_3)$ and $\omega(S_1) = 22$

(2) $S_2 = (c_1, c_2, c_1, c_3, c_1, c_2, c_3, c_2, c_1, c_2, c_3)$ and $\omega(S_2) = 21$

(3) $S_3 = (c_1, c_2, c_1, c_3, c_2, c_1, c_2, c_3, c_2, c_1, c_3)$ and $\omega(S_3) = 21$

(4) $S_4 = (c_1, c_2, c_1, c_3, c_2, c_3, c_1, c_3, c_2, c_1, c_3)$ and $\omega(S_4) = 22$

(5) $S_5 = (c_1, c_2, c_1, c_3, c_1, c_2, c_3, c_1, c_3, c_2, c_1)$ and $\omega(S_5) = 20$

(6) $S_6 = (c_1, c_2, c_1, c_3, c_2, c_3, c_1, c_3, c_2, c_1, c_2)$ and $\omega(S_6) = 21$

(7) $S_7 = (c_2, c_1, c_2, c_3, c_2, c_1, c_3, c_2, c_3, c_1, c_3)$ and $\omega(S_7) = 23$

(8) $S_8 = (c_2, c_1, c_2, c_3, c_2, c_1, c_3, c_1, c_2, c_1, c_3)$ and $\omega(S_8) = 21$

(9) $S_9 = (c_2, c_1, c_2, c_3, c_1, c_2, c_1, c_3, c_1, c_2, c_3)$ and $\omega(S_9) = 21$

(10) $S_{10} = (c_2, c_1, c_2, c_3, c_1, c_3, c_2, c_3, c_1, c_2, c_3)$ and $\omega(S_{10}) = 23$

(11) $S_{11} = (c_2, c_1, c_2, c_3, c_2, c_1, c_3, c_2, c_3, c_1, c_2)$ and $\omega(S_{11}) = 22$

(12) $S_{12} = (c_2, c_1, c_2, c_3, c_1, c_3, c_2, c_3, c_1, c_2, c_1)$ and $\omega(S_{12}) = 21$

(13) $S_{13} = (c_3, c_2, c_3, c_1, c_3, c_2, c_1, c_3, c_1, c_2, c_1)$ and $\omega(S_{13}) = 22$

(14) $S_{14} = (c_3, c_2, c_3, c_1, c_3, c_2, c_1, c_2, c_3, c_2, c_1)$ and $\omega(S_{14}) = 23$

(15) $S_{15} = (c_3, c_2, c_3, c_1, c_2, c_3, c_2, c_1, c_2, c_3, c_1)$ and $\omega(S_{15}) = 23$

(16) $S_{16} = (c_3, c_2, c_3, c_1, c_2, c_1, c_3, c_1, c_2, c_3, c_1)$ and $\omega(S_{16}) = 22$

(17) $S_{17} = (c_3, c_2, c_3, c_1, c_3, c_2, c_1, c_3, c_1, c_2, c_3)$ and $\omega(S_{17}) = 24$

(18) $S_{18} = (c_3, c_2, c_3, c_1, c_2, c_1, c_3, c_1, c_2, c_3, c_2)$ and $\omega(S_{18}) = 23$

(19) $S_{19} = (c_1, c_3, c_1, c_2, c_1, c_3, c_2, c_1, c_2, c_3, c_2)$ and $\omega(S_{19}) = 21$

(20) $S_{20} = (c_1, c_3, c_1, c_2, c_1, c_3, c_2, c_3, c_1, c_3, c_2)$ and $\omega(S_{20}) = 22$

(21) $S_{21} = (c_1, c_3, c_1, c_2, c_3, c_1, c_3, c_2, c_3, c_1, c_2)$ and $\omega(S_{21}) = 22$

(22) $S_{22} = (c_1, c_3, c_1, c_2, c_3, c_2, c_1, c_2, c_3, c_1, c_2)$ and $\omega(S_{22}) = 21$

(23) $S_{23} = (c_1, c_3, c_1, c_2, c_1, c_3, c_2, c_1, c_2, c_3, c_1)$ and $\omega(S_{23}) = 20$

(24) $S_{24} = (c_1, c_3, c_1, c_2, c_3, c_2, c_1, c_2, c_3, c_1, c_3)$ and $\omega(S_{24}) = 22$ From the above list, we note that $\pi'(P_{11}) = 20$ and $\pi^+(P_{11}) = 24$. We strongly believe that the next conjecture holds. Conjecture 6.2 For a path P_n, $n \geq 4$, there is a unique permutation over all proper b-colourings for which $\varphi^+(P_n)$ is obtained, and exactly two permutations for which $\varphi'(P_n)$ is obtained.

The following general result is of importance for all variations of colouring sums discussed thus far. It holds for improper colourings as well. A general colouring which meets some general colouring index is called the ϑ-chromatic number of G and denoted, $\vartheta(G)$.

Theorem 6.3 (Makungu's Theorem[1]) If the set of colours $C = \{c_j : 1 \leq j \leq k\}$ allows a general colouring, $S : f(v_i) = c_l$, $l \in \{1, 2, 3, \ldots, k\}$ of G, such that $\omega(S) = \vartheta'(G) = \min\{\sum_{i=1}^{k} i \cdot \theta(c_i) : \forall S\text{-colourings of } G\}$, then $\vartheta^+(G) = \sum_{i=1}^{k} i \cdot \theta(c_{(k+1)-i})$.

Proof   Since for $a_1 \geq a_2$ it follows that $1 \cdot a_1 + 2 \cdot a_2 \leq 2 \cdot a_1 + 1 \cdot a_2$, it follows through immediate induction that if $a_1 \geq a_2 \geq a_3 \geq \cdots \geq a_k$ then for permuted one-on-one allocations of the elements in $b_i \in \{1, 2, 3, \cdots, k\}$ to form $\sum_{i=1}^{k} a_i b_i$ we have, $\min\{\sum_{i=1}^{k} a_i b_i\} = \sum_{i=1}^{k} i \cdot a_i$ and $\max\{\sum_{i=1}^{k} a_i b_i\} = \sum_{i=1}^{k} i \cdot a_{(k+1)-i}$. Hence, if a ϑ-colouring of G is allowed by $C = \{c_1, c_2, c_3, \ldots, c_k\}$ such that, $\theta(c_1) \geq \theta(c_2) \geq \theta(c_3) \geq \ldots \geq \theta(c_k)$ then, $\vartheta'(G) = \sum_{i=1}^{k} i \cdot \theta(c_i)$ and $\vartheta^+(G) = \sum_{i=1}^{k} i \cdot \theta(c_{(k+1)-i})$. $\square$

Acknowledgements
The authors gratefully acknowledge the contributions of anonymous referees whose critical and constructive comments played a vital role in improving the quality and presentation style of the paper in a significant way.

Funding
The authors received no direct funding for this research.

Author details
Johan Kok[1]
E-mail: kokkiek2@tshwane.gov.za
ORCID ID: http://orcid.org/0000-0003-0106-1676
N.K. Sudev[2]
E-mail: sudevnk@gmail.com
ORCID ID: http://orcid.org/0000-0001-9692-4053
K.P. Chithra[3]
E-mail: chithrasudev@gmail.com
ORCID ID: http://orcid.org/000-0003-3871-1345
[1]Tshwane Metropolitan Police Department, City of Tshwane, Republic of South Africa.
[2]Department of Mathematics, Vidya Academy of Science & Technology, Thalakkottukara, Thrissur, 680501, India.
[3]Naduvath Mana, Nandikkara, Thrissur, 680301, India.

Note
1 The first author dedicates this theorem to Makungu Mathebula and he hopes this young lady will grow up with a deep fondness for mathematics.

References
Alon, N., Grytczuk, J., Hauszczak, M., & Riordan, O. (2002). Non-repetitive colourings of graphs. *Random Structures & Algorithms, 21,* 336–346. doi:10.1002/rsa.10057

Bondy, J. A., & Murty, U. S. R. (1976). *Graph theory with applications.* London: Macmillan Press.

Chartrand, G., & Lesniak, L. (2000). *Graphs and digraphs.* Boca Raton, FL: CRC Press.

Chartrand, G., & Zhang, P. (2009). *Chromatic graph theory.* Boca Raton, FL: CRC Press.

Effatin, B., & Kheddouci, H. (2003). The b-chromatic number of some power graphs. *Discrete Mathematics & Theoretical Computer Science, 6,* 45–54.

Gross, J. T., & Yellen, J. (2006). *Graph theory and its applications.* Boca Raton, FL: CRC Press.

Harary, F. (1969). *Graph theory.* Philippines: Addison Wesley.

Irving, R. W., & Manlove, D. F. (1999). The b-chromatic number of a graph. *Discrete Applied Mathematics, 91,* 127–141.

Kok, J., Sudev, N. K., & Sudev, C. (in press). On the curling number of certain graphs. arXiv: 1506.00813v2.

Kok, J., & Sudev, N. K. (in press). The b-chromatic numbers of certain graphs and digraphs. arXiv: 1511.00680.

Kouider, M., & Mahéo, M. (2002). Some bounds for the b-chromatic number of a graph. *Discrete Mathematics, 256,* 267–277.

Lisna, P. C., & Sunitha, M. S. (2015). b-Chromatic sum of a graph. *Discrete Mathematics, Algorithms and Applications, 7*(3), 1–15. doi:10.1142/S1793830915500408

Škrabuľáková, E. (in press). Thue choice number versus Thue chromatic number of graphs. arXiv:1508.02559v1.

Vaidya, S. K., & Isaac, R. V. (2014). The b-chromatic number of some path related graphs. *International Journal of Computing Science and Mathematics, 4,* 7–12.

Vaidya, S. K., & Isaac, R. V. (2015). The b-chromatic number of some graphs. *International Journal of Mathematics and Soft Computing, 5,* 165–169.

Vivin, J. V., & Vekatachalam, M. (2015). On b-chromatic number of sunlet graph and wheel graph families. *Journal of the Egyptian Mathematical Society, 23,* 215–222.

West, D. B. (2001). *Introduction to graph theory.* Delhi: Pearson Education.

4

On I-convergent sequence spaces defined by a compact operator and a modulus function

Vakeel A. Khan[1]*, Mohd Shafiq[1] and Rami Kamel Ahmad Rababah[1]

*Corresponding author: Vakeel A. Khan, Department of Mathematics, Aligarh Muslim University, Aligarh 202002, India
E-mail: vakhanmaths@gmail.com

Reviewing editor: Lishan Liu, Qufu Normal University, China

Abstract: In this article, we introduce and study I-convergent sequence spaces $S^I(f)$, $S_0^I(f)$, and $S_\infty^I(f)$ with the help of compact operator T on the real space $\mathbb{R}$ and a modulus function f. We study some topological and algebraic properties, and prove some inclusion relations on these spaces.

Subjects: Science; Pure Mathematics; Advanced Mathematics; Mathematics & Statistics

Keywords: compact operator; ideal; filter; I-convergent sequence; I-null sequence; I-bounded sequence; solid and monotone space; symmetric space; modulus function

AMS subject classifications: 41A10; 41A25; 41A36; 40A30

1. Introduction and preliminaries

Let $\mathbb{N}, \mathbb{R}$, and $\mathbb{C}$ be the sets of all natural, real, and complex numbers, respectively. We denote

$$\omega = \{x = (x_k) : x_k \in \mathbb{R} \text{ or} \mathbb{C}\}$$

the space of all real or complex sequences.

ABOUT THE AUTHORS

Vakeel A. Khan received his MPhil and PhD degrees in Mathematics from Aligarh Muslim University, Aligarh, India. Currently, he is a senior assistant professor in the same university. He is a vigorous researcher in the area of Sequence Spaces , and has published a number of research papers in reputed national and international journals, including Numerical Functional Analysis and Optimization (Taylor's and Francis), Information Sciences (Elsevier), Applied Mathematics A Journal of Chinese Universities and Springer-Verlag (China), Rendiconti del Circolo Matematico di Palermo Springer-Verlag (Italy), Studia Mathematica (Poland), Filomat (Serbia), Applied Mathematics & Information Sciences (USA).

Mohd Shafiq did his MSc in Mathematics from University of Jammu, Jammu and Kashmir, India. Currently, he is a PhD scholar at Aligarh Muslim University, Aligarh, India. His research interests are Functional Analysis, sequence spaces, I-convergence, invariant means, zweier sequences, and interval numbers theory.

Rami Kamel Ahmad Rababah is a research scholar in the Department of Mathematics, Aligarh Muslim University, Aligarh, India.

PUBLIC INTEREST STATEMENT

The term sequence has a great role in Analysis. Sequence spaces play an important role in various fields of Real Analysis, Complex Analysis, Functional Analysis, and Topology Convergence of sequences has always remained a subject of interest to the researchers. Several new types of convergence of sequences were studied by the researchers and named them as usual convergence, uniform convergence, strong convergence,week convergence, etc. Later, the idea of statistical convergence came into existence which is the generalization of usual convergence. Statistical convergence has several applications in different fields of Mathematics like Number Theory, Trigonometric Series, Summability Theory, Probability Theory, Measure Theory, Optimization, and Approximation Theory. The notion of Ideal convergence (I-convergence) is a generalization of the statistical convergence and equally considered by the researchers for their research purposes since its inception.

Let ℓ_∞, c, and c_0 be denote the Banach spaces of bounded, convergent, and null sequences of reals, respectively with norm

$$\|x\| = \sup_k |x_k|$$

Any subspace λ of ω is called a sequence space. A sequence space λ with linear topology is called a K-space provided each of maps $p_i \to \mathbb{C}$ defined by $p_i(x) = x_i$ is continuous, for all $i \in \mathbb{N}$. A space λ is called an FK-space provided λ is complete linear metric space. An FK-space whose topology is normable is called a BK-space.

Definition 1.1   Let X and Y be two normed linear spaces and $T : D(T) \to Y$ be a linear operator, where $D(T) \subset X$. Then, the operator T is said to be bounded, if there exists a positive real k such that

$$\| Tx \| \le k \| x \|, \quad \text{for all } x \in D(T)$$

The set of all bounded linear operators $B(X, Y)$ is a normed linear space normed by (see Kreyszig, 1978)

$$\| T \| = \sup_{x \in X, \|x\|=1} \| Tx \|$$

and $B(X, Y)$ is a Banach space if Y is Banach space.

Definition 1.2   Let X and Y be two normed linear spaces. An operator $T : X \to Y$ is said to be a compact linear operator (or completely continuous linear operator), if

(1) T is linear,
(2) T maps every bounded sequence (x_k) in X onto a sequence $T(x_k)$ in Y which has a convergent subsequence.

The set of all compact linear operators $C(X, Y)$ is closed subspace of $B(X, Y)$ and $C(X, Y)$ is a Banach space if Y is Banach space.

Following Basar and Altay (2003) and Şengönül (2009), we introduce the sequence spaces S and S_0 with the help of compact operator T on the real space $\mathbb{R}$ as follows.

$$S = \left\{ x = (x_k) \in \ell_\infty : T(x) \in c \right\}$$

and

$$S_0 = \left\{ x = (x_k) \in \ell_\infty : T(x) \in c_0 \right\}$$

Definition 1.3   A function $f : [0, \infty) \longrightarrow [0, \infty)$ is called a modulus if

(1) $f(t) = 0$ if and only if $t = 0$,
(2) $f(t + u) \le f(t) + f(u)$ for all $t, u \ge 0$,
(3) f is increasing, and
(4) f is continuous from the right at zero.

For any modulus function f, we have the inequalities

$$| f(x) - f(y) | \le f(| x - y |)$$

and

$$f(nx) \le nf(x), \quad \text{for all } x, y \in [0, \infty]$$

A modulus function f is said to satisfy $\Delta_2 -$ Condition for all values of u if there exists a constant $K > 0$ such that $f(Lu) \le KLf(u)$ for all values of $L > 1$.

The idea of modulus was introduced by Nakano (1953).

Ruckle (1967, 1968, 1973) used the idea of a modulus function f to construct the sequence space

$$X(f) = \left\{ x = (x_k) : \sum_{k=1}^{\infty} f(|x_k|) < \infty \right\}$$

This space is an FK-space and Ruckle (1967, 1968, 1973) proved that the intersection of all such $X(f)$ spaces is ϕ, the space of all finite sequences.

The space $X(f)$ is closely related to the space ℓ_1 which is an $X(f)$ space with $f(x) = x$ for all real $x \geq 0$. Thus Ruckle (1967, 1968, 1973) proved that, for any modulus f.

$X(f) \subset \ell_1$ and $X(f)^\alpha = \ell_\infty$

where

$$X(f)^\alpha = \left\{ y = (y_k) \in \omega : \sum_{k=1}^{\infty} f(|y_k x_k|) < \infty \right\}$$

Spaces of the type $X(f)$ are a special case of the spaces structured by Gramsch (1967). From the point of view of local convexity, spaces of the type $X(f)$ are quite pathological. Symmetric sequence spaces, which are locally convex have been frequently studied by Garling (1966), Köthe (1970), and Ruckle (1967, 1968, 1973).

The sequence spaces by the use of modulus function was further investigated by Maddox (1969, 1986), Khan (2005, 2006), Bhardwaj (2003), and many others.

As a generalization of usual convergence, the concept of statistical convergent was first introduced by Fast (1951) and also independently by Buck (1953) and Schoenberg (1959) for real and complex sequences. Later on, it was further investigated from sequence space point of view and linked with the Summability Theory by Fridy (1985), Šalát (1980), Tripathy (1998), Khan (2007), Khan and Sabiha (2012), Khan, Shafiq, and Rababah (2015), and many others.

Definition 1.4 A sequence $x = (x_k) \in \omega$ is said to be statistically convergent to a limit $L \in \mathbb{C}$ if for every $\varepsilon > 0$, we have

$$\lim_{k \to \infty} \frac{1}{k} |\{n \in \mathbb{N} : |x_n - L| \geq \varepsilon, n \leq k\}| = 0$$

where vertical lines denote the cardinality of the enclosed set.

That is, if $\delta(A(\varepsilon)) = 0$, where

$$A(\varepsilon) = \{k \in \mathbb{N} : |x_k - L| > \varepsilon\}$$

The notation of ideal convergence (I-convergence) was introduced and studied by Kostyrko, Mačaj, Salát, and Wilczyński (2000). Later on, it was studied by Šalát, Tripathy, and Ziman (2004, 2005), Tripathy and Hazarika (2009, 2011), Khan and Ebadullah (2011), Khan, Ebadullah, Esi, and Shafiq (2013), and many others.

Now, we recall the following definitions:

Definition 1.5 Let $\mathbb{N}$ be a non-empty set. Then a family of sets $I \subseteq 2^\mathbb{N}$ (power set of $\mathbb{N}$) is said to be an ideal if

(1) I is additive i.e $\forall A, B \in I \Rightarrow A \cup B \in I$

(2) I is hereditary i.e $\forall A \in I$ and $B \subseteq A \Rightarrow B \in I$.

Definition 1.6   A non-empty family of sets $£(I) \subseteq 2^{\mathbb{N}}$ is said to be filter on $\mathbb{N}$ if and only if

(1) $\Phi \notin £(I)$,

(2) $\forall A, B \in £(I)$ we have $A \cap B \in £(I)$,

(3) $\forall A \in £(I)$ and $A \subseteq B \Rightarrow B \in £(I)$.

Definition 1.7   An Ideal $I \subseteq 2^{\mathbb{N}}$ is called non-trivial if $I \neq 2^{\mathbb{N}}$.

Definition 1.8   A non-trivial ideal $I \subseteq 2^{\mathbb{N}}$ is called admissible if

$\{\{x\} : x \in \mathbb{N}\} \subseteq I$

Definition 1.9   A non-trivial ideal I is maximal if there cannot exist any non-trivial ideal $J \neq I$ containing I as a subset.

Remark 1.10   For each ideal I, there is a filter $£(I)$ corresponding to I. i.e $£(I) = \{K \subseteq \mathbb{N} : K^c \in I\}$, where $K^c = \mathbb{N} \setminus K$.

Definition 1.11   A sequence $x = (x_k) \in \omega$ is said to be I-convergent to a number L if for every $\varepsilon > 0$, the set $\{k \in \mathbb{N} : |x_k - L| \geq \varepsilon\} \in I$.

In this case, we write $I - \lim x_k = L$.

Definition 1.12   A sequence $x = (x_k) \in \omega$ is said to be I-null if $L = 0$. In this case, we write $I - \lim x_k = 0$.

Definition 1.13   A sequence $x = (x_k) \in \omega$ is said to be I-Cauchy if for every $\varepsilon > 0$ there exists a number $m = m(\varepsilon)$ such that $\{k \in \mathbb{N} : |x_k - x_m| \geq \varepsilon\} \in I$.

Definition 1.14   A sequence $x = (x_k) \in \omega$ is said to be I-bounded if there exists some $M > 0$ such that $\{k \in \mathbb{N} : |x_k| \geq M\} \in I$.

Definition 1.15   A sequence space E is said to be solid (normal) if $(\alpha_k x_k) \in E$ whenever $(x_k) \in E$ and for any sequence (α_k) of scalars with $|\alpha_k| \leq 1$, for all $k \in \mathbb{N}$.

Definition 1.16   A sequence space E is said to be symmetric if $(x_{\pi(k)}) \in E$ whenever $x_k \in E$. where π is a permutation on $\mathbb{N}$.

Definition 1.17   A sequence space E is said to be sequence algebra if $(x_k) * (y_k) = (x_k . y_k) \in E$ whenever $(x_k), (y_k) \in E$.

Definition 1.18   A sequence space E is said to be convergence free if $(y_k) \in E$ whenever $(x_k) \in E$ and $x_k = 0$ implies $y_k = 0$, for all k.

Definition 1.19   Let $K = \{k_1 < k_2 < k_3 < k_4 < k_5 \ldots\} \subset \mathbb{N}$ and E be a Sequence space. A K-step space of E is a sequence space $\lambda_K^E = \{(x_{k_n}) \in \omega : (x_k) \in E\}$.

Definition 1.20   A canonical pre-image of a sequence $(x_{k_n}) \in \lambda_K^E$ is a sequence $(y_k) \in \omega$ defined by

$$y_k = \begin{cases} x_k, & \text{if } k \in K, \\ 0, & \text{otherwise} \end{cases}$$

A canonical preimage of a step space λ_K^E is a set of preimages all elements in λ_K^E. i.e. y is in the canonical preimage of λ_K^E iff y is the canonical preimage of some $x \in \lambda_K^E$.

Definition 1.21   A sequence space E is said to be monotone if it contains the canonical preimages of its step space.

Definition 1.22   (see, Khan et al., 2015; Kostyrko et al., 2000). If $I = I_f$, the class of all finite subsets of v. Then, I is an admissible ideal in v and I_f convergence coincides with the usual convergence.

Definition 1.23   (see, Khan et al., 2015; Kostyrko et al., 2000). If $I = I_\delta = \{A \subseteq \mathbb{N} : \delta(A) = 0\}$. Then, I is an admissible ideal in $\mathbb{N}$ and we call the I_δ-convergence as the logarithmic statistical convergence.

Definition 1.24   (see, Khan et al., 2015; Kostyrko et al., 2000). If $I = I_d = \{A \subseteq \mathbb{N} : d(A) = 0\}$. Then, I is an admissible ideal in $\mathbb{N}$ and we call the I_d-convergence as the asymptotic statistical convergence.

Remark 1.25   If $I_\delta - \lim x_k = l$, then $I_d - \lim x_k = l$

Definition 1.26   A map $\hbar$ defined on a domain $D \subset X$ i.e $\hbar : D \subset X \to \mathbb{R}$ is said to satisfy Lipschitz condition if $|\hbar(x) - \hbar(y)| \leq K|x - y|$ where K is known as the Lipschitz constant. The class of K-Lipschitz functions defined on D is denoted by $\hbar \in (D, K)$.

Definition 1.27   A convergence field of I-covergence is a set

$F(I) = \{x = (x_k) \in l_\infty : \text{there exists } I - \lim x \in \mathbb{R}\}$

The convergence field $F(I)$ is a closed linear subspace of l_∞ with respect to the supremum norm, $F(I) = l_\infty \cap c^I$ (see Šalát et al., 2004, 2005).

Definition 1.28   Let X be a linear space. A function $g : X \longrightarrow R$ is called paranorm, if for all $x, y \in X$,

$(P_1)\, g(x) = 0$ if $x = \theta$,
$(P_2)\, g(-x) = g(x)$,
$(P_3)\, g(x + y) \leq g(x) + g(y)$,
(P_4) If (λ_n) is a sequence of scalars with $\lambda_n \to \lambda$ $(n \to \infty)$ and $x_n, a \in X$ with $x_n \to a$ $(n \to \infty)$ in the sense that $g(x_n - a) \to 0$ $(n \to \infty)$, then $g(\lambda_n x_n - \lambda a) \to 0$ $(n \to \infty)$.

The notation of paranorm sequence spaces was studied at the initial stage by Nakano (1953). Later on, it was further investigated by Maddox (1969), Tripathy and Hazarika (2009), Khan et al. (2013), and the references therein.

Throughout the article, we use the same techniques as used in Tripathy and Hazarika (2009, 2011).

We used the following lemmas for establishing some results of this article.

LEMMA 1   (see, Tripathy & Hazarika, 2009, 2011). *Every solid space is monotone.*

LEMMA 2   (see, Tripathy & Hazarika, 2009, 2011). *If $I \subseteq 2^N$ and $M \subseteq N$. If $M \notin I$, then $M \cap N \notin I$.*

LEMMA 3   (see, Tripathy & Hazarika, 2009, 2011). *Let $K \in \pounds(I)$ and $M \subseteq N$. If $M \notin I$, then $M \cap K \notin I$.*

Throughout the article, S^I, S_0^I, S_∞^I, $\mathcal{M}_S^I$, and $\mathcal{M}_{S_0}^I$ represent the I-convergent, I-null, I-Bounded, bounded I-convergent, and bounded I-null Sequences spaces defined by a compact operator T on the real space $\mathbb{R}$, respectively.

2. Main results
In this article, we introduce the following classes of sequences.

$$S^I(f) = \left\{ x = (x_k) \in \ell_\infty : \{k \in \mathbb{N} : f\Big(\mid T(x_k) - L \mid \Big) \geq \epsilon\} \in I, \text{ for some } L \in \mathbb{C} \right\} \tag{2.1}$$

$$S_0^I(f) = \left\{ x = (x_k) \in \ell_\infty : \{k \in \mathbb{N} : f\Big(\mid (Tx_k) \mid \Big) \geq \epsilon\} \in I \right\} \tag{2.2}$$

$$S_\infty^I(f) = \left\{ x = (x_k) \in \ell_\infty : \{k \in \mathbb{N} : \exists K < 0 \text{ such that } f\left(\,|\,T(x_k)\,|\,\right) \geq K\} \in I \right\} \qquad (2.3)$$

$$S_\infty(f) = \left\{ x = (x_k) \in \ell_\infty : \sup_k f\left(\,|\,T(x_k)\,|\,\right) < \infty \right\} \qquad (2.4)$$

where f is a modulus function.

We also denote

THEOREM 2.1   Let f be a modulus function. Then, the classes of sequences $S^I(f)$, $S_0^I(f)$, $\mathcal{M}_S^I(f)$, and $\mathcal{M}_{S_0}^I(f)$ are linear spaces.

Proof   We shall prove the result for $S^I(f)$. The proof for the other spaces will follow similarly. For, let $x = (x_k)$, $y = (y_k) \in S^I(f)$ and α, β be scalars. Then, for a given $\epsilon > 0$, we have

$$\left\{ k \in \mathbb{N} : f\left(\,|\,T(x_k) - L_1\,|\,\right) \geq \frac{\epsilon}{2}, \quad \text{for some } L_1 \in \mathbb{C} \right\} \in I \qquad (2.5)$$

$$\left\{ k \in \mathbb{N} : f\left(\,|\,T(x_k) - L_2\,|\,\right) \geq \frac{\epsilon}{2}, \quad \text{for some } L_1 \in \mathbb{C} \right\} \in I \qquad (2.6)$$

Let

$$A_1 = \left\{ k \in \mathbb{N} : f\left(|T(x_k) - L_1|\right) < \frac{\epsilon}{2}, \quad \text{for some } L_1 \in \mathbb{C} \right\} \in \mathcal{L}(I) \qquad (2.7)$$

$$A_2 = \left\{ k \in \mathbb{N} : f\left(|T(y_k) - L_2|\right) < \frac{\epsilon}{2}, \quad \text{for some } L_2 \in \mathbb{C} \right\} \in \mathcal{L}(I) \qquad (2.8)$$

be such that $A_1^c, A_2^c \in I$.

Since f is a modulus function, we have

$$A_3 = \left\{ k \in \mathbb{N} : f\left(|(\alpha T(x_k) + \beta T(y_k) - (\alpha L_1 + \beta L_2)|\right) < \epsilon \right\}$$
$$\supseteq \left[\left\{ k \in \mathbb{N} : f\left(|\alpha||T(x_k) - L_1|\right) < \frac{\epsilon}{2} \right\} \right.$$
$$\left. \cap \left\{ k \in \mathbb{N} : f\left(|\beta||T(y_k) - L_2|\right) < \frac{\epsilon}{2} \right\} \right]$$
$$\supseteq \left[\left\{ k \in \mathbb{N} : f\left(|T(x_k) - L_1|\right) < \frac{\epsilon}{2} \right\} \right.$$
$$\left. \cap \left\{ k \in \mathbb{N} : f\left(|T(y_k) - L_2|\right) < \frac{\epsilon}{2} \right\} \right]$$

Therefore,

$$A_3 = \left\{ k \in \mathbb{N} : f\left(|(\alpha T(x_k) + \beta T(y_k) - (\alpha L_1 + \beta L_2)|\right) < \epsilon \right\}$$
$$\supseteq \left[\left\{ k \in \mathbb{N} : f\left(|T(x_k) - L_1|\right) < \frac{\epsilon}{2} \right\} \right. \qquad (2.9)$$
$$\left. \cap \left\{ k \in \mathbb{N} : f\left(|T(y_k) - L_2|\right) < \frac{\epsilon}{2} \right\} \right]$$

implies that $A_3 \in \pounds(I)$. Thus, $A_3^c = A_1^c \cup A_2^c \in I$. Therefore, $\alpha x_k + \beta y_k \in S^I(f)$, for all scalars α, β, and (x_k), $(y_k) \in S^I(f)$.

Hence, $S^I(f)$ is a linear space.

THEOREM 2.2   The classes of sequences $\mathcal{M}_S^I(f)$ and $\mathcal{M}_{S_0}^I(f)$ are paranormed spaces, paranormed by

$$g(x) = g(x_k) = \sup_k f\left(|T(x_k)|\right)$$

Proof   Let $x = (x_k), y = (y_k) \in \mathcal{M}_S^I(f)$.

(P_1) It is clear that $g(x) = 0$ if $x = \theta$, a zero vector.
(P_2) $g(x) = g(-x)$ is obvious.
(P_3) For $x = (x_k), y = (y_k) \in \mathcal{M}_S^I(f)$, we have

$$g(x + y) = g(x_k + y_k) = \sup_k f\left(|T(x_k + y_k)|\right)$$
$$= \sup_k f\left(|T(x_k) + T(y_k)|\right) \leq \sup_k f\left(|T(x_k)|\right)$$
$$+ \sup_k f\left(|T(y_k)|\right)| = g(x) + g(y)$$

Therefore, $g(x + y) \leq g(x) + g(y)$
(P_4) Let (λ_k) be a sequence of scalars with $(\lambda_k) \to \lambda\,(k \to \infty)$ and $(x_k), L \in \mathcal{M}_S^I(f)$ such that

$$x_k \to L\,(k \to \infty)$$

in the sense that

$$g(x_k - L) \to 0\,(k \to \infty)$$

Then, since the inequality

$$g(x_k) \leq g(x_k - L) + g(L)$$

holds by subadditivity of g, the sequence $\{g(x_k)\}$ is bounded.

Therefore,

$$g[(\lambda_k x_k - \lambda L)] = g[(\lambda_k x_k - \lambda x_k + \lambda x_k - \lambda L)]$$
$$= g[(\lambda_k - \lambda)x_k + \lambda(x_k - L)]$$
$$\leq g[(\lambda_k - \lambda)x_k] + g[\lambda(x_k - L)]$$
$$\leq |(\lambda_k - \lambda)|\,g(x_k) + |\lambda|\,g(x_k - L) \to 0$$

as $(k \to \infty)$. That is to say that scalar multiplication is continuous. Hence, $\mathcal{M}_S^I(f)$ is a paranormed space.

For $\mathcal{M}_{S_0}^I(f)$, the result is similar.

THEOREM 2.3   A sequence $x = (x_k) \in \ell_\infty$ I-converges if and only if for every $\epsilon > 0$, there exists $N_\epsilon \in \mathbb{N}$ such that

$$\left\{k \in \mathbb{N}: f\left(|T(x_k) - T(x_{N_\epsilon})|\right) < \epsilon\right\} \in \pounds(I) \tag{2.10}$$

Proof   Let $x = (x_k) \in \ell_\infty$.

Suppose that $L = I - \lim x$. Then, the set

$$B_\epsilon = \left\{k \in \mathbb{N}: f\left(|T(x_k) - L|\right) < \frac{\epsilon}{2}\right\} \in \pounds(I) \text{ for all } \epsilon < 0$$

Fix an $N_\epsilon \in B_\epsilon$. Then we have,

$$f\left(\,|\,T(x_k) - T(x_{N_\epsilon})\,|\,\right) \le f\left(\,|\,T(x_k) - L\,|\,\right) + f\left(\,|\,T(x_{N_\epsilon}) - L\,|\,\right) < \frac{\epsilon}{2} + \frac{\epsilon}{2} = \epsilon$$

which holds for all $k \in B_\epsilon$. Hence $\left\{ k \in \mathbb{N} : f\left(\,|\,T(x_k) - T(x_{N_\epsilon})\,|\,\right) < \epsilon \right\} \in \pounds(I)$ Conversely, suppose that

$$\left\{ k \in \mathbb{N} : f\left(\,|\,T(x_k) - T(x_{N_\epsilon})\,|\,\right) < \epsilon \right\} \in \pounds(I)$$

That is $\left\{ k \in \mathbb{N} : |\, f\left(|T(x_k)|\right) - f\left(|T(x_{N_\epsilon})|\right) |< \epsilon \right\} \in \pounds(I)$, for all $\epsilon > 0$. Then, the set

$$C_\epsilon = \left\{ k \in \mathbb{N} : f\left(|T(x_k)|\right) \in [f \le (|T(x_{N_\epsilon})|\,) - \epsilon, f(|T(x_{N_\epsilon})|\,) + \epsilon] \right\} \in \pounds(I) \text{ for all } \epsilon < 0$$

Let $J_\epsilon = \left[f(|T(x_{N_\epsilon})|) - \epsilon, f(|T(x_{N_\epsilon})|) + \epsilon \right]$. If we fix an $\epsilon > 0$ then we have $C_\epsilon \in \pounds(I)$ as well as $C_{\frac{\epsilon}{2}} \in \pounds(I)$.

Hence $C_\epsilon \cap C_{\frac{\epsilon}{2}} \in \pounds(I)$. This implies that

$$J = J_\epsilon \cap J_{\frac{\epsilon}{2}} \ne \phi$$

That is

$$\{k \in \mathbb{N} : f\left(|T(x_k)|\right) \in J\} \in \pounds(I)$$

That is

$$diamJ \le diamJ_\epsilon$$

where the diam of J denotes the length of interval J.

In this way, by induction, we get the sequence of closed intervals

$$J_\epsilon = I_0 \supseteq I_1 \supseteq \dots \supseteq I_k \supseteq \dots$$

with the property that $diamI_k \le \frac{1}{2} diamI_{k-1}$ for $(k = 2,3,4,\dots)$ and $\{k \in \mathbb{N} : f\left(|T(x_k)|\right) \in I_k\} \in \pounds(I)$ for $(k = 1,2,3,4,\dots)$. Then, there exists a $\xi \in \cap I_k$ where $k \in \mathbb{N}$ such that $\xi = I - \lim f\left(|T(x_k)|\right)$ showing that $x = (x_k) \in \ell_\infty$ is I-convergent. Hence the result.

THEOREM 2.4   Let f_1 and f_2 be two modulus functions and satisfying Δ_2 − Condition, then

(a) $\mathcal{X}(f_2) \subseteq \mathcal{X}(f_1 f_2)$,
(b) $\mathcal{X}(f_1) \cap (f_2) \subseteq \mathcal{X}(f_1 + f_2)$
   for $\mathcal{X} = S^I, S_o^I, \mathcal{M}_S^I$ and $\mathcal{M}_{S_o}^I$.

Proof   (a) Let $x = (x_k) \in S_o^I(f_2)$ be any arbitrary element. Then, the set

$$\left\{ k \in \mathbb{N} : f_2\left(\,|\,T(x_k)\,|\,\right) \ge \epsilon \right\} \in I \tag{2.11}$$

Let $\epsilon > 0$ and choose δ with $0 < \delta < 1$ such that $f_1(t) < \epsilon$, $0 \le t \le \delta$.

Let us denote

$$y_k = f_2\left(\,|\,T(x_k)\,|\,\right)$$

and consider

$$\lim_k f_1(y_k) = \lim_{y_k \le \delta, k \in \mathbb{N}} f_1(y_k) + \lim_{y_k > \delta, k \in \mathbb{N}} f_1(y_k)$$

Now, since f_1 is an modulus function, we have

$$\lim_{y_k \leq \delta, k \in \mathbb{N}} f_1(y_k) \leq f_1(2) \lim_{y_k \leq \delta, k \in \mathbb{N}} (y_k) \tag{2.12}$$

For $y_k > \delta$, we have

$$y_k < \frac{y_k}{\delta} < 1 + \frac{y_k}{\delta}$$

Now, since f_1 is non-decreasing and modulus, it follows that

$$f_1(y_k) < f_1\left(1 + \frac{y_k}{\delta}\right) < \frac{1}{2}f_1(2) + \frac{1}{2}f_1\left(\frac{2y_k}{\delta}\right)$$

Again, since f_1 satisfies $\Delta_2 -$ Condition, we have

$$f_1(y_k) < \frac{1}{2}K\frac{(y_k)}{\delta}f_1(2) + \frac{1}{2}K\frac{(y_k)}{\delta}f_1(2)$$

Thus, $f_1(y_k) < K\frac{(y_k)}{\delta}f_1(2)$ Hence,

$$\lim_{y_k > \delta, k \in \mathbb{N}} f_1(y_k) \leq \max\{1, K\delta^{-1}f_1(2)\} \lim_{y_k > \delta, k \in \mathbb{N}} (y_k) \tag{2.13}$$

Therefore, from Equations 2.11–2.13, we have $(x_k) \in S_\circ^I(f_1 f_2)$ Thus, $S_\circ^I(f_2) \subseteq S_\circ^I(f_1 f_2)$. Hence, $\mathcal{X}(f_2) \subseteq \mathcal{X}(f_1 f_2)$ for $\mathcal{X} = S_\circ^I$. For $\mathcal{X} = S^I$, $\mathcal{M}_S^I$, and $\mathcal{M}_{S_\circ}^I$ the inclusions can be established similarly.

(b) Let $x = (x_k) \in S_\circ^I(f_1) \cap S_\circ^I(f_2)$. Let $\epsilon > 0$ be given. Then, the sets

$$\left\{k \in \mathbb{N} : f_1\left(|T(x_k)|\right) \geq \epsilon\right\} \in I \tag{2.14}$$

and

$$\left\{k \in \mathbb{N} : f_2\left(|T(x_k)|\right) \geq \epsilon\right\} \in I \tag{2.15}$$

Therefore, from Equations 2.14 and 2.15 the set

$$\left\{k \in \mathbb{N} : (f_1 + f_2)\left(|T(x_k)|\right) \geq \epsilon\right\} \in I$$

Thus, $x = (x_k) \in S_\circ^I(f_1 + f_2)$. Hence, $S_\circ^I(f_1) \cap S_\circ^I(f_2) \subseteq S_\circ^I(f_1 + f_2)$. For $\mathcal{X} = S^I$, $\mathcal{M}_S^I$, and $\mathcal{M}_{S_\circ}^I$ the inclusions are similar.

For $f_2(x) = x$ and $f_1(x) = f(x)$, $\forall x \in [0, \infty)$, we have the following corollary.

COROLLARY 2.5 $\mathcal{X} \subseteq \mathcal{X}(f)$ for $\mathcal{X} = S^I$, $S_\circ^I$, $\mathcal{M}_S^I$ and $\mathcal{M}_{S_\circ}^I$

THEOREM 2.6 For any modulus function f, the spaces $S_\circ^I(f)$ and $\mathcal{M}_{S_\circ}^I(f)$ are solid and monotone.

Proof we prove the result for the space $S_\circ^I(f)$. For $\mathcal{M}_{S_\circ}^I(f)$, the proof can be obtained similarly.

For, let $(x_k) \in S_\circ^I(f)$ be any arbitrary element. Then, the set

$$\{k \in \mathbb{N} : f\left(|T(x_k)|\right) \geq \epsilon\} \in I \tag{2.16}$$

Let (α_k) be a sequence of scalars such that

$| \alpha_k | \leq 1$, for all $k \in \mathbb{N}$

Then the result follows from Equation 2.16 and the following inequality.

$$f\left(\, | \, T(\alpha_k x_k) \, | \, \right) = f\left(\, | \, \alpha_k T(x_k) \, | \, \right) \leq | \, \alpha_k \, | \, f\left(\, | \, T(x_k) \, | \, \right) \leq f\left(\, | \, T(x_k) \, | \, \right), \quad \text{for all } k \in \mathbb{N}$$

That the space $S_o^I(f)$ is monotone follows from the Lemma (I). Hence $S_o^I(f)$ is solid and monotone.

THEOREM 2.7   The spaces $S^I(f)$ and $\mathcal{M}_S^I(f)$ are not neither solid nor monotone.

Proof   Here we give a counter example for the proof of this result.

Counter example. Let $I = I_f$ and $f(x) = x$ for all $x \in [0, \infty)$. Consider the K-step $\mathcal{Z}_K$ of $\mathcal{Z}$ defined as follows.

Let $(x_k) \in \mathcal{Z}$ and let $(y_k) \in \mathcal{Z}_K$ be such that

$$y_k = \begin{cases} x_k, & \text{if } k \text{ is even,} \\ 0, & \text{otherwise} \end{cases}$$

Consider the sequence (x_k) defined as by $x_k = 1$ for all $k \in \mathbb{N}$. Then $(x_k) \in S^I(f)$ and $\mathcal{M}_S^I(f)$ but its K-step preimage does not belong to $S^I(f)$ and $\mathcal{M}_S^I(f)$. Thus, $S^I(f)$ and $\mathcal{M}_S^I(f)$ are not monotone. Hence, $S^I(f)$ and $\mathcal{M}_S^I(f)$ are not solid by Lemma(I).

THEOREM 2.8   If $(x = x_k)$ and $(y = y_k)$ be two sequences with $T(x \cdot y) = T(x)T(y)$. Then, the spaces $S^I(f)$ and $S_o^I(f)$ are sequence algebra.

Proof   Let $(x = x_k)$ and $(y = y_k)$ be two elements of $S_o^I(f)$ with $T(x \cdot y) = T(x)T(y)$.

Then, the sets

$$\left\{ k \in \mathbb{N} : f\left(\, | \, T(x_k) \, | \, \right) \geq \epsilon \right\} \in I \qquad (2.17)$$

and

$$\left\{ k \in \mathbb{N} : f\left(\, | \, T(y_k) \, | \, \right) \geq \epsilon \right\} \in I \qquad (2.18)$$

Therefore,

$$\left\{ k \in \mathbb{N} : f\left(\, | \, T(x_k) . T(y_k) \, | \, \right) \geq \epsilon \right\} \in I$$

Thus, $(x_k).(y_k) \in S_o^I(f)$.

Hence, $S_o^I(f)$ is sequence algebra. For $S^I(f)$, the result can be proved similarly.

THEOREM 2.9   Let f be a modulus function. Then, $S_o^I(f) \subset S^I(f) \subset S_\infty^I(f)$.

Proof   The inclusion $S_o^I(f) \subset S^I(f)$ is obvious.

Next, let $(x_k) \in S^I(f)$. Then there exists some L such that

$$\{ k \in \mathbb{N} : f(\, | \, T(x_k) - L \, | \,) \geq \epsilon \} \in I$$

We have

$$f\left(\mid T(x_k)\mid\right) \leq \frac{1}{2}f\left(\mid T(x_k) - L\mid\right) + f\left(\frac{1}{2}\mid L\mid\right)$$

Taking supremum over k on both sides, we get $(x_k) \in S_\infty^I(f)$

Hence, $S_o^I(f) \subset S^I(f) \subset S_\infty^I(f)$

THEOREM 2.10 If $f(x) = x$ for all $x \in [0, \infty]$. Then, the function $\hbar : \mathcal{M}_S^I(f) \to \mathbb{R}$ defined by $\hbar(x) = I - \lim f(\mid T(x_k)\mid)$, where $\mathcal{M}_S^I(f) = S_\infty(f) \cap S^I(f)$ is a Lipschitz function and hence uniformly continuous.

Proof Clearly, the function $\hbar$ is well defined. Let $x = (x_k)$, $y = (y_k) \in \mathcal{M}_S^I(f)$, $x \neq y$.

Then, the sets

$$A_x = \left\{k \in \mathbb{N} : f\left(\mid T(x) - \hbar(x)\mid\right) \geq \parallel x - y \parallel_*\right\} \in I$$

$$A_y = \left\{k \in \mathbb{N} : f\left(\mid T(y) - \hbar(y)\mid\right) \geq \parallel x - y \parallel_*\right\} \in I$$

where

$$\parallel x - y \parallel_* = \sup_k f\left(\mid T(x_k) - T(y_k)\mid\right)$$

Thus, the sets

$$B_x = \left\{k \in \mathbb{N} : \mid T(x) - \hbar(x)\mid < \parallel x - y \parallel_*\right\} \in \pounds(I)$$

$$B_y = \left\{k \in \mathbb{N} : \mid T(y) - \hbar(y)\mid < \parallel x - y \parallel_*\right\} \in \pounds(I)$$

Hence, $B = B_x \cap B_y \in \pounds(I)$, so that $B \neq \emptyset$ Now, taking $k \in B$, we have

$$\mid \hbar(x) - \hbar(y)\mid \leq \mid \hbar(x) - T(x_k)\mid + \mid T(x_k) - T(y_k)\mid + \mid T(y_k) - \hbar(y)\mid \leq 3\parallel x - y \parallel_*$$

Therefore, $\hbar$ is Lipschitz function and hence uniformly continuous.

THEOREM 2.11 If $f(x) = x$ for all $x \in [0, \infty]$ and if $x - (x_k)$, $y = (y_k) \subset \mathcal{M}_S^I(f)$ with $T(x \cdot y) - T(x)T(y)$. Then $(x \cdot y) \in \mathcal{M}_S^I(f)$ and $\hbar(xy) = \hbar(x)\hbar(y)$ where $\hbar : \mathcal{M}_S^I(f) \to \mathbb{R}$ is defined by $\hbar(x) = I - \lim f(\mid T(x_k)\mid)$.

Proof For $\epsilon > 0$, the sets

$$B_x = \{k \in \mathbb{N} : |T(x_k) - \hbar(x)| < \epsilon\} \in \pounds(I) \tag{2.19}$$

$$B_y = \{k \in \mathbb{N} : |T(y_k) - \hbar(y)| < \epsilon\} \in \pounds(I) \tag{2.20}$$

where $\parallel x - y \parallel_* = \epsilon$

Now,

$$\begin{aligned}|T(x_k y_k) - \hbar(x)\hbar(y)| &= |T(x_k)T(y_k) - T(x_k)\hbar(y) + T(x_k)\hbar(y) - \hbar(x)\hbar(y)| \\ &\leq |T(x_k)||y_k - \hbar(y)| + |\hbar(y)||x_k - \hbar(x)|\end{aligned} \tag{2.21}$$

As $\mathcal{M}_S^I(f) \subseteq S_\infty(f)$, there exists an $M \in \mathbb{R}$ such that $|T(x_k)| < M$ and $|\hbar(y)| < M$.

Therefore, from Equations 2.19–2.21, we have

$$|T(x_k y_k) - \hbar(x)\hbar(y)| \leq M\epsilon + M\epsilon = 2M\epsilon$$

for all $k \in B_x \cap B_y \in \pounds(I)$.

Hence $(x \cdot y) \in \mathcal{M}_S^I(f)$ and $\hbar(xy) = \hbar(x)\hbar(y)$.

Acknowledgements

The authors would like to record their gratitude to the reviewer for his careful reading and making some useful corrections which improved the presentation of the paper.

Funding

The authors received no direct funding for this research.

Author details

Vakeel A. Khan[1]
E-mail: vakhanmaths@gmail.com
Mohd Shafiq[1]
E-mail: shafiqmaths7@gmail.com
Rami Kamel Ahmad Rababah[1]
E-mail: rami2013r@gmail.com

[1] Department of Mathematics, Aligarh Muslim University, Aligarh 202002, India.

References

Basar, F., & Altay, B. (2003). On the spaces of sequences of p-bounded variation and related matrix mapping. *Ukrainian Mathematical Journal, 55*, 136–147.

Bhardwaj, V. K. (2003). A generalization of a sequence space of Ruckle. *Bulletin of Calcutta Mathematical Society, 95*, 411–420.

Buck, R. C. (1953). Generalized asymptotic density. *American Journal of Mathematics, 75*, 335–346.

Fast, H. (1951). Sur la convergence statistique [About statistical convergence]. *Colloquium Mathematicum, 2*, 241–244.

Fridy, J. A. (1985). On statistical convergence. *Analysis, 5*, 301–313.

Garling, D. J. H. (1966). On symmetric sequence spaces. *Proceedings of the London Mathematical Society, 16*, 85–106.

Gramsch, B. (1967). Die Klasse metrisher linearer Raume $L(\varphi)$ [The class of metric linear spaces $L(\varphi)$]. *Mathematische Annalen, 171*, 61–78.

Khan, V. A. (2005). On a sequence space defined by modulus function. *Southeast Asian Bulletin of Mathematics, 29*, 1–7.

Khan, V. A. (2006). Difference sequence spaces defined by a sequence modulii. *Southeast Asian Bulletin of Mathematics, 30*, 1061–1067.

Khan, V. A. (2007). Statistically pre-Cauchy sequences and Orlicz function. *Southeast Asian Bulletin of Mathematics, 6*, 1107–1112.

Khan, V. A., & Ebadullah, K. (2011). On some I-convergent sequence spaces defined by a modullus function. *Theory and Applications of Mathematics and Computer Science, 1*, 22–30.

Khan, V. A., Ebadullah, K., Esi, A., & Shafiq, M. (2013). On some Zweier I-convergent sequence spaces defined by a modulus function. *Africa Mathematika, Journal of the African Mathematical Society* (Springer Verlag Berlin Heidelberg), *26*, 115–125.

Khan, V. A., Shafiq, M., & Rababah, R. K. A. (2015). On BVs I-convergent sequence spaces defined by an Orlicz function. *Theory and Applications of Mathematics and Computer Science, 5*, 62–70.

Khan, V. A., & Tabassum, S. (2012). Statistically pre-Cauchy double sequences. *Southeast Asian Bulletin of Mathematics, 36*, 249–254.

Kostyrko, P., Šalát, T., & Wilczyński, W. (2000). I-convergence. *Raal Analysis Analysis Exchange, 26*, 669–686.

Köthe, G. (1970). *Topological vector spaces* (Vol. 1). Berlin: Springer.

Kreyszig, E. (1978). *Introductory functional analysis with applications*. New York, NY: Wiley.

Maddox, I. J. (1969). Some properties of paranormed sequence spaces. *Journal of the London Mathematical Society, 1*, 316–322.

Maddox, I. J. (1986). Sequence spaces defined by a modulus. *Mathematical Proceedings of the Cambridge Philosophical Society, 100*, 161–166.

Nakano, H. (1953). Concave modulars. *Journal of the Mathematical Society of Japan, 5*, 29–49.

Ruckle, W. H. (1967). Symmetric coordinate spaces and symmetric bases. *Canadian Journal of Mathematics, 19*, 828–838.

Ruckle, W. H. (1968). On perfect symmetric BK-spaces. *Mathematische Annalen, 175*, 121–126.

Ruckle, W. H. (1973). FK-spaces in which the sequence of coordinate vectors is bounded. *Canadian Journal of Mathematics, 25*, 973–975.

Šalát, T. (1980). On statistical convergent sequences of real numbers. *Mathematica Slovaca, 30*, 139–150.

Šalát, T., Tripathy, B. C., & Ziman, M. (2004). On some properties of I-convergence. *Tatra Mountains Mathematical Publications, 28*, 279–286.

Šalát, T., Tripathy, B. C., & Ziman, M. (2005). On I-convergence field. *Italian Journal of Pure and Applied Mathematics, 17*, 45–54.

Schoenberg, I. J. (1959). The integrability of certain functions and related summability methods. *American Mathematical Monthly, 66*, 361–375.

Sengönül, M. (2009). The Zweier sequence space. *Demonstratio Mathematica, XL*, 181–196.

Tripathy, B. C. (1998). On statistical convergence. *Proceedings of the Estonian Academy of Sciences. Physics. Mathematics Analysis, 47*, 299–303.

Tripathy, B. C., & Hazarika, B. (2009). Paranorm I-convergent sequence spaces. *Mathematica Slovaca, 59*, 485–494.

Tripathy, B. C., & Hazarika, B. (2011). Some I-convergent sequence spaces defined by Orlicz function. *Acta Mathematicae Applicatae Sinica, 27*, 149–154.

5

On the convergence of extended Newton-type method for solving variational inclusions

M.H. Rashid[1*]

*Corresponding author: M.H. Rashid, Faculty of Science, Department of Mathematics, University of Rajshahi, Rajshahi-6205, Bangladesh
E-mail: harun_math@ru.ac.bd
Reviewing editor: Lishan Liu, Qufu Normal University, China

Abstract: In this paper, we introduce and study the extended Newton-type method for solving the variational inclusion $0 \in f(x) + g(x) + F(x)$, where $f : \Omega \subseteq X \to Y$ is Fréchet differentiable in a neighborhood Ω of a point $\bar{x}$ in X, $g : \Omega \subseteq X \to Y$ is linear and differentiable at point $\bar{x}$, and F is a set-valued mapping with closed graph acting in Banach spaces X and Y. Semilocal and local convergence of the extended Newton-type method are analyzed.

Subjects: Mathematics & Statistics; Advanced Mathematics; Pure Mathematics

Keywords: variational inclusions; semilocal convergence; Lipschitz–like mappings; extended Newton-type method; divided difference

AMS Subject Classifications: 49J53, 47H04, 65K10

1. Introduction

In this study, we are concerned with the problem of approximating a solution of a variational inclusions. Let X and Y be Banach spaces. We consider here a variational inclusions problem to find a point $\bar{x} \in \Omega$ satisfying

ABOUT THE AUTHOR

Mohammed Harunor Rashid is an Associate Professor in the Department of Mathematics, University of Rajshahi, Bangladesh, where he teaches calculus, geometry, real analysis, numerical analysis, vector and tensor analysis, operations research, functional analysis and other courses in both undergraduate and graduate level. He has completed his PhD from Zhejiang University, China. His research interests focus on fuzzy mathematics, nonlinear numerical functional analysis and optimization. He has published his research contributions in some internationally renowned journals whose publishers are Springer, Taylor & Francis, Yokohama and other journals.

PUBLIC INTEREST STATEMENT

This study is concerned with the problem of approximating a point which satisfies the following variational inclusion

$$f(x) + g(x) + F(x) \ni 0 \qquad (1)$$

where X and Y be Banach spaces, $f : X \to Y$ is Frechet differentiable, g: X→Y admits first order divided difference and $F : X \rightrightarrows 2^Y$ is a set-valued mapping with closed graph.

To solve (1), for an initial point near to a solution, the sequences generated by the Newton-type method are not uniquely defined and not every generated sequence is convergent. The convergence result, obtained by Alexis and Pietrus (2008) or Rashid, Wang and Li (2012), guarantees the existence of convergent sequence. Therefore, from the viewpoint of practical computations, these kind of Newton-type methods are not convenient in practical applications. To overcome this drawback, we propose an extended Newton-type method (ENM) for solving (1) and analyze its semilocal and local convergence results. I believe this finding would appeal to the readership of Cogent Mathematics.

$$0 \in f(\bar{x}) + g(\bar{x}) + F(\bar{x}), \tag{1.1}$$

where $f : \Omega \subseteq X \to Y$ is differentiable in a neighborhood Ω of a point $\bar{x}$ in X, $g : \Omega \subseteq X \to Y$ is linear and differentiable at $\bar{x}$ but may not differentiable in a neighborhood Ω of $\bar{x}$, and $F : X \rightrightarrows 2^Y$ is a set-valued mapping with closed graph.

When $g = 0$, the inclusion (1.1) reduce to a generalized equation of the form

$$0 \in f(\bar{x}) + F(\bar{x}). \tag{1.2}$$

Several iterative methods have been presented for solving (1.2). Dontchev (1996b) established a quadratically convergent Newton-type method under a pseudo-Lipschitz property for set-valued mapping when ∇f is Lipschitz on a neighborhood of a solution x^* of (1.2) and subsequently he (1996c) proved the stability of the method. When ∇f is Hölder on a neighborhood of x^*, Piétrus (2000a) obtained superlinear convergence by following Dontchev's method and later he (2000b) proved the stability of this method in this mild differentiability context. For solving (1.2), Hilout, Alexis, and Piétrus (2006) considered the following sequence

$$\begin{cases} x_0 \text{ and } x_1 \text{ are given starting points} \\ y_k = \alpha x_k + (1 - \alpha)x_{k-1}; \quad \alpha \text{ is fixed in } (0,1) \\ 0 \in f(x_k) + [y_k, x_k; f](x_{k+1} - x_k) + F(x_{k+1}), \end{cases}$$

where $[y_k, x_k; f]$ is the first-order divided difference of f on the points y_k and x_k. This operator will be defined in Section 2. They proved the convergence of this method is superlinear when f is only continuous and differentiable at x^*. Furthermore, it should be mentioned that Argyros (2004) has studied local as well as semilocal convergence analysis for two-point Newton-like methods in a Banach space setting under very general Lipschitz type conditions for solving (Argyros, 2004) in the case when F=0.

Alexis and Piétrus (2008) introduced a method for solving the variational inclusions (1.1), which can be defined as follows:

$$0 \in f(x_k) + g(x_k) + (\nabla f(x_k) + [2x_{k+1} - x_k, x_k; g])(x_{k+1} - x_k) + F(x_{k+1}), \tag{1.3}$$

where $\nabla f(x)$ denotes the Fréchet derivative of f at x and $[x, y; g]$, the first order divided difference of g on the points x and y; and proved the convergence is superlinear and quadratic when ∇f is Lipschitz continuous. Rashid, Wang, and Li (2012) established local convergence results for the method (1.3) under the weaker conditions than Alexis and Pietrus (2008). In particular, Rashid et al. (2012) extended the results by fixing a gap in the proof of corresponding ones (Alexis & Piétrus, 2008, Theorem 1).

Although the method (1.3) guarantees the existence of a convergent sequence $\{x_k\}$, the points $x_1, x_2, \ldots$ of the sequence $\{x_k\}$ are not converges separately. Therefore, for a starting point near to a solution, the sequences generated by the method (1.3) are not uniquely defined. For instance, the convergence result, obtained by Alexis and Piétrus (2008) or Rashid et al. (2012), guarantees the existence of a convergent sequence. Hence, in view of numerical computation, these kind of methods are not convenient in practical application. This drawback motivates us to introduce a method 'so-called' extended Newton-type method. The difference between the method (1.3) and our proposed method is that the extended Newton-type method generates a convergent sequence $\{x_k\}$ whose each point $x_1, x_2, \ldots$ converge individually but this does not happen for the method (1.3). Thus, we propose the following extended Newton-type method:

Algorithm 1.1. *Let $\eta \geq 1$ and $x_0 \in X$. For $k = 0, 1, 2, \ldots$, having x_k, determine x_{k+1} as follows:*
Choose $d_k \in D(x_k)$ such that

$$\|d_k\| \leq \eta \operatorname{dist}(0, D(x_k)),$$

and set $x_{k+1} = x_k + d_k$, where the set $D(x)$ is defined for any $x \in X$ by

$$D(x) = \left\{ d \in X : 0 \in f(x) + g(x) + (\nabla f(x) + [2d + x, x; g])d + F(x + d) \right\}.$$

Remark 1.1 If $g = 0$, then the set D will be replaced by the set

$$D'(x) := \left\{ d \in X : 0 \in f(x) + \nabla f(x)d + F(x + d) \right\}$$

Then the Algorithm 1.1 reduces to the same algorithm corresponding one given by Rashid, Yu, Li, and Wu (2013).

There have been studied many fruitful works on semilocal convergence analysis for the Gauss–Newton method in the case when $F = 0$ and $g = 0$ [see Dedieu and Kim (2002), Dedieu and Shub (2000), Xu and Li (2008), for more details] or when $F = C$ and $g = 0$ [see Li and Ng (2007), for details]. Rashid et al. (2013) have studied semilocal convergence analysis for the Gauss–Newton-type method to solve the generalized Equation (1.2). However, in our best knowledge, there is no study on semilocal convergence analysis discovered for the general case (1.1), even for the method (1.3).

Our purpose here is to analyze the semilocal convergence of the extended Newton-type method defined by Algorithm 1.1. The main tool is the Lipschitz-like property of set-valued mappings, which was introduced by Aubin (1984) in the context of nonsmooth analysis and studied by many mathematicians [see for example, Alexis and Piétrus (2008), Argyros and Hilout (2008), Dontchev (1996a), Hilout et al. (2006), Piétrus (2000b)] and the references therein. The main results are the convergence criteria, established in Section 3, which, based on the attraction region around the initial point, provide some sufficient conditions ensuring the convergence to a solution of any sequence generated by Algorithm 1.1. As a result, local convergence results for the extended Newton-type method are obtained.

This paper is organized as follows: In Section 2, we recall some necessary notations, notions, some preliminary results and also recall a fixed-point theorem which has been proved by Dontchev and Hager (1994). This fixed-point theorem is the main tool to prove the existence of the sequence generated by Algorithm 1.1. In Section 3, we consider the extended Newton-type method as well as the concept of Lipschitz-like property to show the existence and the convergence of the sequence generated by Algorithm 1.1. In the last section, we give a summary of the major results to close our paper.

2. Notations and preliminaries

In this section, we give some notations and collect some results that will be helpful to prove our main results. Throughout this paper, we suppose that X and Y are two real or complex Banach spaces. Let $x \in X$ and $r > 0$. The closed ball centered at x with radius r is denoted by $\mathbb{B}_r(x)$. Let $F : X \rightrightarrows 2^Y$ be a set-valued mapping with $\operatorname{dom} F \neq \emptyset$. The domain $\operatorname{dom} F$, the inverse F^{-1} and the graph $\operatorname{gph} F$ of F are, respectively, defined by

$$\operatorname{dom} F := \{ x \in X : F(x) \neq \emptyset \},$$
$$F^{-1}(y) := \{ x \in X : y \in F(x) \} \text{ for each } y \in Y$$

and

$gphF := \{(x,y) \in X \times Y : y \in F(x)\}.$

Let $A \subseteq X$. The distance function of A is defined by

$dist(x,A) := \inf\{\|x-a\| : a \in A\}$ for each $x \in X$,

while the excess from the set A to the set $C \subseteq X$ is defined by

$e(C,A) := \sup\{dist\,(x,A) : x \in C\}.$

All the norms are denoted by $\|\cdot\|$ and the space of linear operators from X to Y is denoted by $\mathcal{L}(X,Y)$.

Now, we recall a few definitions, some results and then state the Banach fixed point theorem. We begin with the definition of the first-order divided difference operators. The following definition, is given by Argyros (2007), introduces the notion of divided differences of nonlinear operators.

Definition 2.1 An operator belonging to the space $\mathcal{L}(X,Y)$ is called the first order divided difference of the operator $g: X \to Y$ on the points x and y in X ($x \neq y$) if the following properties hold:

(a) $[x,y;g](y-x) = g(y) - g(x)$ for $x \neq y$;

(b) if g is Fréchet differentiable at $x \in X$, then $[x,y;g] = \nabla g(x)$.

Recall from Rashid et al. (2013) the notions of pseudo-Lipschitz and Lipschitz-like set-valued mappings. These notions were introduced by Aubin [see, Aubin (1984), Aubin and Frankowska (1990), for more details] and have been studied extensively.

Definition 2.2 Let $\Gamma: Y \rightrightarrows 2^X$ be a set-valued mapping and let $(\bar{y},\bar{x}) \in gph\Gamma$. Let $r_{\bar{x}} > 0$, $r_{\bar{y}} > 0$ and $M > 0$. Then Γ is said to be

(a) Lipschitz-like on $\mathbb{B}_{r_{\bar{y}}}(\bar{y})$ relative to $\mathbb{B}_{r_{\bar{x}}}(\bar{x})$ with constant M if the following inequality holds:

$e(\Gamma(y_1) \cap \mathbb{B}_{r_{\bar{x}}}(\bar{x}), \Gamma(y_2)) \leq M\|y_1 - y_2\|$ for any $y_1, y_2 \in \mathbb{B}_{r_{\bar{y}}}(\bar{y})$

(b) pseudo-Lipschitz around $(\bar{y},\bar{x})$ if there exist constants $r'_{\bar{y}} > 0$, $r'_{\bar{x}} > 0$ and $M' > 0$ such that Γ is Lipschitz-like on $\mathbb{B}_{r'_{\bar{y}}}(\bar{y})$ relative to $\mathbb{B}_{r'_{\bar{x}}}(\bar{x})$ with constant M'.

Remark 2.1 Γ is Lipschitz-like on $\mathbb{B}_{r_{\bar{y}}}(\bar{y})$ relative to $\mathbb{B}_{r_{\bar{x}}}(\bar{x})$ with constant M is equivalent to the following statement: if for every $y_1, y_2 \in \mathbb{B}_{r_{\bar{y}}}(\bar{y})$ and for every $x_1 \in \Gamma(y_1) \cap \mathbb{B}_{r_{\bar{x}}}(\bar{x})$, there exists $x_2 \in \Gamma(y_2)$ such that

$\|x_1 - x_2\| \leq M\|y_1 - y_2\|.$

The following lemma is useful and it has been taken from [Rashid et al., 2013, Lemma 2.1].

LEMMA 2.1 *Let $\Gamma: Y \rightrightarrows 2^X$ be a set-valued mapping and let $(\bar{y},\bar{x}) \in gph\ \Gamma$. Assume that Γ is Lipschitz-like on $\mathbb{B}_{r_{\bar{y}}}(\bar{y})$ relative to $\mathbb{B}_{r_{\bar{x}}}(\bar{x})$ with constant M. Then*

$dist\,(x, \Gamma(y)) \leq M\,dist\,(y, \Gamma^{-1}(x))$

holds for every $x \in \mathbb{B}_{r_{\bar{x}}}(\bar{x})$ and $y \in \mathbb{B}_{\frac{r_{\bar{y}}}{3}}(\bar{y})$ satisfying $dist(y, \Gamma^{-1}(x)) \leq \dfrac{r_{\bar{y}}}{3}$

We end this section with the following lemma. This lemma is a fixed-point statement which has been proved by Dontchev and Hager (1994) and employing the standard iterative concept for contracting mapping. This lemma is used to prove the existence of the sequence generated by Algorithm 1.1.

LEMMA 2.2   Let $\Phi:X \rightrightarrows 2^X$ be a set-valued mapping. Let $\eta_0 \in X, r > 0$ and $0 < \lambda < 1$ be such that

$$\text{dist}(\eta_0, \Phi(\eta_0)) < r(1-\lambda) \tag{2.1}$$

and

$$e(\Phi(x_1) \cap \mathbb{B}_r(\eta_0), \Phi(x_2)) \leq \lambda \|x_1 - x_2\| \quad \text{for any } x_1, x_2 \in \mathbb{B}_r(\eta_0). \tag{2.2}$$

Then Φ has a fixed point in $\mathbb{B}_r(\eta_0)$, that is, there exists $x \in \mathbb{B}_r(\eta_0)$ such that $x \in \Phi(x)$. If Φ is additionally single-valued, then the fixed point of Φ in $\mathbb{B}_r(\eta_0)$ is unique.

The previous lemma is a generalization of a fixed-point theorem which has been given by Ioffe and Tikhomirov (1979), where in assertion (b) the excess e is replaced by Hausdorff distance.

3. Convergence analysis of extended Newton-type method

Throughout this section, we suppose that $f: \Omega \subseteq X \rightarrow Y$ is a Fréchet differentiable function on a neighborhood Ω of $\bar{x}$ with its derivative denoted by ∇f, $g: \Omega \subseteq X \rightarrow Y$ is linear and differentiable at $\bar{x}$, and let $F: X \rightrightarrows 2^Y$ be set-valued mapping with closed graph. We prove the existence and convergence of the sequences generated by extended Newton-type method, defined by the Algorithm 1.1, on a neighborhood Ω of a point $\bar{x}$.

Let $x \in X$ and define the mapping Q_x by

$$Q_x(\cdot) := f(x) + g(\cdot) + \nabla f(x)(\cdot - x) + F(\cdot).$$

Then

$$D(x) = \left\{ d \in X : 0 \in Q_x(x+d) \right\}, \text{ since } g(x+d) = g(x) + [2d+x, x; g]d.$$

We remark that

$$g(x) + [2d+x, x; g]d = g(x) - \frac{1}{2}[2d+x, x; g](x - (2d+x))$$

$$= g(x) - \frac{1}{2}(g(x) - g(2d+x)) = \frac{1}{2}g(x) + \frac{1}{2}g(2d+x)$$

$$= \frac{1}{2}(g(2x+2d)) = g(x+d).$$

Furthermore, we have the following equivalence

$$z \in Q_x^{-1}(y) \Longleftrightarrow y \in f(x) + g(z) + \nabla f(x)(z-x) + F(z), \text{ for any } z \in X \text{ and } y \in Y. \tag{3.1}$$

In particular,

$$\bar{x} \in Q_{\bar{x}}^{-1}(\bar{y}) \quad \text{for each } (\bar{x}, \bar{y}) \in \text{gph}(f+g+F). \tag{3.2}$$

Let $(\bar{x}, \bar{y}) \in \text{gph}(f+g+F)$ and let $r_{\bar{x}} > 0$, $r_{\bar{y}} > 0$. Throughout the whole paper, we assume that $\mathbb{B}_{r_{\bar{x}}}(\bar{x}) \subseteq \Omega \cap \text{dom} F$, the function g is Fréchet differentiable at $\bar{x}$ and admits a first-order divided difference satisfying the following condition:

there exist $v > 0$ such that for all x, y, u and $v \in \mathbb{B}_{r_{\bar{x}}}(\bar{x})$ $(x \neq y, u \neq v)$,

$$\|[x, y ; g] - [u, v ; g]\| \leq v(\|x - u\| + \|y - v\|)$$

and the mapping $Q_{\bar{x}}^{-1}(\cdot)$ is Lipschitz-like on $\mathbb{B}_{r_{\bar{y}}}(\bar{y})$ relative to $\mathbb{B}_{r_{\bar{x}}}(\bar{x})$ with constant M, that is,

$$e(Q_{\bar{x}}^{-1}(y_1) \cap \mathbb{B}_{r_{\bar{x}}}(\bar{x}), Q_{\bar{x}}^{-1}(y_2)) \leq M\|y_1 - y_2\| \quad \text{for any} \cong_1, y_2 \in \mathbb{B}_{r_{\bar{y}}}(\bar{y}). \tag{3.3}$$

Let $\varepsilon > 0$ and write

$$\bar{r} := \min\left\{ r_{\bar{y}} - 2\varepsilon r_{\bar{x}}, \frac{r_{\bar{x}}(1 - M\varepsilon)}{4M} \right\}. \tag{3.4}$$

Then

$$\bar{r} < 0 \iff \varepsilon < \min\left\{ \frac{r_{\bar{y}}}{2r_{\bar{x}}}, \frac{1}{M} \right\}. \tag{3.5}$$

The following lemma plays a crucial role for convergence analysis of the extended Newton-type method. The proof is a refinement of the one for (Rashid et al., 2013, Lemma 3.1).

LEMMA 3.1 *Suppose that $Q_{\bar{x}}^{-1}(\cdot)$ is Lipschitz-like on $\mathbb{B}_{\bar{r}}(\bar{y})$ relative to $\mathbb{B}_{r_{\bar{x}}}(\bar{x})$ with constant M and that*

$$\sup_{x \in \mathbb{B}_{\frac{r_{\bar{x}}}{2}}(\bar{x})} \|\nabla f(x) - \nabla f(\bar{x})\| \leq \varepsilon < \min\left\{ \frac{r_{\bar{y}}}{2r_{\bar{x}}}, \frac{1}{M} \right\}. \tag{3.6}$$

Let $x \in \mathbb{B}_{\frac{r_{\bar{x}}}{2}}(\bar{x})$. Then $Q_x^{-1}(\cdot)$ is Lipschitz-like on $\mathbb{B}_{\bar{r}}(\bar{y})$ relative to $\mathbb{B}_{\frac{r_{\bar{x}}}{2}}(\bar{x})$ with constant $\dfrac{M}{1 - M\varepsilon}$, that is,

$$e(Q_x^{-1}(y_1) \cap \mathbb{B}_{\frac{r_{\bar{x}}}{2}}(\bar{x}), Q_x^{-1}(y_2)) \leq \frac{M}{1 - M\varepsilon}\|y_1 - y_2\| \quad \text{for any } y_1, y_2 \in \mathbb{B}_{\bar{r}}(\bar{y}).$$

Proof Noted that (3.5) and (3.6) imply $\bar{r} > 0$. Now let

$$y_1, y_2 \in \mathbb{B}_{\bar{r}}(\bar{y}) \quad \text{and} \quad x' \in Q_x^{-1}(y_1) \cap \mathbb{B}_{\frac{r_{\bar{x}}}{2}}(\bar{x}). \tag{3.7}$$

It suffices to show that there exist $x'' \in Q_x^{-1}(y_2)$ such that

$$\|x' - x''\| \leq \frac{M}{1 - M\varepsilon}\|y_1 - y_2\|.$$

To this end , we shall verify that there exists a sequence $\{x_k\} \subset \mathbb{B}_{r_{\bar{x}}}(\bar{x})$ such that

$$y_2 \in f(x) + g(x_k) + \nabla f(x)(x_{k-1} - x) + \nabla f(\bar{x})(x_k - x_{k-1}) + F(x_k), \tag{3.8}$$

and

$$\|x_k - x_{k-1}\| \leq M\|y_1 - y_2\|(M\varepsilon)^{k-2} \tag{3.9}$$

hold for each $k = 2, 3, 4, \ldots$. We proceed by mathematical induction on k. Write

$$z_i := y_i - f(x) - \nabla f(x)(x_1 - x) + f(\bar{x}) + \nabla f(\bar{x})(x_1 - \bar{x}) \quad \text{for each } i = 1, 2.$$

Note by (3.7) that

$$\|x - x'\| \leq \|x - \bar{x}\| + \|\bar{x} - x'\| \leq r_{\bar{x}}.$$

It follows, from (3.7) and the relation $\bar{r} \leq r_{\bar{y}} - 2\varepsilon r_{\bar{x}}$ by (3.4), that

$$\|z_i - \bar{y}\| \leq \|y_i - \bar{y}\| + \|f(x) - f(\bar{x}) - \nabla f(\bar{x})(x - \bar{x})\| + \|(\nabla f(x) - \nabla f(\bar{x}))(x - x')\|$$
$$\leq \bar{r} + \varepsilon(\|x - \bar{x}\| + \|x - x'\|)$$
$$\leq \bar{r} + \varepsilon \left(\frac{r_{\bar{x}}}{2} + r_{\bar{x}} \right) \leq r_{\bar{y}}.$$

That is $z_i \in \mathbb{B}_{r_{\bar{y}}}(\bar{y})$ for each $i = 1,2$. Denote $x_1 := x'$. Then $x_1 \in Q_x^{-1}(y_1)$ by (3.7) and it follows from (3.1) that

$$y_1 \in f(x) + g(x_1) + \nabla f(x)(x_1 - x) + F(x_1)$$

which can be rewritten as

$$y_1 + f(\bar{x}) + \nabla f(\bar{x})(x_1 - \bar{x})) \in f(x) + g(x_1) + \nabla f(x)(x_1 - x) + F(x_1) + f(\bar{x}) + \nabla f(\bar{x})(x_1 - \bar{x}).$$

This, by the definition of z_1, means that

$$z_1 \in f(\bar{x}) + g(x_1) + \nabla f(\bar{x})(x_1 - \bar{x}) + F(x_1).$$

Hence, $x_1 \in Q_{\bar{x}}^{-1}(z_1)$ by (3.1). This, together with (3.7), implies that

$$x_1 \in Q_{\bar{x}}^{-1}(z_1) \cap \mathbb{B}_{r_{\bar{x}}}(\bar{x}).$$

By the assumed Lipschitz-like property of $Q_{\bar{x}}^{-1}(\cdot)$ and noting that $z_1, z_2 \in \mathbb{B}_{r_{\bar{y}}}(\bar{y})$, it follows from (3.3) that there exists $x_2 \in Q_{\bar{x}}^{-1}(z_2)$ such that

$$\|x_2 - x_1\| \leq M\|z_1 - z_2\| = M\|y_1 - y_2\|.$$

Moreover, by the definition of z_2 and $x_1 = x'$, we have

$$x_2 \in Q_{\bar{x}}^{-1}(z_2) = Q_{\bar{x}}^{-1}\left(y_2 - f(x) - \nabla f(x)(x_1 - x) + f(\bar{x}) + \nabla f(\bar{x})(x_1 - \bar{x})\right)$$

which together with (3.1) implies that

$$y_2 \in f(x) + g(x_2) + \nabla f(x)(x_1 - x) + \nabla f(\bar{x})(x_2 - x_1) + F(x_2).$$

This shows that (3.8) and (3.9) are true with constructed x_1, x_2.

We assume that $x_1, x_2, \ldots, x_n$ are constructed such that (3.8) and (3.9) are true for $k = 2,3, \ldots, n$. We need to construct x_{n+1} such that (3.8) and (3.9) are also true for $k = n+1$. For this purpose, write

$$z_i^n := y_2 - f(x) - \nabla f(x)(x_{n+i-1} - x) + f(\bar{x}) + \nabla f(\bar{x})(x_{n+i-1} - \bar{x}) \quad \text{for each } i = 0,1.$$

Then, by the inductional assumption,

$$\|z_0^n - z_1^n\| = \|(\nabla f(\bar{x}) - \nabla f(x))(x_n - x_{n-1})\|$$
$$\leq \varepsilon \|x_n - x_{n-1}\| \leq \|y_1 - y_2\|(M\varepsilon)^{n-1}. \tag{3.10}$$

Since $\|x_1 - \bar{x}\| \leq \dfrac{r_{\bar{x}}}{2}$ by (3.7) and $\|y_1 - y_2\| \leq 2\bar{r}$ by (3.7), it follows from (3.9) that

$$\|x_n - \bar{x}\| \leq \sum_{k=2}^{n} \|x_k - x_{k-1}\| + \|x_1 - \bar{x}\|$$

$$\leq 2M\bar{r} \sum_{k=2}^{n} (M\varepsilon)^{k-2} + \frac{r_{\bar{x}}}{2}.$$

$$\leq \frac{2M\bar{r}}{1 - M\varepsilon} + \frac{r_{\bar{x}}}{2}.$$

By (3.4), we have $\bar{r} \leq \dfrac{r_{\bar{x}}(1 - M\varepsilon)}{4M}$ and so

$$\|x_n - \bar{x}\| \leq r_{\bar{x}}. \tag{3.11}$$

Consequently,

$$\|x_n - x\| \leq \|x_n - \bar{x}\| + \|\bar{x} - x\| \leq \frac{3}{2} r_{\bar{x}}. \tag{3.12}$$

Furthermore, using (3.7) and (3.12), one has that, for each $i = 0, 1$,

$$\|z_i^n - \bar{y}\|$$
$$\leq \|y_2 - \bar{y}\| + \|f(x) - f(\bar{x}) - \nabla f(\bar{x})(x - \bar{x})\| + \|(\nabla f(x) - \nabla f(\bar{x}))(x - x_{n+i-1})\|$$
$$\leq \bar{r} + \varepsilon \left(\|x - \bar{x}\| + \|x - x_{n+i-1}\| \right) \leq \bar{r} + \varepsilon \left(\frac{r_{\bar{x}}}{2} + \frac{3r_{\bar{x}}}{2} \right)$$
$$= \bar{r} + 2\varepsilon r_{\bar{x}}.$$

It follows from the definition of $\bar{r}$ in (3.4) that $z_i^n \in \mathbb{B}_{r_{\bar{y}}}(\bar{y})$ for each $i = 0, 1$. Since assumption (3.8) holds for $k = n$, we have

$$y_2 \in f(x) + g(x_n) + \nabla f(x)(x_{n-1} - x) + \nabla f(\bar{x})(x_n - x_{n-1}) + F(x_n)$$

which can be written as

$$y_2 + f(\bar{x}) + \nabla f(\bar{x})(x_{n-1} - \bar{x}) \in f(x) + \nabla f(x)(x_{n-1} - x) + f(\bar{x}) + g(x_n)$$
$$+ \nabla f(\bar{x})(x_n - x_{n-1}) + F(x_n) + \nabla f(\bar{x})(x_{n-1} - \bar{x});$$

i.e. $z_0^n \in f(\bar{x}) + g(x_n) + \nabla f(\bar{x})(x_n - \bar{x}) + F(x_n)$ by the definition of z_0^n. This together with (3.1) and (3.11) yields that

$$x_n \in Q_{\bar{x}}^{-1}(z_0^n) \cap \mathbb{B}_{r_{\bar{x}}}(\bar{x}).$$

Using (3.3) again, there exists an element $x_{n+1} \in Q_{\bar{x}}^{-1}(z_1^n)$ such that

$$\|x_{n+1} - x_n\| \leq M\|z_0^n - z_1^n\| \leq M\|y_1 - y_2\|(M\varepsilon)^{n-1}, \tag{3.13}$$

where the last inequality holds by (3.10). By the definition of z_1^n, we have

$$x_{n+1} \in Q_{\bar{x}}^{-1}(z_1^n) = Q_{\bar{x}}^{-1}(y_2 - f(x) - \nabla f(x)(x_n - x) + f(\bar{x}) + \nabla f(\bar{x})(x_n - \bar{x})),$$

which together with (3.1) implies

$$y_2 \in f(x) + g(x_{n+1}) + \nabla f(x)(x_n - x) + \nabla f(\bar{x})(x_{n+1} - x_n) + F(x_{n+1}).$$

This together with (3.13) completes the induction step and ensure the existence of a sequence $\{x_n\}$ satisfying (3.8) and (3.9).

Since $M\varepsilon < 1$, we see from (3.9) that $\{x_k\}$ is a Cauchy sequence and hence it is convergent, to say x'', that is, $x'' := \lim_{k\to\infty} x_k$. Note that F has closed graph. Then, taking limit in (3.8), we get $y_2 \in f(x) + g(x'') + \nabla f(x)(x'' - x) + F(x'')$, that is,

$$x'' \in Q_x^{-1}(y_2) = [f(x) + g(\cdot) + \nabla f(x)(\cdot - x) + F(\cdot)]^{-1}(y_2).$$

Therefore, we obtain

$$\|x' - x''\| \leq \lim_{n\to\infty} \sup \sum_{k=2}^{n} \|x_k - x_{k-1}\|$$

$$\leq \lim_{n\to\infty} \sup \sum_{k=2}^{n} (M\varepsilon)^{k-2} M \|y_1 - y_2\|$$

$$\leq \frac{M}{1 - M\varepsilon} \|y_1 - y_2\|.$$

That is,

$$e(Q_x^{-1}(y_1) \cap \mathbb{B}_{\frac{r_{\bar{x}}}{2}}(\bar{x}), Q_x^{-1}(y_2)) \leq \frac{M}{1 - M\varepsilon} \|y_1 - y_2\|.$$

This completes the proof of the Lemma 3.1. $\square$

For our convenience, we define the mapping $Z_x : X \to Y$, for each $x \in X$, by

$$Z_x(\cdot) := f(\bar{x}) + g(\cdot) + \nabla f(\bar{x})(\cdot - \bar{x}) - f(x) - g(x) - \big(\nabla f(x) + [2(\cdot) - x, x; g]\big)(\cdot - x)$$

and the set-valued mapping $\Phi_x : X \rightrightarrows 2^X$ by

$$\Phi_x(\cdot) := Q_{\bar{x}}^{-1}[Z_x(\cdot)]. \tag{3.14}$$

Then for any $x', x'' \in X$, we have

$$\|Z_x(x') - Z_x(x'')\| = \|g(x') - g(x'') - [2x' - x, x; g](x' - x) + [2x'' - x, x; g](x'' - x)$$
$$+ (\nabla f(\bar{x}) - \nabla f(x))(x' - x'')\|$$
$$\leq \|g(x') - g(x'') - [2x'' - x, x; g](x' - x'')\| + \|([2x'' - x, x; g]$$
$$- [2x' - x, x; g])(x' - x)\| + \|\nabla f(\bar{x}) - \nabla f(x)\| \|x' - x''\|$$
$$\leq \big(\|[x'', x'; g] - [2x'' - x, x; g]\| + \|\nabla f(\bar{x}) - \nabla f(x)\|\big) \|x' - x''\|$$
$$+ \|[2x'' - x, x; g] - [2x' - x, x; g]\| \|x' - x\|. \tag{3.15}$$

Our first main theorem is as follows, which provides some sufficient conditions ensuring the convergence of the extended Newton-type method with initial point x_0.

THEOREM 3.1 *Suppose that $\eta > 1$ and that $Q_{\bar{x}}^{-1}(\cdot)$ is Lipschitz-like on $\mathbb{B}_{r_{\bar{y}}}(\bar{y})$ relative to $\mathbb{B}_{r_{\bar{x}}}(\bar{x})$ with constant M. Let*

$$\varepsilon \geq \sup_{x, x' \in \mathbb{B}_{\frac{r_{\bar{x}}}{2}}(\bar{x})} \|\nabla f(x) - \nabla f(x')\| \tag{3.16}$$

and let $\bar{r}$ be defined in (3.4). Let $v > 0$ and $\delta > 0$ be such that

(a) $\delta \leq \min \left\{ \dfrac{r_{\bar{x}}}{4}, \dfrac{\bar{r}}{3(\varepsilon + 3v)}, 1, \dfrac{3 - 5M\varepsilon}{80Mv}, \dfrac{r_{\bar{y}}}{17(\varepsilon + 3v)} \right\},$

(b) $6\eta M(\varepsilon + 3v) \leq 1 - M\varepsilon,$

(c) $\|\bar{y}\| < (\varepsilon + 3v)\delta.$

Suppose that

$$\lim_{x \to \bar{x}} \text{dist}(\bar{y}, f(x) + g(x) + F(x)) = 0. \tag{3.17}$$

Then there exists some $\hat{\delta} > 0$ such that any sequence $\{x_n\}$ generated by Algorithm 1.1 with initial point in $\mathbb{B}_{\hat{\delta}}(\bar{x})$ converges to a solution x^* of (1.1), that is, x^* satisfies $0 \in f(x^*) + g(x^*) + F(x^*)$.

Proof Letting that $q := \dfrac{\eta M(\varepsilon + 3v)}{1 - M\varepsilon}$. Then by the relation $6\eta M(\varepsilon + 3v) \le 1 - M\varepsilon$ from assumption (b), we obtain

$$q := \frac{\eta M(\varepsilon + 3v)}{1 - M\varepsilon} \le \frac{1}{6}.$$

Take $0 < \hat{\delta} \le \delta$ such that

$$\text{dist}(0, f(x_0) + g(x_0) + F(x_0)) \le (\varepsilon + 3v)\delta \quad \text{for each } x_0 \in \mathbb{B}_{\hat{\delta}}(\bar{x}) \tag{3.18}$$

(Noting that such $\hat{\delta}$ exists by (3.17) and assumption (c)). Let $x_0 \in \mathbb{B}_{\hat{\delta}}(\bar{x})$. We will proceed by mathematical induction to show that Algorithm 1.1 generates at least one sequence and any sequence $\{x_n\}$ generated by Algorithm 1.1 satisfies the following assertions:

$$\|x_n - \bar{x}\| \le 2\delta \tag{3.19}$$

and

$$\|x_{n+1} - x_n\| \le q^{n+1}\delta \tag{3.20}$$

hold for each $n = 0, 1, 2, \dots$. For this purpose, we define

$$r_x := \frac{5}{2} \left(M(\varepsilon + 3v\|x - \bar{x}\|)\|x - \bar{x}\| + M\|\bar{y}\| \right) \quad \text{for each } x \in X. \tag{3.21}$$

Then, thanks to the fact that $6\eta M(\varepsilon + 3v) \le 1 - M\varepsilon < 1$ by assumption (b) and $\|\bar{y}\| < (\varepsilon + 3v)\delta$ by assumption (c). Since $\eta > 1$, (3.21) yields that

$$r_x < 5M(\varepsilon + 6v\delta)\delta + M(\varepsilon + 3v)\delta$$
$$\le 6M\varepsilon\delta + 33v\delta^2 < 6M\varepsilon\delta + 33v\delta$$
$$< 11M\varepsilon\delta + 33v\delta = 11M(\varepsilon + 3v)\delta$$
$$< \frac{11}{6\eta} \le 2\delta \quad \text{for each } x \in \mathbb{B}_{2\delta}(\bar{x}). \tag{3.22}$$

Note that (3.19) is trivial for $n = 0$. To show (3.20) holds for $n = 0$, first we need to show that x_1 exists. To complete this, we have to prove that $D(x_0) \ne \emptyset$ by applying Lemma 2.2 to the map Φ_{x_0} with $\eta_0 = \bar{x}$. Let us check that both assertions (2.1) and (2.2) of Lemma 2.2 hold with $r := r_{x_0}$ and $\lambda := \dfrac{3}{5}$. Noting that $\bar{x} \in Q_{\bar{x}}^{-1}(\bar{y}) \cap \mathbb{B}_{2\delta}(\bar{x})$ by (3.2) and according to the definition of the excess e and the mapping Φ_{x_0} in (3.14), we obtain

$$\text{dist}(\bar{x}, \Phi_{x_0}(\bar{x})) \le e(Q_{\bar{x}}^{-1}(\bar{y}) \cap \mathbb{B}_{r_{x_0}}(\bar{x}), \Phi_{x_0}(\bar{x})) \le e(Q_{\bar{x}}^{-1}(\bar{y}) \cap \mathbb{B}_{2\delta}(\bar{x}), \Phi_{x_0}(\bar{x}))$$
$$\le e(Q_{\bar{x}}^{-1}(\bar{y}) \cap \mathbb{B}_{r_x}(\bar{x}), Q_{\bar{x}}^{-1}[Z_{x_0}(\bar{x})]) \tag{3.23}$$

(noting that $\mathbb{B}_{2\delta}(\bar{x}) \subseteq \mathbb{B}_{r_x}$). By the choice of ε, we have

$$\|Z_{x_0}(x) - \bar{y}\| = \|f(\bar{x}) + g(x) + \nabla f(\bar{x})(x - \bar{x}) - f(x_0) - g(x_0)$$
$$- (\nabla f(x_0) + [2x - x_0, x_0; g])(x - x_0) - \bar{y}\|$$
$$\leq \|f(\bar{x}) - f(x_0) - \nabla f(x_0)(\bar{x} - x_0)\| + \|\nabla f(x_0) - \nabla f(\bar{x})(\bar{x} - x)\|$$
$$+ \|g(x) - g(x_0) - [2x - x_0, x_0; g](x - x_0)\| + \|\bar{y}\|$$
$$\leq \varepsilon(\|\bar{x} - x_0\| + \|\bar{x} - x\|) + \|[x_0, x; g] - [2x - x_0, x_0; g]\|\|x - x_0\|$$
$$+ \|\bar{y}\|$$
$$\leq \varepsilon(\|\bar{x} - x_0\| + \|\bar{x} - x\|) + v(2\|x_0 - x\| + \|x - x_0\|)\|x - x_0\|$$
$$+ \|\bar{y}\|. \tag{3.24}$$

Note that $\|x_0 - \bar{x}\| \leq \hat{\delta} \leq \delta, 17(\varepsilon + 3v)\delta \leq r_{\bar{y}}$ by assumption (a) and $\|\bar{y}\| < (\varepsilon + 3v)\delta$ by assumption (c), it follows from (3.24) that, for each $x \in \mathbb{B}_{2\delta}(\bar{x})$,

$$\|Z_{x_0}(x) - \bar{y}\| \leq 3\varepsilon\delta + 27v\delta^2 + (\varepsilon + 3v)\delta < 3\varepsilon\delta + 27v\delta + (\varepsilon + 3v)\delta$$
$$\leq 9(\varepsilon + 3v)\delta + (\varepsilon + 3v)\delta = 10(\varepsilon + 3v)\delta$$
$$\leq r_{\bar{y}}. \tag{3.25}$$

In particular, letting $x = \bar{x}$ in (3.24). Then we have that

$$\|Z_{x_0}(\bar{x}) - \bar{y}\| \leq \varepsilon\|\bar{x} - x_0\| + v(2\|x_0 - \bar{x}\| + \|\bar{x} - x_0\|)\|\bar{x} - x_0\| + \|\bar{y}\|$$
$$= (\varepsilon + 3v\|\bar{x} - x_0\|)\|\bar{x} - x_0\| + \|\bar{y}\|$$
$$\leq (\varepsilon + 3v\delta)\delta + \|\bar{y}\| < (\varepsilon + 3v)\delta + \|\bar{y}\|$$
$$\leq 2(\varepsilon + 3v)\delta \leq r_{\bar{y}} \tag{3.26}$$

and hence $Z_{x_0}(\bar{x}) \in \mathbb{B}_{r_{\bar{y}}}(\bar{y})$.

Hence, by (3.21), (3.23), (3.26) and the assumed Lipschitz-like property, we have

$$\text{dist}(\bar{x}, \Phi_{x_0}(\bar{x})) \leq M\|\bar{y} - Z_{x_0}(\bar{x})\|$$
$$\leq M(\varepsilon + 3v\|\bar{x} - x_0\|)\|\bar{x} - x_0\| + M\|\bar{y}\|$$
$$= \left(1 - \frac{3}{5}\right)r_{x_0} = (1 - \lambda)r;$$

that is, the assertion (2.1) of Lemma 2.2 is satisfied.

Now, we show that the assertion (2.2) of Lemma 2.2 holds. To end this, let $x', x'' \in \mathbb{B}_{r_{x_0}}(\bar{x})$. Then we have that $x', x'' \in \mathbb{B}_{r_{x_0}}(\bar{x}) \subseteq \mathbb{B}_{2\delta}(\bar{x}) \subseteq \mathbb{B}_{r_{\bar{x}}}(\bar{x})$ by (3.22) and assumption (a), and $Z_{x_0}(x'), Z_{x_0}(x'') \in \mathbb{B}_{r_{\bar{y}}}(\bar{y})$ by (3.25). This together with the assumed Lipschitz-like property implies that

$$e(\Phi_{x_0}(x') \cap \mathbb{B}_{r_{x_0}}(\bar{x}), \Phi_{x_0}(x'')) \leq e(\Phi_{x_0}(x') \cap \mathbb{B}_{r_{\bar{x}}}(\bar{x}), \Phi_{x_0}(x''))$$
$$= e(Q_{\bar{x}}^{-1}[Z_{x_0}(x')] \cap \mathbb{B}_{r_{\bar{x}}}(\bar{x}), Q_{\bar{x}}^{-1}[Z_{x_0}(x'')])$$
$$\leq M\|Z_{x_0}(x') - Z_{x_0}(x'')\|. \tag{3.27}$$

Using (3.15) and the choice of x_0, we have

$$\|Z_{x_0}(x') - Z_{x_0}(x'')\| \leq \left(\|[x'', x'; g] - [2x'' - x_0, x_0; g]\| + \|\nabla f(\bar{x}) - \nabla f(x_0)\|\right)\|x' - x''\|$$
$$+ \|[2x'' - x_0, x_0; g] - [2x' - x_0, x_0; g]\|\|x' - x_0\|$$
$$\leq \left(v(\|x_0 - x''\| + \|x' - x_0\|) + \varepsilon\right)\|x' - x''\|$$
$$+ 2v\|x'' - x'\|\|x' - x_0\|$$
$$\leq (\varepsilon + 12v\delta)\|x' - x''\| \leq (\varepsilon + 16v\delta)\|x' - x''\|.$$

It follows, from $\delta \leq \dfrac{3-5M\varepsilon}{80Mv}$ as in assumption (a) together with (3.27) that

$$e(\Phi_{x_0}(x')\cap \mathbb{B}_{r_{x_0}}(\bar{x}),\Phi_{x_0}(x''))\leq M(\varepsilon+16v\delta)\|x'-x''\|\leq \tfrac{3}{5}\|x'-x''\|=\lambda\|x'-x''\|.$$

This yields that the assertion (2.2) of Lemma 2.2 is satisfied. Since both assertions of Lemma 2.2 are fulfilled, we can say that the Lemma 2.2 is applicable and hence we can conclude that there exists $\hat{x}_1\in\mathbb{B}_{r_{x_0}}(\bar{x})$ satisfying $\hat{x}_1\in\Phi_{x_0}(\hat{x}_1)$. This means that $0\in f(x_0)+g(x_0)+(\nabla f(x_0)+[2\hat{x}_1-x_0,x_0;g])(\hat{x}_1-x_0)+F(\hat{x}_1)$, i.e. $D(x_0)\neq\emptyset$. Since $\eta>1$ and $D(x_0)\neq\emptyset$, we can choose $d_0\in D(x_0)$ such that

$$\|d_0\|\leq\eta\,dist(0,D(x_0)).$$

By Algorithm 1.1, $x_1:=x_0+d_0$ is defined. Furthermore, by the definition of $D(x_0)$, we can write

$$D(x_0):=\left\{d_0\in X:0\in f(x_0)+g(x_0)+(\nabla f(x_0)+[2d_0+x_0,x_0;g])d_0+F(x_0+d_0)\right\}$$
$$=\left\{d_0\in X:0\in f(x_0)+g(x_0+d_0)+\nabla f(x_0)d_0+F(x_0+d_0)\right\}$$
$$=\left\{d_0\in X:x_0+d_0\in Q_{x_0}^{-1}(0)\right\}$$

and so

$$dist(0,D(x_0))=dist(x_0,Q_{x_0}^{-1}(0)). \tag{3.28}$$

Now, we show that (3.20) holds also for $n=0$. Note by (3.16) that

$$\varepsilon\geq\sup_{x\in\mathbb{B}_{\frac{r_{\bar{x}}}{2}}(\bar{x})}\|\nabla f(x)-\nabla f(\bar{x})\|$$

and note also that $\bar{r}>0$ by assumption (a). Therefore, (3.5) satisfies (3.6). Hence, by the assumed Lipschitz-like property of $Q_{\bar{x}}^{-1}(\cdot)$, it follows from Lemma 3.1 that the mapping $Q_x^{-1}(\cdot)$ is Lipschitz-like on $\mathbb{B}_{\bar{r}}(\bar{y})$ relative to $\mathbb{B}_{\frac{r_{\bar{x}}}{2}}(\bar{x})$ with constant $\dfrac{M}{1-M\varepsilon}$ for each $x\in\mathbb{B}_{\frac{r_{\bar{x}}}{2}}(\bar{x})$. In particular, $Q_{x_0}^{-1}(\cdot)$ is Lipschitz-like on $\mathbb{B}_{\bar{r}}(\bar{y})$ relative to $\mathbb{B}_{\frac{r_{\bar{x}}}{2}}(\bar{x})$ with constant $\dfrac{M}{1-M\varepsilon}$ as $x_0\in\mathbb{B}_\delta(\bar{x})\subset\mathbb{B}_\delta(\bar{x})\subset\mathbb{B}_{\frac{r_{\bar{x}}}{2}}(\bar{x})$ by assumption (a) and by the choice of $\hat{\delta}$. Furthermore, assumptions (a) and (c) imply that

$$\|\bar{y}\|<(\varepsilon+3v)\delta\leq\dfrac{\bar{r}}{3}. \tag{3.29}$$

and hence (3.18) implies that

$$dist(0,Q_{x_0}(x_0))=dist\big(0,f(x_0)+g(x_0)+F(x_0)\big)\leq(\varepsilon+3v)\delta$$
$$\leq\dfrac{\bar{r}}{3} \tag{3.30}$$

It is noted earlier that $x_0\in\mathbb{B}_{\frac{r_{\bar{x}}}{2}}(\bar{x})$ and $0\in\mathbb{B}_{\frac{\bar{r}}{3}}(\bar{y})$ by (3.29). Thus, we can apply Lemma 2.1 and utilizing it, we get

$$dist\big(x_0,Q_{x_0}^{-1}(0)\big)\leq\dfrac{M}{1-M\varepsilon}\,dist\big(0,Q_{x_0}(x_0)\big).$$

This together with (3.28) gives that

$$dist\big(0,D(x_0)\big)=dist\big(x_0,Q_{x_0}^{-1}(0)\big)\leq\dfrac{M}{1-M\varepsilon}\,dist\big(0,Q_{x_0}(x_0)\big). \tag{3.31}$$

According to Algorithm 1.1 and using (3.10) and (3.31), we have

$$\|d_0\| \le \eta \operatorname{dist}(0,D(x_0)) \le \frac{\eta M}{1-M\varepsilon} \operatorname{dist}(0,Q_{x_0}(x_0))$$
$$\le \frac{\eta M(\varepsilon+3v)\delta}{1-M\varepsilon} = q\delta.$$

This implies that

$$\|x_1 - x_0\| = \|d_0\| \le q\delta$$

and therefore, (3.20) is hold for $n=0$.

We assume that $x_1, x_2, \ldots, x_k$ are constructed so that (3.19) and (3.20) are hold for $n=0,1,2,\ldots,k-1$. We will show that there exists x_{k+1} such that (3.19) and (3.20) are also hold for $n=k$. Since (3.19) and (3.20) are true for each $n \le k-1$, we have the following inequality

$$\|x_k - \bar{x}\| \le \sum_{i=0}^{k-1} \|d_i\| + \|x_0 - \bar{x}\| \le \delta \sum_{i=0}^{k-1} q^{i+1} + \delta \le \frac{\delta q}{1-q} + \delta \le 2\delta.$$

This shows that (3.19) holds for $n=k$. Now with almost the same argument as we did for the case when $n=0$, we can show that (3.20) hold for $n=k$. The proof is complete. $\qquad\square$

In particular, in the case when $\bar{x}$ is a solution of (1.2), i.e. $\bar{y}=0$, Theorem 3.1 is reduced to the following corollary, which gives the local convergent result for the extended Newton-type method.

COROLLARY 3.1   *Suppose that $\eta > 1$ and $\bar{x}$ satisfies $0 \in f(\bar{x})+g(\bar{x})+F(\bar{x})$. Let $Q_{\bar{x}}^{-1}(\cdot)$ be pseudo-Lipschitz around $(0,\bar{x})$. Let $\tilde{r} > 0$, $v > 0$ and suppose that ∇f is continuous on $\mathbb{B}_{\tilde{r}}(\bar{x})$ and that*

$$\lim_{x \to \bar{x}} \operatorname{dist}(0, f(x)+g(x)+F(x)) = 0.$$

Then there exists some δ such that any sequence $\{x_n\}$ generated by Algorithm 1.1 with initial point in $\mathbb{B}_\delta(\bar{x})$ converges to a solution x^* of (1.1), that is, x^* satisfies that $0 \in f(x^*)+g(x^*)+F(x^*)$.

Proof   Let $Q_{\bar{x}}^{-1}(\cdot)$ is pseudo-Lipschitz around $(0,\bar{x})$. Then there exist constants r_0, $\hat{r}_{\bar{x}}$ and M satisfy the following condition:

$$e(Q_{\bar{x}}^{-1}(y_1) \cap \mathbb{B}_{\hat{r}_{\bar{x}}}(\bar{x}), Q_{\bar{x}}^{-1}(y_2)) \le M\|y_1 - y_2\|, \quad \text{for every } y_1, y_2 \in \mathbb{B}_{r_0}(0). \tag{3.32}$$

Thus, by the definition of Lipschitz-like property, we can say that $Q_{\bar{x}}^{-1}(\cdot)$ is Lipschitz-like on $\mathbb{B}_{r_0}(0)$ relative to $\mathbb{B}_{\hat{r}_{\bar{x}}}$ with constant M which satisfy (3.32). Then, for each $0 < \tilde{r} \le \hat{r}_{\bar{x}}$, one has that

$$e(Q_{\bar{x}}^{-1}(y_1) \cap \mathbb{B}_{\tilde{r}}(\bar{x}), Q_{\bar{x}}^{-1}(y_2)) \le M\|y_1 - y_2\|, \quad \text{for every } y_1, y_2 \in \mathbb{B}_{r_0}(0),$$

that is, $Q_{\bar{x}}^{-1}(\cdot)$ is Lipschitz-like on $\mathbb{B}_{r_0}(0)$ relative to $\mathbb{B}_{\tilde{r}}(\bar{x})$ with constant M. Let $\varepsilon \in (0,1)$ be such that $M((6\eta+1)\varepsilon+3v) \le 1$. By the continuity of ∇f, we can choose $r_{\bar{x}} \in (0,\hat{r}_{\bar{x}})$ such that $\frac{r_{\bar{x}}}{2} \le \tilde{r}$, $r_0 - 2\varepsilon r_{\bar{x}} > 0$ and

$$\varepsilon \ge \sup_{x,x' \in \mathbb{B}_{\frac{r_{\bar{x}}}{2}}(\bar{x})} \|\nabla f(x) - \nabla f(x')\|.$$

Then

$$\bar{r} = \min\left\{ r_0 - 2\varepsilon r_{\bar{x}}, \frac{r_{\bar{x}}(1-M\varepsilon)}{4M} \right\} < 0$$

and

$$\min \left\{ \frac{r_{\bar{x}}}{4}, \frac{\bar{r}}{3(\varepsilon+3v)}, \frac{r_0}{17(\varepsilon+3v)}, \frac{3-5M\varepsilon}{80Mv} \right\} > 0. \tag{3.33}$$

By (3.33), we can choose $0 < \delta \leq 1$ such that

$$\delta \leq \min \left\{ \frac{r_{\bar{x}}}{4}, \frac{\bar{r}}{3(\varepsilon+3v)}, 1, \frac{r_0}{17(\varepsilon+3v)}, \frac{3-5M\varepsilon}{80Mv} \right\}.$$

Thus it is routine to check that inequalities (a)–(c) of Theorem 3.1 are satisfied. Therefore, Theorem 3.1 is applicable to complete the proof.   □

In the following theorem, we show that if ∇f is Lipschitz continuous around $\bar{x}$, then the sequence generated by Algorithm 1.1 converges quadratically.

THEOREM 3.2   Let $\eta > 1$ and suppose that $Q_{\bar{x}}^{-1}(\cdot)$ is Lipschitz-like on $\mathbb{B}_{\bar{r}}(\bar{y})$ relative to $\mathbb{B}_{r_{\bar{x}}}(\bar{x})$ with constant M and that ∇f is Lipschitz continuous on $\mathbb{B}_{\frac{r_{\bar{x}}}{2}}(\bar{x})$ with Lipschitz constant L. Let

$$\bar{r} := \min \left\{ r_{\bar{y}} - 2Lr_{\bar{x}}^2, \frac{r_{\bar{x}}(1-MLr_{\bar{x}})}{4M} \right\}.$$

Let $v > 0, \delta > 0$ be such that

(a) $\delta \leq \min \left\{ \frac{r_{\bar{x}}}{4}, 6\bar{r}, 1, \frac{4r_{\bar{y}}}{33(L+6v)} \right\}$,

(b) $(M+1)(L+6v)(\eta\delta + 2r_{\bar{x}}) \leq 2$,

(c) $\|\bar{y}\| < \frac{(L+6v)\delta^2}{4}$.

Suppose that

$$\lim_{x \to \bar{x}} \mathrm{dist}(\bar{y}, f(x) + g(x) + F(x)) = 0. \tag{3.34}$$

Then there exist some $\hat{\delta} > 0$ such that any sequence $\{x_n\}$ generated by Algorithm 1.1 with initial point in $\mathbb{B}_{\hat{\delta}}(\bar{x})$ converges quadratically to a solution x^* of (1.1).

Proof   Setting

$$q := \frac{\eta M(L+6v)\delta}{2(1-MLr_{\bar{x}})} \tag{3.35}$$

Then, thanks to the assumption (b) for allowing us to write the fact that

$$M(L+6v)\eta\delta + 2MLr_{\bar{x}} < (M+1)(L+6v)(\eta\delta + 2r_{\bar{x}}) \leq 2.$$

It follows from (3.35) that

$$q := \frac{\eta M(L+6v)\delta}{2(1-MLr_{\bar{x}})} \leq 1. \tag{3.36}$$

Taking $0 < \hat{\delta} \leq \delta$ such that

$$\mathrm{dist}(0, f(x_0) + g(x_0) + F(x_0)) \leq \frac{(L+6v)\delta^2}{4} \quad \text{for each } x_0 \in \mathbb{B}_{\hat{\delta}}(\bar{x}). \tag{3.37}$$

It is noting that such $\hat{\delta}$ exists by (3.34) and assumption (c). Let $x_0 \in \mathbb{B}_{\hat{\delta}}(\bar{x})$. To complete the proof of this theorem, we use almost similar argument that we used for completing the proof of Theorem 3.1.

We show that Algorithm 1.1 generates at least one sequence and such sequence $\{x_n\}$ generated by Algorithm 1.1 satisfies the following assertions:

$$\|x_n - \bar{x}\| \leq 2\delta; \tag{3.38}$$

and

$$\|d_n\| \leq q\left(\frac{1}{2}\right)^{2^n}\delta \tag{3.39}$$

hold for each $n = 0, 1, 2, \ldots.$ Let

$$r_x := \frac{3}{2}\left(M(L+6v)\|x-\bar{x}\|^2 + 2M\|\bar{y}\|\right) \quad \text{for each } x \in X. \tag{3.40}$$

Owing to the fact $\delta \leq \dfrac{r_{\bar{x}}}{4}$ in assumption (a) and $\eta > 1$, by assumption (b) we can write as follows

$$9(M+1)(L+6v)\delta = (M+1)(L+6v)(\delta + 8\delta)$$
$$\leq (M+1)(L+6v)(\eta\delta + 2r_{\bar{x}})$$
$$\leq 2.$$

This gives

$$M(L+6v)\delta \leq \frac{2}{9} \quad \text{and} \quad (L+6v)\delta \leq \frac{2}{9} \tag{3.41}$$

and hence by $\delta \leq 6\bar{r}$ in assumption (a) together with second inequality of (3.41), we get

$$\|\bar{y}\| < \frac{(L+6v)\delta^2}{4} \leq \frac{2}{9\cdot 4}\cdot 6\bar{r} = \frac{\bar{r}}{3}; \tag{3.42}$$

thanks to assumption (c). Utilizing the first inequality from (3.41) and assumption (c) together, we obtain from (3.40) that

$$r_x < \frac{3}{2}\left(4M(L+6v)\delta^2 + \frac{M(L+6v)\delta^2}{2}\right)$$
$$= \frac{27}{4}M(L+6v)\delta^2 \leq 2\delta \quad \text{for each } x \in \mathbb{B}_{2\delta}(\bar{x}). \tag{3.43}$$

Note that (3.38) is trivial for $n = 0$. In order to show that (3.39) is hold for $n = 0$, we need to prove $D(x_0) \neq \emptyset$. The nonemptyness of $D(x_0)$ will ensure us to deduce the existence of the point x_1. To complete this, we will apply Lemma 2.2 to the map Φ_{x_0} with $\eta_0 = \bar{x}$. Let us check that both assertions (2.1) and (2.2) of Lemma 2.2 hold with $r := r_{x_0}$ and $\lambda := \frac{2}{3}$. Noting that $x \in Q_{\bar{x}}^{-1}(y) \cap \mathbb{B}_{2\delta}(\bar{x})$ by (3.2) and according to the definition of the excess e and the mapping Φ_{x_0} by (3.14), we obtain

$$\text{dist}(\bar{x}, \Phi_{x_0}(\bar{x})) \leq e(Q_{\bar{x}}^{-1}(\bar{y}) \cap \mathbb{B}_{r_{x_0}}(\bar{x}), \Phi_{x_0}(\bar{x})) \leq e(Q_{\bar{x}}^{-1}(\bar{y}) \cap \mathbb{B}_{2\delta}(\bar{x}), \Phi_{x_0}(\bar{x}))$$
$$\leq e(Q_{\bar{x}}^{-1}(\bar{y}) \cap \mathbb{B}_{r_{\bar{x}}}(\bar{x}), Q_{\bar{x}}^{-1}[Z_{x_0}(\bar{x})]). \tag{3.44}$$

By the assumed Lipschitz continuity of ∇f and for each $x \in \mathbb{B}_{2\delta}(\bar{x}) \subseteq \mathbb{B}_{\frac{r_{\bar{x}}}{2}}(\bar{x})$, we obtain that

$$\|Z_{x_0}(x)-\bar{y}\| = \|f(\bar{x})+g(x)+\nabla f(\bar{x})(x-\bar{x})-f(x_0)-g(x_0)$$
$$-(\nabla f(x_0)+[2x-x_0,x_0;g])(x-x_0)-\bar{y}\|$$
$$\leq \|f(\bar{x})-f(x_0)-\nabla f(x_0)(\bar{x}-x_0)\|+\|\nabla f(x_0)-\nabla f(\bar{x})(\bar{x}-x)\|$$
$$+\|g(x)-g(x_0)-[2x-x_0,x_0;g](x-x_0)\|+\|\bar{y}\|$$
$$\leq \frac{L}{2}\|\bar{x}-x_0\|^2+L\|x_0-\bar{x}\|\|\bar{x}-x\|+\|[x_0,x;g]$$
$$-[2x-x_0,x_0;g]\|\|x-x_0\|+\|\bar{y}\|$$
$$\leq \frac{L}{2}\|\bar{x}-x_0\|^2+L\|x_0-\bar{x}\|\|\bar{x}-x\|+v\Big(2\|x_0-x\|$$
$$+\|x-x_0\|\Big)\|x-x_0\|+\|\bar{y}\|$$
$$\leq \frac{L}{2}(\delta^2+4\delta^2)+3v(3\delta)^2+\|\bar{y}\|=\frac{5L\delta^2}{2}+27v\delta^2+\|\bar{y}\|$$
$$\leq \frac{9}{2}(L+6v)\delta^2+\|\bar{y}\|. \tag{3.45}$$

It follows, from the facts $33(L+6v)\delta\leq 4r_{\bar{y}}$, $\delta\leq 1$ and $\|\bar{y}\|<\dfrac{(L+6v)\delta^2}{4}$, respectively, in assumptions (a) and (c), that

$$\|Z_{x_0}(x)-\bar{y}\| \leq \frac{9}{2}(L+6v)\delta^2+\frac{(L+6v)\delta^2}{4}=\frac{19}{4}(L+6v)\delta^2$$
$$\leq \frac{19}{4}(L+6v)\delta\leq r_{\bar{y}}. \tag{3.46}$$

This shows that $Z_{x_0}(x)\in \mathbb{B}_{r_{\bar{y}}}(\bar{y})$. In particular, let $x=\bar{x}$ in (3.45). Then it is easily shown that

$$Z_{x_0}(\bar{x})\in\mathbb{B}_{r_{\bar{y}}}(\bar{y}) \quad \text{and} \quad \|Z_{x_0}(\bar{x})-\bar{y}\|\leq\frac{(L+6v)}{2}\|\bar{x}-x_0\|^2+\|\bar{y}\|. \tag{3.47}$$

Using assumed Lipschitz-like property and (3.47) in (3.44), we have

$$\text{dist}(\bar{x},\Phi_{x_0}(\bar{x})) \leq M\|\bar{y}-Z_{x_0}(\bar{x})\|\leq\frac{M(L+6v)}{2}\|\bar{x}-x_0\|^2+M\|\bar{y}\|$$
$$=\Big(1-\frac{2}{3}\Big)r_{x_0}=(1-\lambda)r,$$

that is, the assertion (2.1) of Lemma 2.2 is satisfied.

Now, we show that assertion (2.2) of Lemma 2.2 holds. To end this, let $x',x''\in\mathbb{B}_{r_{x_0}}(\bar{x})$. Then we have that $x',x''\in\mathbb{B}_{r_{x_0}}(\bar{x})\subseteq\mathbb{B}_{2\delta}(\bar{x})\subseteq\mathbb{B}_{r_{\bar{x}}}(\bar{x})$ by (3.43) and $Z_{x_0}(x'),Z_{x_0}(x'')\in\mathbb{B}_{r_{\bar{y}}}(\bar{y})$ by (3.46). This together with the assumed Lipschitz-like property implies that

$$e(\Phi_{x_0}(x')\cap\mathbb{B}_{r_{x_0}}(\bar{x}),\Phi_{x_0}(x''))\leq e(\Phi_{x_0}(x')\cap\mathbb{B}_{r_{\bar{x}}}(\bar{x}),\Phi_{x_0}(x''))$$
$$=e(Q_{\bar{x}}^{-1}[Z_{x_0}(x')]\cap\mathbb{B}_{\delta}(\bar{x}),Q_{\bar{x}}^{-1}[Z_{x_0}(x'')])$$
$$\leq M\|Z_{x_0}(x')-Z_{x_0}(x'')\|.$$

By the choice of x_0, (3.15) yields that

$$\|Z_{x_0}(x')-Z_{x_0}(x'')\| \leq \Big(\|[x'',x';g]-[2x''-x_0,x_0;g]\|+\|\nabla f(\bar{x})-\nabla f(x_0)\|\Big)\|x'-x''\|$$
$$+\|[2x''-x_0,x_0;g]-[2x'-x_0,x_0;g]\|\|x'-x_0\|$$
$$\leq \Big(v\big(\|x_0-x''\|+\|x'-x_0\|\big)+L\|\bar{x}-x_0\|\Big)\|x'-x''\|$$
$$+2v\|x''-x'\|\|x'-x_0\|$$
$$\leq (L+12v)\delta\|x'-x''\|\leq 2(L+6v)\delta\|x'-x''\|.$$

Combining above inequality and first inequality from (3.41), we obtain that

$$e(\Phi_{x_0}(x') \cap \mathbb{B}_{r_{x_0}}(\bar{x}), \Phi_{x_0}(x'')) \leq 2M(L+6v)\delta\|x'-x''\| \leq 3M(L+6v)\delta\|x'-x''\|$$

$$\leq \frac{2}{3}\|x'-x''\| = \lambda\|x'-x''\|.$$

This means that the assertion (2.2) of Lemma 2.2 is also satisfied. Since both assertions of Lemma 2.2 are fulfilled, we can conclude that Lemma 2.2 is applicable to deduce the existence of a point $\hat{x}_1 \in \mathbb{B}_{r_{x_0}}(\bar{x})$ such that $\hat{x}_1 \in \Phi_{x_0}(\hat{x}_1)$. This implies that $0 \in f(x_0)+g(x_0)+(\nabla f(x_0)+[2\hat{x}_1-x_0,x_0;g])(\hat{x}_1-x_0)+F(\hat{x}_1)$, that is $D(x_0) \neq \emptyset$. Since $\eta > 1$ and $D(x_0) \neq \emptyset$, we can choose $d_0 \in D(x_0)$ such that

$$\|d_0\| \leq \eta \, dist(0, D(x_0)).$$

By Algorithm 1.1, $x_1 := x_0+d_0$ is defined. Furthermore, by the definition of $D(x_0)$, we can write

$$D(x_0): = \left\{ d_0 \in X : 0 \in f(x_0)+g(x_0)+(\nabla f(x_0)+[2d_0+x_0,x_0;g])d_0+F(x_0+d_0) \right\}$$

$$= \left\{ d_0 \in X : 0 \in f(x_0)+g(x_0+d_0)+\nabla f(x_0)d_0+F(x_0+d_0) \right\}$$

$$= \left\{ d_0 \in X : x_0+d_0 \in Q_{x_0}^{-1}(0) \right\}$$

and so

$$dist(0, D(x_0)) = dist(x_0, Q_{x_0}^{-1}(0)). \tag{3.48}$$

Now we are ready to show that (3.39) is hold for $n=0$. Since ∇f is Lipschitz continuous on $\mathbb{B}_{\frac{r_{\bar{x}}}{2}}(\bar{x})$ with Lipschitz constant L, we have

$$Lr_{\bar{x}} \geq \sup_{x', x'' \in \mathbb{B}_{\frac{r_{\bar{x}}}{2}}(\bar{x})} \|\nabla f(x')-\nabla f(x'')\| \geq \sup_{x \in \mathbb{B}_{\frac{r_{\bar{x}}}{2}}(\bar{x})} \|\nabla f(x)-\nabla f(\bar{x})\|. \tag{3.49}$$

Note that $\bar{r} > 0$ by assumption (a). Therefore, (3.5) and (3.49) imply that assumption (3.6) is satisfied with $\varepsilon := Lr_{\bar{x}}$. Since $Q_{\bar{x}}^{-1}(\cdot)$ is Lipschitz-like on $\mathbb{B}_{\bar{r}}(\bar{y})$ relative to $\mathbb{B}_{r_{\bar{x}}}(\bar{x})$, it follows from Lemma 3.1 that for each $x \in \mathbb{B}_{\frac{r_{\bar{x}}}{2}}(\bar{x})$ the mapping $Q_x^{-1}(\cdot)$ is Lipschitz-like on $\mathbb{B}_{\bar{r}}(\bar{y})$ relative to $\mathbb{B}_{\frac{r_{\bar{x}}}{2}}(\bar{x})$ with constant $\frac{M}{1-MLr_{\bar{x}}}$. In particular, $Q_{x_0}^{-1}(\cdot)$ is Lipschitz-like on $\mathbb{B}_{\bar{r}}(\bar{y})$ relative to $\mathbb{B}_{\frac{r_{\bar{x}}}{2}}(\bar{x})$ with constant $\frac{M}{1-MLr_{\bar{x}}}$ as $x_0 \in \mathbb{B}_{\hat{\delta}}(\bar{x}) \subseteq \mathbb{B}_{2\delta}(\bar{x}) \subseteq \mathbb{B}_{\frac{r_{\bar{x}}}{2}}(\bar{x})$ by assumption (a). Furthermore, (3.37) implies that

$$dist(0, Q_{x_0}(x_0)) - dist(0, f(x_0)+g(x_0)+F(x_0)) \leq \frac{\bar{r}}{3}.$$

It is noted earlier that $x_0 \in \mathbb{B}_{\frac{r_{\bar{x}}}{2}}(\bar{x})$ and $0 \in \mathbb{B}_{\frac{\bar{r}}{3}}(\bar{y})$ by (3.42). Thus, Lemma 2.1 is applied to get the following inequality:

$$dist(x_0, Q_{x_0}^{-1}(0)) \leq \frac{M \, dist(0, Q_{x_0}(x_0))}{1-MLr_{\bar{x}}} = \frac{M \, dist(0, f(x_0)+g(x_0)+F(x_0))}{1-MLr_{\bar{x}}}.$$

Furthermore, by (3.48), we have that

$$dist(0, D(x_0)) = dist(x_0, Q_{x_0}^{-1}(0)) \leq \frac{M \, dist(0, f(x_0)+g(x_0)+F(x_0))}{1-MLr_{\bar{x}}}. \tag{3.50}$$

According to Algorithm 1.1 and using (3.36), (3.37), and (3.50), we have

$$\|d_0\| \le \eta \, \text{dist}(0, D(x_0))$$
$$\le \frac{\eta M \, \text{dist}(0, f(x_0) + g(x_0) + F(x_0))}{(1 - ML r_{\bar{x}})}$$
$$\le \frac{\eta M (L + 6v)\delta^2}{4(1 - ML r_{\bar{x}})} = q\left(\frac{1}{2}\right)\delta.$$

This yields that

$$\|x_1 - x_0\| = \|d_0\| \le q\left(\frac{1}{2}\right)\delta$$

and therefore, (3.39) is true for $n = 0$. We assume that $x_1, x_2, \ldots, x_k$ are constructed and (3.38), and (3.39) are true for $n = 0, 1, 2, \ldots, k-1$. We show that there exists x_{k+1} such that (3.38) and (3.39) are also hold for $n = k$. Since (3.38) and (3.39) are true for each $n \le k-1$, we have the following inequality:

$$\|x_k - \bar{x}\| \le \sum_{i=0}^{k-1} \|d_i\| + \|x_0 - \bar{x}\| \le q\delta \sum_{i=0}^{k-1} \left(\frac{1}{2}\right)^{2^i} + \delta \le 2\delta.$$

This shows that (3.38) holds for $n = k$. Now with almost the same argument as we did for the case when $n = 0$, we can show that (3.39) holds for $n = k$. The proof is complete. □

Consider the special case when $\bar{x}$ is a solution of (1.1)(that is, $\bar{y} = 0$) in Theorem 3.2. We have the following corollary, which gives the local quadratic convergence result for the extended Newton-type method. The proof of this corollary is similar to that we did for Corollary 3.1.

COROLLARY 3.2 *Suppose that $\bar{x}$ satisfies $0 \in f(\bar{x}) + g(\bar{x}) + F(\bar{x})$ and that $Q_{\bar{x}}^{-1}(\cdot)$ is pseudo-Lipschitz around $(0, \bar{x})$. Let $\eta > 1$, $v > 0$, $\tilde{r} > 0$ and suppose that ∇f is Lipschitz continuous on $\mathbb{B}_{\tilde{r}}(\bar{x})$ with Lipschitz constant L. Suppose that*

$$\lim_{x \to \bar{x}} \text{dist}(0, f(x) + g(x) + F(x)) = 0.$$

Then there exist some $\hat{\delta} > 0$ such that any sequence $\{x_n\}$ generated by Algorithm 1.1 with initial point in $\mathbb{B}_{\hat{\delta}}(\bar{x})$ converges quadratically to a solution x^* of (1.1).

4. Concluding remarks
When $\eta > 1$, we have established semi-local and local convergence results for the extended Newton-type method under the assumptions that $Q_{\bar{x}}^{-1}(\cdot)$ is Lipschitz-like and ∇f is continuous. In particular, if ∇f is additionally Lipschitz continuous, we further show that the extended Newton-type method is quadratically convergent. For the case where $\eta = 1$, the question, whether the results are true for the extended Newton-type method, is a little bit complicated. However, from the proof of the main theorems, one sees that all the results obtained in the present paper remain true provided that, for any $x \in \Omega \subseteq X$, the following implication holds:

$$D(x) \ne \emptyset \Longrightarrow \exists \, \bar{d} \in D(x) \text{ such that} \|\bar{d}\| = \min_{d \in D(x)} \|d\|.$$

To see the detail proof of the above implication, one can refer to (Rashid et al., 2013).

Funding
The authors received no direct funding for this research.

[1] Faculty of Science, Department of Mathematics, University of Rajshahi, Rajshahi-6205, Bangladesh.

Author details
M.H. Rashid[1]
E-mail: harun_math@ru.ac.bd

References

Alexis, C.J., & Pietrus, A. (2008). On the convergence of some methods for variational inclusions. *Revista de la Real Academia de Ciencias Exactas, Físicas y Naturales. Serie A, Matemáticas, 102,* 355–361.

Argyros, I. K. (2004). A unifying local-semilocal convergence analysis and applications for two-point Newton-like methods in Banach space. *Journal of Mathematical Analysis and Applications, 298,* 374–397.

Argyros, I. K. (2007). *Computational theory of iterative methods.* Studies of computer mathematics. Amsterdam: Elsevier.

Argyros, I. K., & Hilout, S. (2008). Local convergence of Newton-like methods for generalized equations. *Applied Mathematics and Computer, 197,* 507–514.

Aubin, J. P. (1984). Lipschitz behavior of solutions to convex minimization problems. *Mathematics of Operations Research, 9,* 87–111.

Aubin, J. P., & Frankowska, H. (1990). *Set-valued analysis.* Boston: Birkhäuser.

Dedieu, J. P., & Kim, M. H. (2002). Newton's method for analytic systems of equations with constant rank derivatives. *Journal of Complexity, 18,* 187–209.

Dedieu, J. P., & Shub, M. (2000). Newton's method for overdetermined systems of equations. *Mathematics of Computation, 69,* 1099–1115.

Dontchev, A. L. (1996a). The Graves theorem revisited. *Journal of Convex Analysis, 3,* 45–53.

Dontchev, A. L. (1996b). Local convergence of the Newton method for generalized equation. *C.R.A.S Paris Series-I, 322,* 327–331.

Dontchev, A. L. (1996c). Uniform convergence of the Newton method for Aubin continuous maps. *Serdica Mathematical Journal, 22,* 385–398.

Dontchev, A. L., & Hager, W. W. (1994). An inverse mapping theorem for set-valued maps. *Proceedings of the American Mathematical Soceity, 121,* 481–498.

Hilout, S., Alexis, C. J., & Pietrus, A. (2006). A semilocal convergence of the secant-type method for solving a generalized equations. *Positivity, 10,* 673–700.

Ioffe, A. D., & Tikhomirov, V. M. (1979). *Theory of extremal problems.* Studies in mathematics and its applications. Amsterdam: North-Holland.

Li, C., & Ng, K. F. (2007). Majorizing functions and convergence of the Gauss-Newton method for convex composite optimization. *SIAM Journal on Optimization, 18,* 613–642.

Pietrus, A. (2000a). Generalized equations under mild differentiability conditions. *Revista de la Real Academia de Ciencias Exactas, Fisicas y Naturales – Serie A: Matematicas, 94*(1), 15–18.

Pietrus, A. (2000b). Does Newton's method for set-valued maps converges uniformly in mild differentiability context? *Revista Colombiana de Matemáticas, 32,* 49–56.

Rashid, M. H., Wang, J. H., & Li, C. (2012). Convergence analysis of a method for variational inclusions. *Applicable Analysis, 91,* 1943–1956. Retrieved from: http://dx.doi.org/10.1080/00036811.2011.618127

Rashid, M. H., Yu, S. H., Li, C., & Wu, S. Y. (2013). Convergence analysis of the Gauss–Newton-type method for Lipschitz-like mappings. *Journal of Optimization Theory and Applications, 158,* 216–233. doi:10.1007/s10957-012-0206-3.

Xu, X. B., & Li, C. (2008). Convergence criterion of Newton's method for singular systems with constant rank derivatives. *Journal of Mathematical Analysis and Applications, 345,* 689–701.

6

Boundedness on Orlicz space of Toeplitz type operators related to multiplier operator and mean oscillation

Ouyang Difei[1*]

*Corresponding author: Ouyang Difei, College of Mathematics, Changsha University of Science and Technology, Changsha 410077, Hunan, P.R. China

E-mail: ouyangdifei1965@163.com

Reviewing editor: Lishan Liu, Qufu Normal University, China

Abstract: In this paper, the boundedness for certain Toeplitz type operator related to the multiplier operator from Lebesgue space to Orlicz space is obtained.

Subjects: Science; Advanced Mathematics; Pure Mathematics

Keywords: Toeplitz type operator; multiplier operator; BMO space; Orlicz space

MR Subject classifications: 42B20; 42B25

1. Introduction and preliminaries

Let b be a locally integrable function on R^n and T be an integral operator. For a suitable function f, the commutator generated by b and T is defined by $[b,T]f = bT(f) - T(bf)$. It is well known that one important role of commutators is to characterize function spaces, which is originated by Coifman, Rochberg, and Weisss (1976). They characterized BMO space via the Lp boundedness of the commutator for singular integral operator. Since then, similar results of other operators have also been obtained (see Chanillo, 1982; Janson, 1978, Paluszynski, 1995). Now, with the development of singular integral operators (see Garcia-Cuerva & Rubio de Francia, 1985; Stein, 1993), their commutators have been well studied. In Coifman et al. (1976), Wang and Liu (2009a, 2009b), the authors proved that the commutators generated by the singular integral operators and BMO functions are bounded on $L^p(R^n)$ for $1 < p < \infty$. Chanillo (1982) proved a similar result when singular integral operators are replaced by the fractional integral operators. Janson (1978) proved the boundedness for the commutators generated by the singular integral operators and BMO functions from Lebesgue spaces to Orlicz spaces. Lu and Mo (2009), some multiplier operators are introduced and the boundedness for the operators are obtained (see Kurtz & Wheeden, 1979; Muckenhoupt, Wheeden, & Young, 1987; Wang & Liu, 2009a, 2009b; You, 1988; Zhang & Chen, 2005, 2006). Krantz and Li (2001), Lu and Mo (2009), some Toeplitz type operators related to the singular integral operators are introduced, and the boundedness for the operators generated by BMO and Lipschitz functions are obtained. Motivated by these, in this paper, we will prove the boundedness properties of the Toeplitz type operator associated to the multiplier operator from Lebesgue space to Orlicz space.

ABOUT THE AUTHOR

Ouyang Difei has been studying the topics from 2000 year. The issue will provide some new thinking and methods for the domain. It is natural and important for the domain.

PUBLIC INTEREST STATEMENT

It is one of the core problems in analysis mathematics of studying the boundedness of integral operators on function spaces. The Toeplitz type operators are important ones, which are the non-trival and natural generalizations of the commututor operator. It is an advanced and hot research topics in harmonic analysis to study the boundedness of Toeplitz type operators on function spaces. In this paper, the boundedness for certain Toeplitz type operator related to the multiplier operator from Lebesgue space to Orlicz space is obtained.

First, let us introduce some notations. Throughout this paper, Q will denote a cube of R^n with sides parallel to the axes. For any locally integrable function f, the sharp function of f is defined by

$$f^{\#}(x) = \sup_{Q \ni x} \frac{1}{|Q|} \int_Q |f(y) - f_Q| dy$$

where, and in what follows, $f_Q = |Q|^{-1} \int_Q f(x) dx$. It is well-known that (see Garcia-Cuerva & Rubio de Francia, 1985)

$$f^{\#}(x) \approx \sup_{Q \ni x} \inf_{c \in C} \frac{1}{|Q|} \int_Q |f(y) - c| dy$$

Let M be the Hardy–Littlewood maximal operator defined by

$$M(f)(x) = \sup_{Q \ni x} \frac{1}{|Q|} \int_Q |f(y)| dy$$

We write that $M_p f = \left(M(f^p) \right)^{1/p}$ for $0 < p < \infty$. For $1 \le r < \infty$ and $0 < \eta < n$, let

$$M_{\eta, r}(f)(x) = \sup_{Q \ni x} \left(\frac{1}{|Q|^{1 - r\eta/n}} \int_Q |f(y)|^r dy \right)^{1/r}$$

We say that f belongs to $BMO(R^n)$ if $f^{\#}$ belongs to $L^\infty(R^n)$ and $||f||_{BMO} = ||f^{\#}||_{L^\infty}$. More generally, let ρ be a non-decreasing positive function on $[0, +\infty)$ and define $BMO_\rho(R^n)$ as the space of all functions f such that

$$\frac{1}{|Q(x,r)|} \int_{Q(x,r)} |f(y) - f_Q| dy \le C\rho(r)$$

For $\beta > 0$, the Lipschitz space $Lip_\beta(R^n)$ is the space of functions f such that

$$||f||_{Lip_\beta} = \sup_{x \ne y} |f(x) - f(y)| / |x - y|^\beta < \infty$$

For f, m_f denotes the distribution function of f, that is $m_f(t) = |\{x \in R^n : |f(x)| < t\}|$.

Let ρ be a non-decreasing convex function on $[0, +\infty)$ with $\rho(0) = 0$. ρ^{-1} denotes the inverse function of ρ. The Orlicz space $L_\rho(R^n)$ is defined by the set of functions f such that $\int_{R^n} \rho(\lambda |f(x)|) dx < \infty$ for some $\lambda > 0$. The norm is given by

$$||f||_{L_\rho} = \inf_{\lambda > 0} \lambda^{-1} \left(1 + \int_{R^n} \rho(\lambda |f(x)|) dx \right)$$

2. Results
In this paper, we will study the multilinear operator as following (see Kurtz & Wheeden, 1979).

A bounded measurable function k defined on $R^n \setminus \{0\}$ is called a multiplier. The multiplier operator T associated with k is defined by

$$T(f)(x) = k(x)\hat{f}(x), \quad \text{for } f \in S(R^n)$$

where $\hat{f}$ denotes the Fourier transform of f and $S(R^n)$ is the Schwartz test function class. Now, we recall the definition of the class $M(s, l)$. Denote by $|x| \sim t$ the fact that the value of x lies in the annulus $\{x \in R^n : ct < |x| < Ct\}$, where $0 < c < C < \infty$ are values specified in each instance.

Definition 1   Let $l \ge 0$ be a real number and $1 \le s \le 2$. we say that the multiplier k satisfies the condition $M(s, l)$, if

$$\left(\int_{|\xi| \sim R} |D^\alpha k(\xi)|^s d\xi \right)^{\frac{1}{s}} < CR^{n/s - |\alpha|}$$

for all $R > 0$ and multi-indices α with $|\alpha| \le l$, when l is a positive integer, and, in addition, if

$$\left(\int_{|\xi| \sim R} |D^\alpha k(\xi) - D^\alpha k(\xi - z)|^s d\xi \right)^{\frac{1}{s}} \le C \left(\frac{|z|}{R} \right)^\gamma R^{\frac{n}{s} - |\alpha|}$$

for all $|z| < R/2$ and all multi-indices α with $|\alpha| = [l]$, the integer part of l, i.e. $[l]$ is the greatest integer less than or equal to l, and $l = [l] + \gamma$ when l is not an integer.

Denote $D(R^n) = \{\phi \in S(R^n): \text{supp}(\phi) \text{ is compact}\}$ and $\hat{D}_0(R^n) = \{\phi \in S(R^n): \hat{\phi} \in D(R^n) \text{ and } \hat{\phi}$ vanishes in a neighbourhood of the origin$\}$. The following boundedness property of T on $L^p(R^n)$ is proved by Strömberg and Torkinsky (see Kurtz & Wheeden, 1979).

Definition 2   For a real number $\tilde{l} \ge 0$ and $1 \le \tilde{s} < \infty$, we say that K verifies the condition $\tilde{M}(\tilde{s}, \tilde{l})$, and write $K \in \tilde{M}(\tilde{s}, \tilde{l})$, if

$$\left(\int_{|x| \sim R} |D^{\tilde{\alpha}} K(x)|^{\tilde{s}} dx \right)^{\frac{1}{\tilde{s}}} \le C R^{n/\tilde{s} - n - |\tilde{\alpha}|}, \quad R > 0$$

for all multi-indices $|\tilde{\alpha}| \le \tilde{l}$ and, in addition, if

$$\left(\int_{|x| \sim R} |D^{\tilde{\alpha}} K(x) - D^{\tilde{\alpha}} K(x - z)|^{\tilde{s}} dx \right)^{\frac{1}{\tilde{s}}} \le C \left(\frac{|z|}{R} \right)^v R^{\frac{n}{\tilde{s}} - n - u}, \quad \text{if } 0 < v < 1$$

$$\left(\int_{|x| \sim R} |D^{\tilde{\alpha}} K(x) - D^{\tilde{\alpha}} K(x - z)|^{\tilde{s}} dx \right)^{\frac{1}{\tilde{s}}} \le C \left(\frac{|z|}{R} \right) \left(\log \frac{R}{|z|} \right) R^{\frac{n}{\tilde{s}} - n - u}, \quad \text{if } v = 1$$

for all $|z| > \frac{R}{2}, R < 0$, and all multi-indices $\tilde{\alpha}$ with $|\tilde{\alpha}| = u$, where u denotes the largest integer strictly less than $\tilde{l}$ with $\tilde{l} = u + v$.

LEMMA 1   (*see Kurtz & Wheeden, 1979*)   Let $k \in M(s, l), 1 \le s \le 2$, and $l > \frac{n}{s}$. Then the associated mapping T, defined a priori for $f \in \hat{D}_0(R^n)$, $T(f)(x) = (f * K)(x)$, extends to a bounded mapping from $L^p(R^n)$ into itself for $1 < p < \infty$ and $K(x) = \check{k}(x)$.

LEMMA 2   (*see Kurtz & Wheeden, 1979*)   Suppose $k \in M(s, l)$, $1 \le s \le 2$. Given $1 \le \tilde{s} < \infty$, let $r \ge 1$ be such that $\frac{1}{r} = \max\{\frac{1}{s}, 1 - \frac{1}{\tilde{s}}\}$. Then $K \in \tilde{M}(\tilde{s}, \tilde{l})$, where $\tilde{l} = l - \frac{n}{r}$.

LEMMA 3   (*see Kurtz & Wheeden, 1979*)   Let $1 \le s < \infty$, suppose that   is a positive real number with $l > n/r, 1/r = \max\{1/s, 1 - 1/\tilde{s}\}$, and $k \in M(s, l)$. Then there is a positive constant a, such that

$$\left(\int_{Q_k} |K(x - z) - K(x_Q - z)|^{\tilde{s}} dz \right)^{1/\tilde{s}} \le C 2^{-ka} (2^k h)^{-n/\tilde{s}'}$$

Now we can define the Toeplitz type operator associated to the multiplier operator as following.

Definition 3   Let b be a locally integrable function on R^n and T be the multiplier operator. By Lemma 1, $T(f)(x) = (K * f)(x)$ for $K(x) = \check{k}(x)$. The Toeplitz type operator associated to T is defined by

$$T^b = \sum_{k=1}^{m} T^{k,1} M^b T^{k,2}$$

where $T^{k,1}$ are T or $\pm I$ (the identity operator), $T^{k,2}$ are the bounded linear operators on $L^p(R^n)$ for $1 < p < \infty$ and $k = 1, \dots, m, M^b(f) = bf$.

Note that the commutator $[b, T](f) = bT(f) - T(bf)$ is a particular operator of the Toeplitz type operator T^b. It is well known that commutators are of great interest in harmonic analysis and have been widely studied by many authors (see Janson, 1978; Janson & Peetre, 1988; Krantz & Li, 2001; Paluszynski, 1995; Pérez & Pradolini, 2001; Pérez & Trujillo-Gonzalez, 2002). The main purpose of this paper is to prove the boundedness properties for the Toeplitz type operator T^b from Lebesgue spaces to Orlicz spaces.

We shall prove the following results in Section 4.

THEOREM 1   Let T be the multiplier operator as Definition 3. Suppose that $Q = Q(x_0, d)$ is a cube with $\text{supp}\, f \subset (2Q)^c$ and $x, \tilde{x} \in Q$.

(1) If $b \in BMO(R^n)$, then

$$|T^{(b-b_Q)\chi_{(2Q)^c}}(f)(x) - T^{(b-b_Q)\chi_{(2Q)^c}}(f)(x_0)| \le C||b||_{BMO} \sum_{k=1}^{m} M_r(T^{k,2}(f))(\tilde{x}) \text{ for any } r > 1;$$

(II)(2) If $0 < \beta \le 1$ and $b \in Lip_\beta(R^n)$, then

$$|T^{(b-b_Q)\chi_{(2Q)^c}}(f)(x) - T^{(b-b_Q)\chi_{(2Q)^c}}(f)(x_0)| \le C||b||_{Lip_\beta} \sum_{k=1}^{m} M_{\beta,r}(T^{k,2}(f))(\tilde{x}) \text{ for any } r > 1.$$

THEOREM 2   Let $0 < \beta \le 1, 1 < p < n/\beta$ and φ, ψ be two non-decreasing positive functions on $[0, +\infty)$ with $(\psi^l)^{-1}(t) = t^{1/p}\varphi^l(t^{-1/n})$. Suppose that ψ is convex, $\psi(0) = 0$, $\psi(2t) \le C\psi(t)$. Let T be the multiplier operator as Definition 3. If $T^1(g) = 0$ for any $g \in L^u(R^n)(1 < u < \infty)$, then T^b is bounded from $L^p(R^n)$ to $L_{\psi^l}(R^n)$ if $b \in BMO(R^n)$.

COROLLARY 1   Let $0 < \beta \le 1, 1 < p < n/\beta$ and T be the multiplier operator as Definition 3. If $T^1(g) = 0$ for any $g \in l^u(R^n)(1 < u < \infty)$, then T^b is bounded on $L^p(R^n)$ if $b \in BMO(R^n)$.

COROLLARY 2   Let $1 < p < s < \infty$ and T be the multiplier operator as Definition 3. If $T^1(g) = 0$ for any $g \in L^u(R^n)(1 < u < \infty)$, then T^b is bounded from $L^p(R^n)$ to $L^s(R^n)$ if $b \in BMO(R^n)$.

3. Some lemma
We need the following preliminary lemmas.

LEMMA 4   (see Kurtz & Wheeden, 1979)   Let T be the multiplier operator as Definition 3. Then T is bounded on $L^p(R^n)$ for $1 < p < \infty$.

LEMMA 5   (see Garcia-Cuerva & Rubio de Francia, 1985)   Let $0 < p < \infty$. Then, for any smooth function f for which the left-hand side is finite,

$$\int_{R^n} M(f)(x)^p dx \le C \int_{R^n} f^\#(x)^p dx$$

LEMMA 6   (see Chanillo, 1982)   Suppose that $0 < \eta < n, 1 \le r < p < n/\eta$ and $1/s = 1/p - \eta/n$. Then $||M_{\eta,r}(f)||_{L^s} \le C||f||_{L^p}$.

LEMMA 7   (see Janson, 1978)   Let ρ be a non-decreasing positive function on $[0, +\infty)$ and η be an infinitely differentiable function on R^n with compact support such that $\int_{R^n} \eta(x)dx = 1$. Denote that $b_t(x) = \int_{R^n} b(x - ty)\eta(y)dy$. Then $||b - b_t||_{BMO} \le C\rho(t)||b||_{BMO_\rho}$.

LEMMA 8   *(see Janson, 1978)*   *Let $0 < \beta < 1$ or $\beta = 1$ and ρ be a non-decreasing positive function on $[0, +\infty)$. Then $||b_t||_{Lip_\beta} \leq Ct^{-\beta}\rho(t)||b||_{BMO_\rho}$.*

LEMMA 9   *(see Janson, 1978)*   *Suppose $1 \leq p_2 < p < p_1 < \infty$, ρ is a non-increasing function on R^+, B is a linear operator such that $m_{B(f)}(t^{1/p_1}\rho(t)) \leq Ct^{-1}$ if $||f||_{L^{p_1}} \leq 1$ and $m_{B(f)}(t^{1/p_2}\rho(t)) \leq Ct^{-1}$ if $||f||_{L^{p_2}} \leq 1$. Then $\int_0^\infty m_{B(f)}(t^{1/p}\rho(t))dt \leq C$ if $||f||_{L^p} \leq (p/p_1)^{1/p}$.*

4. Proofs of theorems

Now we are in position to prove our results.

Proof of Theorem 1   For $\text{supp}f \subset (2Q)^c$ and $x, \tilde{x} \in Q$, note that $|x - y| \sim |x_0 - y|$ for $x \in Q$ and $y \in R^n \setminus 2Q$. We have

$$|T^{(b-b_Q)\chi_{(2Q)^c}}(f)(x) - T^{(b-b_Q)\chi_{(2Q)^c}}(f)(x_0)|$$

$$\leq \sum_{k=1}^m |T^{k,1}M^{(b-b_Q)\chi_{(2Q)^c}}T^{k,2}(f)(x) - T^{k,1}M^{(b-b_Q)\chi_{(2Q)^c}}T^{k,2}(f)(x_0)|$$

(I) By the Hölder's inequality and Lemma 3, we obtain, for $1 < s, t < \infty$ with $1/r + 1/s + 1/t = 1$,

$$|T^{k,1}M^{(b-b_Q)\chi_{(2Q)^c}}T^{k,2}(f)(x) - T^{k,1}M^{(b-b_Q)\chi_{(2Q)^c}}T^{k,2}(f)(x_0)|$$

$$\leq \int_{(2Q)^c} |b(y) - b_Q||K(x - y) - K(x_0 - y)||T^{k,2}(f)(y)|dy$$

$$\leq \sum_{j=1}^\infty \int_{2^j d \leq |y - x_0| < 2^{j+1}d} |b(y) - b_Q||K(x - y) - K(x_0 - y)||T^{k,2}(f)(y)|dy$$

$$\leq C \sum_{j=1}^\infty \left(\int_{2^{j+1}Q} |b(y) - b_Q|^s dy\right)^{1/s} \left(\int_{2^{j+1}Q} |T^{k,2}(f)(y)|^r dy\right)^{1/r}$$

$$\times \left(\int_{2^j d \leq |y - x_0| < 2^{j+1}d} |K(x - y) - K(x_0 - y)|^t dy\right)^{1/t} dx$$

$$\leq C||b||_{BMO} \sum_{j=1}^\infty j2^{-aj} \left(\frac{1}{|2^{j+1}Q|}\int_{2^{j+1}Q} |T^{k,2}(f)(y)|^r dy\right)^{1/r}$$

$$\leq C||b||_{BMO}M_r(T^{k,2}(f))(\tilde{x})$$

thus

$$|T^{(b-b_Q)\chi_{(2Q)^c}}(f)(x) - T^{(b-b_Q)\chi_{(2Q)^c}}(f)(x_0)|$$

$$\leq \sum_{k=1}^m |T^{k,1}M^{(b-b_Q)\chi_{(2Q)^c}}T^{k,2}(f)(x) - T^{k,1}M^{(b-b_Q)\chi_{(2Q)^c}}T^{k,2}(f)(x_0)|$$

$$\leq C||b||_{BMO}\sum_{k=1}^m M_r(T^{k,2}(f))(\tilde{x})$$

(II) Note that, for $b \in Lip_\beta(R^n)$,

$$|b(x) - b_Q| \leq \frac{1}{|Q|} \int_Q ||b||_{Lip_\beta} |x - y|^\beta dy \leq C||b||_{Lip_\beta} (|x - x_0| + d)^\beta$$

similar to the proof of (I), we obtain, for $1/r + 1/r' = 1$,

$$|T^{k,1} M^{(b-b_Q)\chi_{(2Q)^c}} T^{k,2}(f)(x) - T^{k,1} M^{(b-b_Q)\chi_{(2Q)^c}} T^{k,2}(f)(x_0)|$$

$$\leq \sum_{j=1}^\infty \int_{2^j d \leq |y - x_0| < 2^{j+1} d} |b(y) - b_Q| |K(x-y) - K(x_0-y)| |T^{k,2}(f)(y)| dy$$

$$\leq C||b||_{Lip_\beta} \sum_{j=1}^\infty |2^{j+1}Q|^{\beta/n} \left(\int_{2^j d \leq |y - x_0| < 2^{j+1} d} |K(x,y) - K(x_0,y)|^{r'} dy \right)^{1/r'}$$

$$\times \left(\int_{2^{j+1}Q} |T^{k,2}(f)(y)|^r dy \right)^{1/r}$$

$$\leq C||b||_{Lip_\beta} \sum_{j=1}^\infty 2^{-j} \left(\frac{1}{|2^{j+1}Q|^{1-r\beta/n}} \int_{2^{j+1}Q} |T^{k,2}(f)(y)|^r dy \right)^{1/r}$$

$$\leq C||b||_{Lip_\beta} M_{\beta,r}(T^{k,2}(f))(\tilde{x})$$

thus

$$|T^{(b-b_Q)\chi_{(2Q)^c}}(f)(x) - T^{(b-b_Q)\chi_{(2Q)^c}}(f)(x_0)|$$

$$\leq \sum_{k=1}^m |T^{k,1} M^{(b-b_Q)\chi_{(2Q)^c}} T^{k,2}(f)(x) - T^{k,1} M^{(b-b_Q)\chi_{(2Q)^c}} T^{k,2}(f)(x_0)|$$

$$\leq C||b||_{Lip_\beta} \sum_{k=1}^m M_{\beta,r}(T^{k,2}(f))(\tilde{x})$$

These complete the proof.

Proof of Theorem 2   Without loss of generality, we may assume $T^{k,1}$ are $T(k = 1, \ldots, m)$. We prove the theorem in several steps. First, we prove, if $b \in BMO(R^n)$,

$$(T^b(f))^\# \leq C||b||_{BMO} \sum_{k=1}^m M_r(T^{k,2}(f)) \tag{1}$$

for any $1 < r < \infty$.

Fix a cube $Q = Q(x_0, d)$ and $\tilde{x} \in Q$. By $T^1(g) = 0$, we have $T^b(f) = T^{b-b_Q}(f)$, thus

$$T^b(f) = T^{b-b_Q}(f) = T^{(b-b_Q)\chi_{2Q}}(f) + T^{(b-b_Q)\chi_{(2Q)^c}}(f) := I_1(x) + I_2(x)$$

and

$$\frac{1}{|Q|} \int_Q |T^b(f)(x) - I_2(x_0)| dx \leq \frac{1}{|Q|} \int_Q |I_1(x)| dx + \frac{1}{|Q|} \int_Q |I_2(x) - I_2(x_0)| dx = I_1 + I_2$$

For I_1, choose $1 < s < r$, by Hölder's inequality and the boundedness of T (see Lemma 4), we obtain

$$\frac{1}{|Q|} \int_Q |T^{k,1} M^{(b-b_Q)\chi_{2Q}} T^{k,2}(f)(x)| dx$$

$$\leq \left(\frac{1}{|Q|} \int_{R^n} |T^{k,1} M_{(b-b_Q)\chi_{2Q}} T^{k,2}(f)(x)|^s dx \right)^{1/s}$$

$$\leq C|Q|^{-1/s} \left(\int_{R^n} |M^{(b-b_Q)\chi_{2Q}} T^{k,2}(f)(x)|^s dx \right)^{1/s}$$

$$\leq C|Q|^{-1/s} \left(\int_{2Q} |T^{k,2}(f)(x)|^r dx \right)^{1/r} \left(\int_{2Q} |b(x)-b_Q|^{rs/(r-s)} dx \right)^{(r-s)/rs}$$

$$\leq C||b||_{BMO} \left(\frac{1}{|Q|} \int_{2Q} |T^{k,2}(f)(x)|^r dx \right)^{1/r}$$

$$\leq C||b||_{BMO} M_r(T^{k,2}(f))(\tilde{x})$$

thus

$$I_1 \leq \sum_{k=1}^m \frac{1}{|Q|} \int_Q |T^{k,1} M^{(b-b_Q)\chi_{2Q}} T^{k,2}(f)(x)| dx$$

$$\leq C||b||_{BMO} \sum_{k=1}^m M_r(T^{k,2}(f))(\tilde{x})$$

For I_2, by using Theorem 1,

$$I_2 \leq C||b||_{BMO} \sum_{k=1}^m M_r(T^{k,2}(f))(\tilde{x})$$

We now put these estimates together and take the supremum over all Q such that $\tilde{x} \in Q$, we obtain

$$(T^b(f))^{\#}(\tilde{x}) \leq C||b||_{BMO} \sum_{k=1}^m M_r(T^{k,2}(f))(\tilde{x})$$

Thus, taking r such that $1 < r < p$, we obtain, by Lemma 5,

$$||T^b(f)||_{L^p} \leq ||M(T^b(f))||_{L^p} \leq C||(T^b(f))^{\#}||_{L^p}$$

$$\leq C||b||_{BMO} \sum_{k=1}^m ||M_r(T^{k,2}(f))||_{L^p}$$

$$\leq C||b||_{BMO} \sum_{k=1}^m ||T^{k,2}(f)||_{L^p} \tag{2}$$

$$\leq C||b||_{BMO} ||f||_{L^p}$$

Secondly, we prove that, if $b \in Lip_\beta(R^n)$,

$$(T^b(f))^{\#} \leq C||b||_{Lip_\beta} \sum_{k=1}^m M_{\beta,r}(T^{k,2}(f)) \tag{3}$$

for any r with $1 < r < n/\beta$. In fact, similar to the proof of (1) and by Theorem 1, we obtain

$$\frac{1}{|Q|}\int_Q \left|T^b(f)(x) - I_2(x_0)\right| dx$$

$$\leq \sum_{k=1}^{m}\left(\frac{1}{|Q|}\int_{R^n}|T^{k,1}M^{(b-b_Q)\chi_{2Q}}T^{k,2}(f)(x)|^r dx\right)^{1/r}$$

$$+\frac{1}{|Q|}\int_Q |I_2(x) - I_2(x_0)| dx$$

$$\leq C\sum_{k=1}^{m}|Q|^{-1/r}\left(\int_{2Q}(|b(x)-b_Q||T^{k,2}(f)(x)|)^r dx\right)^{1/r}$$

$$+\sum_{k=1}^{m}\frac{1}{|Q|}\int_Q \left|T^{k,1}M^{(b-b_Q)\chi_{(2Q)^c}}T^{k,2}(f)(x) - T^{k,1}M^{(b-b_Q)\chi_{(2Q)^c}}T^{k,2}(f)(x_0)\right| dx$$

$$\leq C\sum_{k=1}^{m}|Q|^{-1/r}||b||_{Lip_\beta}|2Q|^{\beta/n}|Q|^{1/r-\beta/n}\left(\frac{1}{|Q|^{1-r\beta/n}}\int_{2Q}|T^{k,2}(f)(x)|^r dx\right)^{1/r}$$

$$+C||b||_{Lip_\beta}\sum_{k=1}^{m}M_{\beta,r}(T^{k,2}(f))(\tilde{x})$$

$$\leq C||b||_{Lip_\beta}\sum_{k=1}^{m}M_{\beta,r}(T^{k,2}(f))(\tilde{x})$$

Thus, (3) holds. We take $1<r<p<n/\beta, 1/q=1/p-\beta/n$ and obtain, by Lemma 6,

$$||T^b(f)||_{L^q}\leq ||M(T^b(f))||_{L^q}\leq C||(T^b(f))^{\#}||_{L^q}$$

$$\leq C||b||_{Lip_\beta}\sum_{k=1}^{m}||M_{\beta,r}(T^{k,2}(f))||_{L^q}$$

$$\leq C||b||_{Lip_\beta}\sum_{k=1}^{m}||T^{k,2}(f)||_{L^p} \tag{4}$$

$$\leq C||b||_{Lip_\beta}||f||_{L^p}$$

Now we verify that T^b satisfies the conditions of Lemma 9. In fact, for any $1<p_i<n/\beta$, $1/q_i=1/p_i-\beta/n(i=1,2)$ and $||f||_{L^{p_i}}\leq 1$, note that $T^b(f)(x)=T^{b-b_s}(f)(x)+T^{b_s}(f)(x)$, $b-b_s\in BMO(R^n)$ and $b_s\in Lip_\beta(R^n)$, by (2) and Lemma 7, we obtain

$$||T^{b-b_s}(f)||_{L^{p_i}}\leq C||b-b_s||_{BMO}||f||_{L^{p_i}}$$
$$\leq C||b-b_s||_{BMO}\leq C||b||_{BMO_\varphi}\varphi(s)$$

and by (4) and Lemma 8, we obtain

$$||T^{b_s}(f)||_{L^{q_i}}\leq C||b||_{Lip_\beta}||f||_{L^{p_i}}\leq Cs^{-\beta}\varphi(s)||b||_{BMO_\varphi}$$

Thus, for $s=t^{-1/n}$ and $i=1,2$,

$$m_{T^b(f)}(\psi^{-1}(t))\leq m_{T^b(f)}(t^{1/p_i}\varphi(t^{-1/n}))$$

$$\leq m_{T^{b-b_s}(f)}(t^{1/p_i}\varphi(t^{-1/n})/2)+m_{T^{b_s}(f)}(t^{1/p_i}\varphi(t^{-1/n})/2)$$

$$\leq C\left[\left(\frac{\varphi(s)}{t^{1/p_i}\varphi(s)}\right)^{p_i}+\left(\frac{s^{-\beta}\varphi(s)}{t^{1/p_i}\varphi(s)}\right)^{q_i}\right]=Ct^{-1}$$

Taking $1<p_2<p<p_1<n/\beta$ and by Lemma 9, we obtain, for $||f||_{L^p}\leq (p/p_1)^{1/p}$,

$$\int_{R^n} \psi(|T^b(f)(x)|)dx = \int_0^\infty m_{T^b(f)}(\psi^{-1}(t))dt \leq C$$

then, $||T^b(f)||_{L_\psi} \leq C$.

This completes the proof of the theorem.

Acknowledgements

The author would like to express her gratitude to the referees for their valuable comments and suggestions.

Funding

This paper is supported by the Scientific Research Fund of Hunan Province Land and Resources Departments (No. 2013-28).

Author details

Ouyang Difei[1]

E-mail: ouyangdifei1965@163.com

[1] College of Mathematics, Changsha University of Science and Technology, Changsha 410077, Hunan, P.R. China.

References

Chanillo, S. (1982). A note on commutators. *Indiana University Mathematics Journal, 31*, 7–16.

Coifman, R., Rochberg, R., & Weiss, G. (1976). Factorization theorems for Hardy spaces in several variables. *Annals of Mathematics, 103*, 611–635.

Garcia-Cuerva, J., & Rubio de Francia, J. L. (1985). *Weighted norm inequalities and related topics. Mathematics studies* (Vol. 116). North-Holland: Amsterdam.

Janson, S. (1978). Mean oscillation and commutators of singular integral operators. *Arkiv för Matematik, 16*, 263–270.

Janson, S., & Peetre, J. (1988). Paracommutators boundedness and Schatten–von Neumann properties. *Transactions of the American Mathematical Society, 305*, 467–504.

Krantz, S., & Li, S. (2001). Boundedness and compactness of integral operators on spaces of homogeneous type and applications. *Journal of Mathematical Analysis and Applications, 258*, 629–641.

Kurtz, D. S., & Wheeden, R. L. (1979). Results on weighted norm inequalities for multiplies. *Transactions of the American Mathematical Society, 255*, 343–362.

Lu, S. Z., & Mo, H. X. (2009). Toeplitz type operators on Lebesgue spaces. *Acta Mathematica Scientia, 29*, 140–150.

Muckenhoupt, B., Wheeden, R. L., & Young, W. S. (1987). Sufficient conditions for *Lp* multipliers with general weights. *Transactions of the American Mathematical Society, 300*, 463–502.

Paluszynski, M. (1995). Characterization of the Besov spaces via the commutator operator of Coifman, Rochberg and Weiss. *Indiana University Mathematics Journal, 44*, 1–17.

Pérez, C., & Pradolini, G. (2001). Sharp weighted endpoint estimates for commutators of singular integral operators. *Michigan Mathematical Journal, 49*, 23–37.

Pérez, C., & Trujillo-Gonzalez, R. (2002). Sharp weighted estimates for multilinear commutators. *Journal London Mathematical Society, 65*, 672–692.

Stein, E. M. (1993). *Harmonic analysis: Real variable methods, orthogonality and oscillatory integrals*. Princeton, NJ: Princeton University Press.

Wang, K. W., & Liu, L. Z. (2009a). Boundedness for multilinear commutator of multiplier operator on Hardy spaces. *Scientia Series A: Mathematical Science, 17*, 19–26.

Wang, K. W., & Liu, L. Z. (2009b). Sharp inequality for multilinear commutator of multiplier operator. *Acta Mathematica Vietnamica, 34*, 233–244.

You, Z. (1988). Results of commutators obtained norm inequalities. *Advanced in Mathematics, 17*, 79–84 (in Chinese).

Zhang, P., & Chen, J. C. (2005). *Acta Mathematica Sinica, 21*, 765–772.

Zhang, P., & Chen, J. C. (2006). Boundedness properties for commutators of multiplies. *Acta Mathematica Sinica (Chinese Series), 49*, 1387–1396.

7

Some new results on inner product quasilinear spaces

Hacer Bozkurt[1]* and Yılmaz Yılmaz[2]

*Corresponding author: Hacer Bozkurt, Department of Mathematics, Batman University, 72100 Batman, Turkey
E-mail: hacer.bozkurt@batman.edu.tr
Reviewing editor: Hari M. Srivastava, University of Victoria, Canada

Abstract: In this article, we research on the properties of the floor of an element taken from an inner product quasilinear space. We prove some theorems related to this new concept. Further, we try to explore some new results in quasilinear functional analysis. Also, some examples have been given which provide an important information about the properties of floor of an inner product quasilinear space.

Subjects: Advanced Mathematics; Analysis-Mathematics; Functional Analysis; Mathematics & Statistics; Pure Mathematics; Science

Keywords: quasilinear space; inner product quasilinear space; Hilbert quasilinear space; orthogonality; orthonormality

AMS subject classifications: 46C05; 46C07; 46C15; 46C50; 97H50

1. Introduction

Aseev (1986) introduced the theory of quasilinear space (briefly, QLGs) which is generalization of classical linear spaces. He used the partial order relation when he defined the quasilinear spaces and so he can give consistent counterparts of results in linear spaces. Further, he also described the convergence of sequences and norm in quasilinear space. This work has inspired a lot of authors to introduce new results on multivalued mappings, fuzzy quasilinear operators and set-valued analysis (Lakshmikantham, Gnana Bhaskar, & Vasundhara Devi, 2006; Rojas-Medar, Jiménez-Gamerob, Chalco-Canoa, & Viera-Brandão, 2005).

ABOUT THE AUTHORS

Hacer Bozkurt received MSc from Sakarya University, and is currently a PhD scholar at İnönü University. Her research interests are functional analysis, nonlinear functional analysis and interval analysis.

Yılmaz Yılmaz received MSc and PhD degrees in İnönü University, Malatya, Turkey. Currently he is a professor at İnönü University, Malatya, Turkey. His research interests are Functional analysis, sequence spaces, nonlinear functional analysis, Bifurcation theory.

PUBLIC INTEREST STATEMENT

The theory of quasilinear spaces was introduced by Aseev (1986). Aseev used the partial order relation when he defined quasilinear spaces and so he can give consistent counterparts of results in linear spaces. As known, the theory of inner product space and Hilbert spaces play a fundamental role in functional analysis and its applications. We know that any inner product space is a normed space and any normed space is a particular class of normed quasilinear space. Hence, this relation and Aseev's work motivated us to examine quasilinear counterpart of inner product space in classical analysis. Thus, we introduce the concept of inner product quasilinear space. In this paper, we give some results related to floors of inner product quasilinear spaces. Also, some examples have been given which provide an important contribution to understand the structure of inner product quasilinear spaces.

We see from the definition of quasilinear space which given in Aseev (1986), the inverse of some elements of in quasilinear spaces may not be available. Yılmaz, Çakan, and Aytekin (2012), these elements are called as singular elements of quasilinear space. At the same time the others which have an inverse are referred to as regular elements. Then, in Çakan (2016), she noticed that the base of each singular elements of a combination of regular elements of the quasilinear space. Therefore, she defined the concept of the floor of an element in quasilinear space in Çakan (2016) which is very convenient for some analysis of quasilinear spaces. This work has motivated us to introduce some results about the floors of inner product quasilinear spaces, briefly, IPQLS.

In this paper, motivated by the work of Assev (1986) and Çakan (2016), we research some properties of floors of inner product quasilinear spaces and prove some theorems related to floor of a subset of an inner product quasilinear space. Further, we try to extend the results in quasilinear functional analysis. Our consequences gives us some information about the properties of floor of an inner product quasilinear space.

Let us give some notation and preliminary results given by Aseev (1986).

Definition 1.1   A set X is called a quasilinear space (QLS, for short), if a partial order relation "$\leq$", an algebraic sum operation, and an operation of multiplication by real numbers are defined in it in such way that the following conditions hold for any elements $x, y, z, v \in X$ and any real numbers $\alpha, \beta \in \mathbb{R}$:

(1) $x \leq x$;

(2) $x \leq z$ if $x \leq y$ and $y \leq z$,

(3) $x = y$ if $x \leq y$ and $y \leq x$,

(4) $x + y = y + x$,

(5) $x + (y + z) = (x + y) + z$,

(6) there exists an element $\theta \in X$ such that $x + \theta = x$,

(7) $\alpha \cdot (\beta \cdot x) = (\alpha \cdot \beta) \cdot x$,

(8) $\alpha \cdot (x + y) = \alpha \cdot x + \alpha \cdot y$,

(9) $1 \cdot x = x$,

(10) $0 \cdot x = \theta$,

(11) $(\alpha + \beta) \cdot x \leq \alpha \cdot x + \beta \cdot x$,

(12) $x + z \leq y + v$ if $x \leq y$ and $z \leq v$,

(13) $\alpha \cdot x \leq \alpha \cdot y$ if $x \leq y$.

A linear space is a quasilinear space with the partial order relation "=". The most popular example which is not a linear space is the set of all closed intervals of real numbers with the inclusion relation "$\subseteq$", algebraic sum operation

$$A + B = \{a + b : a \in A, \quad b \in B\}$$

and the real scalar multiplication

$\lambda \cdot A = \{\lambda \cdot a\colon a \in A\}.$

We denote this set by $\Omega_c(\mathbb{R})$. Another one is $\Omega(\mathbb{R})$, the set of all compact subsets of real numbers. By a slight modification of algebraic sum operation (with closure) such as

$$A + B = \overline{\{a + b\colon a \in A, \quad b \in B\}}$$

and by the same real scalar multiplication defined above and by the inclusion relation we get the nonlinear QLS, $\Omega_c(E)$ and $\Omega(E)$, the space of all nonempty closed bounded and convex closed bounded subsets of some normed linear space E, respectively.

LEMMA 1.1 *Suppose that any element x in a QLS X has an inverse element x' $\in$ X. Then the partial order in X is determined by equality, the distributivity conditions hold, and consequently, X is a linear space (Aseev, 1986).*

Suppose that X is a QLS and Y $\subseteq$ X. Then Y is called a subspace of X whenever Y is a QLS with the same partial order and the restriction to Y of the operations on X. One can easily prove the following theorem using the condition of to be a QLS. It is quite similar to its linear space analogue (Yılmaz et al., 2012).

THEOREM 1.1 *Y is a subspace of a QLS X if and only if $\alpha \cdot x + \beta \cdot y \in Y$ for every x, y $\in$ Y and $\alpha, \beta \in \mathbb{R}$ (Yılmaz et al., 2012).*

Let X be a QLS. An $x \in X$ is said to be symmetric if $(-1) \cdot x = -x = x$, and X_d denotes the set of all such elements. θ denotes the zero's, additive unit of X and it is minimal, i.e. $x = \theta$ if $x \leq \theta$. An element x' is called inverse of x if $x + x' = \theta$. The inverse is unique whenever it exists and $x' = -x$ in this case. Sometimes x' may not be exist but $-x$ is always meaningful in QLSs. An element x possessing an inverse is called regular, otherwise is called singular. For a singular element x we should note that $x - x \neq 0$. Now, X_r and X_s stand for the sets of all regular and singular elements in X, respectively. Further, X_r, X_d and $X_s \cup \{0\}$ are subspaces of X and they are called regular, symmetric and singular subspaces of X, respectively (Yılmaz et al., 2012).

Proposition 1.1 In a quasilinear space X every regular element is minimal (Yılmaz et al., 2012).

Definition 1.2 Let X be a QLS. A function $\|\cdot\|_X\colon X \longrightarrow \mathbb{R}$ is called a norm if the following conditions hold (Aseev, 1986):

(14) $\|x\|_X > 0$ if $x \neq 0$,

(15) $\|x + y\|_X \leq \|x\|_X + \|y\|_X$,

(16) $\|\alpha \cdot x\|_X = |\alpha| \|x\|_X$,

(17) if $x \leq y$, then $\|x\|_X \leq \|y\|_X$,

(18) if for any $\varepsilon > 0$ there exists an element $x_\varepsilon \in X$ such that, $x \leq y + x_\varepsilon$ and $\|x_\varepsilon\|_X \leq \varepsilon$ then $x \leq y$.

A quasilinear space X with a norm defined on it is called normed quasilinear space (NQLS, for short). It follows from Lemma 1.1 that if any $x \in X$ has an inverse element $x' \in X$, then the concept of NQLS coincides with the concept of a real normed linear space.

Let X be a NQLS. Hausdorff or norm metric on X is defined by the equality

$$h_X(x, y) = \inf \{r \geq 0\colon x \leq y + a_1^r, y \leq x + a_2^r, \|a_i^r\| \leq r\}.$$

Since $x \leq y + (x - y)$ and $y \leq x + (y - x)$, the quantity $h_X(x, y)$ is well-defined for any elements $x, y \in X$, and

$$h_X(x, y) \leq \|x - y\|_X. \tag{1}$$

It is not hard to see that this function satisfies all of the metric axioms.

LEMMA 1.2   *The operations of algebraic sum and multiplication by real numbers are continuous with respect to the Hausdorff metric. The norm is continuous function respect to the Hausdorff metric (Aseev, 1986).*

Example 1.1   Let E be a Banach space. A norm on $\Omega(E)$ is defined by

$$\|A\|_{\Omega(E)} = \sup_{a \in A} \|a\|_E.$$

Then $\Omega(E)$ and $\Omega_c(E)$ are normed quasilinear spaces. In this case, the Hausdorff metric is defined as usual:

$$h_{\Omega_c(E)}(A, B) = \inf\{r \geq 0 : A \subset B + S_r(\theta), B \subset A + S_r(\theta)\},$$

where $S_r(\theta)$ denotes a closed ball of radius r about $\theta \in X$ (Aseev, 1986).

Definition 1.3   Let X be a QLS, $M \subseteq X$ and $x \in M$. The set of

$$F_x^M = \{z \in M_r : z \leq x\}$$

is called *floor in M of x* . In the case of $M = X$ it is called only *floor of x* and written briefly F_x instead of F_x^X (Çakan, 2016).

Floor of an element x in linear spaces is $\{x\}$. Therefore, it is nothing to discuss the notion of floor of an element in a linear space.

Definition 1.4   Let X be a QLS and $M \subseteq X$. Then the union set

$$\bigcup_{x \in M} F_x^M$$

is called *floor of M* and is denoted by $\mathcal{F}_M$. In the case of $M = X$, $\mathcal{F}_X$ is called floor of the qls X.

On the other hand, the set

$$\mathcal{F}_M^X = \bigcup_{x \in M} F_x^X$$

is called *floor in X of M* and is denoted by $\mathcal{F}_M^X$ (Çakan, 2016).

Definition 1.5   Let X be a quasilinear space. X is called solid-floored quasilinear space whenever

$$y = \sup\{x \in X_r : x \leq y\}$$

for every $y \in X$. Otherwise, X is called nonsolid-floored quasilinear space (Çakan, 2016).

Example 1.2 $\Omega(\mathbb{R})$ and $\Omega_c(\mathbb{R})$ are solid-floored quasilinear space. But singular subspace of $\Omega_c(\mathbb{R})$ is a nonsolid-floored quasilinear space.

Definition 1.6 Let X be a QLS. Consolidation of floor of X is the smallest solid-floored QLS $\hat{X}$ containing X, that is, if there exists another solid-floored QLS Y containing X then $\hat{X} \subseteq Y$.

Clearly, $\hat{X} = X$ for some solid-floored QLS X. Further, $\widehat{\Omega_C(\mathbb{R}^n)}_s = \Omega_C(\mathbb{R}^n)$. For a QLS X, the set

$$F_y^{\hat{X}} = \left\{ z \in \left(\hat{X}\right)_r : z \leq y \right\}.$$

is the *floor* of X in $\hat{X}$.

Let us give an extended definition of inner product. This definition and some prerequisites are given by Y. Yılmaz. We can see following inner product as (set-valued) inner product on QLSs.

Definition 1.7 Let X be a quasilinear space. A mapping $\langle\,,\,\rangle : X \times X \to \Omega(\mathbb{R})$ is called an inner product on X if for any $x, y, z \in X$ and $\alpha \in \mathbb{R}$ the following conditions are satisfied:

(19) if $x, y \in X_r$ then $\langle x, y \rangle \in \Omega_C(\mathbb{R})_r \equiv \mathbb{R}$,

(20) $\langle x + y, z \rangle \subseteq \langle x, z \rangle + \langle y, z \rangle$,

(21) $\langle \alpha \cdot x, y \rangle = \alpha \cdot \langle x, y \rangle$,

(22) $\langle x, y \rangle = \langle y, x \rangle$,

(23) $\langle x, x \rangle \geq 0$ for $x \in X_r$ and $\langle x, x \rangle = 0 \Leftrightarrow x = 0$,

(24) $\|\langle x, y \rangle\|_{\Omega(\mathbb{R})} = \sup \left\{ \|\langle a, b \rangle\|_{\Omega(\mathbb{R})} : a \in F_x^{\hat{X}}, b \in F_y^{\hat{X}} \right\}$,

(25) if $x \leq y$ and $u \leq v$ then $\langle x, u \rangle \subseteq \langle y, v \rangle$,

(26) if for any $\varepsilon > 0$ there exists an element $x_\varepsilon \in X$ such that $x \leq y + x_\varepsilon$ and $\langle x_\varepsilon, x_\varepsilon \rangle \subseteq S_\varepsilon(\theta)$ then $x \leq y$.

A quasilinear space with an inner product is called an inner product quasilinear space, briefly, IPQLS.

Example 1.3 One can see easily $\Omega_C(\mathbb{R})$, the space of closed real intervals, is an IPQLS with inner product defined by

$$\langle A, B \rangle = \{ab : a \in A, b \in B\}.$$

Every IPQLS X is a normed QLS with the norm defined by

$$\|x\| = \sqrt{\|\langle x, x \rangle\|_{\Omega(\mathbb{R})}}$$

for every $x \in X$. This norm is called inner product norm. Classical norm of $\Omega_C(\mathbb{R})$ (see Aseev, 1986) is generated by the above inner product.

Proposition 1.2 $x_n \to x$ and $y_n \to y$ in an IPQLS then $\langle x_n, y_n \rangle \to \langle x, y \rangle$.

An IPQLS is called *Hilbert QLS*, if it is complete according to the Inner product (norm) metric. $\Omega_C(\mathbb{R})$ is a Hilbert QLS.

Definition 1.8 (Orthogonality) An element x of an IPQLS X is said to be orthogonal to an element $y \in X$ if

$$\|\langle x, y \rangle\|_{\Omega(\mathbb{R})} = 0.$$

We also say that x and y are orthogonal and we write $x \perp y$. Similarly, for subsets $m, n \subseteq X$ we write $x \perp m$ if $x \perp z$ for all $z \in m$ and $m \perp n$ if $a \perp b$ for all $a \in m$ and $b \in n$.

An orthonormal set $M \subset X$ is an orthogonal set in X whose elements have norm 1, that is, for all $x,\ y \in M$

$$\| < x, y > \|_{\Omega(\mathbb{R})} = \left\{ \begin{array}{ll} 0, & x \neq y \\ 1, & x = y \end{array} \right.$$

Definition 1.9 Let A be a nonempty subset of an inner product quasilinear space X. An element $x \in X$ is said to be orthogonal to A, denoted by $x \perp A$, if $\|\langle x, y \rangle\|_{\Omega(\mathbb{R})} = 0$ for every $y \in A$. The set of all elements of X orthogonal to A, denoted by $A^{\perp}$, is called the orthogonal complement of A and is indicated by

$$A^{\perp} = \{x \in X : \|\langle x, y \rangle\|_{\Omega(\mathbb{R})} = 0, \quad y \in A\}.$$

For any subset A of an IPQLS X, $A^{\perp}$ is a closed subspace of X.

2. Main results

In this section, we try to explore some properties of floor of an element in an inner product quasilinear space. We note that the concept of floor is unneeded in linear spaces. Because, the floor of a linear space is equal to itself.

In general, $(\lambda + \mu) \cdot A = \lambda \cdot A + \mu \cdot A$ equality is not satisfy in a quasilinear space for every $\lambda, \mu \in \mathbb{R}$. For example; Let $A = [-2, 1] \subseteq \Omega_C(\mathbb{R})$ and $\lambda = 1, \mu = -1$, we have

$$(1 + (-1)) \cdot [-2, 1] = 0 \cdot [-2, 1] = \{0\}$$

But

$$1 \cdot [-2, 1] + (-1) \cdot [-2, 1] = [-2, 1] + [-1, 2] = [-3, 3].$$

Here, we see that $\{0\} \neq [-3, 3]$. But, we can say that $(\lambda + \mu) \cdot A \subset \lambda \cdot A + \mu \cdot A$ inequality is provided for any $A \in \Omega_C(\mathbb{R}^n)$.

Definition 2.1 Let X be a quasilinear space. X is called **homogenized quasilinear space** if for every $x \in X$ and $\alpha\beta \geq 0$ the following condition is satisfied:

$$(\alpha + \beta) \cdot x = \alpha \cdot x + \beta \cdot x.$$

Clearly, every linear space is a homogenized quasilinear space. But the reverse is not true.

Let X be a normed linear space. Then $\Omega_C(X)$ is a homogenized quasilinear space but $\Omega(X)$ is non-homogenized quasilinear space.

Proposition 2.1 Let X be a homogenized IPQLS and $x \in X$. Then F_x is convex subset of X.

Proof Let X be a homogenized IPQLS. From Definition 1.3, we get

$$F_x = \{a \in X_r : a \leq x\}$$

for a $x \in X$. So we have

$a \le x$ and $b \le x$

for every $a, b \in F_x$. From the condition (13), we get

$\lambda \cdot a \le \lambda \cdot x$ and $(1 - \lambda) \cdot b \le (1 - \lambda) \cdot x$

for all $0 \le \lambda \le 1$. Hence,

$\lambda \cdot a + (1 - \lambda) \cdot b \le \lambda \cdot x + (1 - \lambda) \cdot x.$

Since, X is a homogenized IPQLS,

$\lambda \cdot x + (1 - \lambda) \cdot x = (\lambda + 1 - \lambda) \cdot x = x$

for every $0 \le \lambda \le 1$. So, we obtain

$\lambda \cdot a + (1 - \lambda) \cdot b \le x.$

Hence $\lambda \cdot a + (1 - \lambda) \cdot b \in F_x$. This completes the proof.   □

Remark 2.1 Floor of an element of an IPQLS X is convex if and only if this IPQLS X is homogenized. If X is not homogenized inner product quasilinear space in the above proposition, then F_x is not convex since $(\alpha + \beta) \cdot x \ne \alpha \cdot x + \beta \cdot x$.

Proposition 2.2 Let X be an IPQLS and $A, B \subseteq X$. Then, we have

(a) $\{0\} \in F_A^\perp$,

(b) $F_{\{0\}} = \{0\}$,

(c) if $A \subseteq B$, then we get $F_A \subseteq F_B$ and $F_A^\perp \subseteq F_B^\perp$.

The proof of proposition is similar to the classical linear counterpart.

THEOREM 2.1 *If M is a convex subspace of Hilbert QLS X, then F_M is complete and convex subspace of Hilbert QLS X.*

Proof Let $a, b \in F_M$. Then, in view of Definition 1.3, there exist a $x \in M$ such that $a \le x$ and there exist a $y \in M$ such that $b \le y$. From (12) and (13), we have

$\alpha \cdot a + (1 - \alpha) \cdot b \le \alpha \cdot x + (1 - \alpha) \cdot y.$

Since M is convex, we find a $z \in M$ such that

$\alpha \cdot a + (1 - \alpha) \cdot b \le z.$

This proves that $\alpha \cdot a + (1 - \alpha) \cdot b \in F_M$.

Let $(a_n) \in F_M$ and $a_n \to a$ for some $a \in X$. Then for any $\epsilon > 0$ there exists an $N \in \mathbb{N}$ such that the following condition holds for any $n > N$:

$$a_n \le a + a_{1n}^\epsilon, \quad a \le a_n + a_{2n}^\epsilon, \quad \|a_{in}^\epsilon\| \le \frac{\epsilon}{2}. \tag{2}$$

On the other hand, if $(a_n) \in F_M$ then there exist a $(x_n) \in M$ such that $a_n \le x_n$ for every $n \in \mathbb{N}$. From here and above inequality, we get

$$a \le x_n + a_{2n}^\epsilon, \quad \|a_{2n}^\epsilon\| \le \epsilon$$

for every $n \in \mathbb{N}$. By the (18), we have $a \le x_n$ for every $n \in \mathbb{N}$. Now, we show that a is a regular element of X. By Lemma 1.2, we know $-a_n \to -a$ when $a_n \to a$. So, for any $\epsilon > 0$ there exists an $n' \in \mathbb{N}$ such that the following condition holds for any $n > n'$:

$$-a_n \le -a + b_{1n}^\epsilon, \quad -a \le -a_n + b_{2n}^\epsilon, \quad \|b_{in}^\epsilon\| \le \frac{\epsilon}{2}. \tag{3}$$

Because of $(a_n) \in F_M$, $a_n - a_n = 0$. By Lemma 1.2, (2) and (3), we get

$$a_n - a_n \le a - a + a_{1n}^\epsilon + b_{1n}^\epsilon, \quad a - a \le a_n - a_n + a_{2n}^\epsilon + b_{2n}^\epsilon, \quad \|a_{in}^\epsilon + b_{in}^\epsilon\| \le \epsilon$$

and

$$0 \le a - a + a_{1n}^\epsilon + b_{1n}^\epsilon, \quad a - a \le 0 + a_{2n}^\epsilon + b_{2n}^\epsilon, \quad \|a_{in}^\epsilon + b_{in}^\epsilon\| \le \epsilon.$$

From here, we have $0 = a - a$ since X is a Hilbert QLS. This shows that a is a regular element of X. Thus, since $(x_n) \in M$ for all $n \in \mathbb{N}$, we obtain $a \in F_M$. This proves that the set F_M is complete.   □

COROLLARY 2.1   Let X be a Hilbert QLS and M is a convex subspace of X. Then F_M is a complete subspace of X even if M is not complete.

Proposition 2.3   If X is an IPQLS and $x \in X$, then F_x is a closed.

Proof   Let $(b_n) \in F_x$ and $b_n \to b$ for some $b \in X$. Then for all $\epsilon > 0$ there exists an $n_0 \in \mathbb{N}$ such that the following condition holds for any $n > n_0$:
$$b_n \le b + c_{1n}^\epsilon, \quad b \le b_n + c_{2n}^\epsilon, \quad \|c_{in}^\epsilon\| \le \epsilon.$$

Since $(b_n) \in F_x$, $b_n \le x$ for every $n \in \mathbb{N}$. So, we have

$$b \le x$$

since $b \le b_n + c_{2n}^\epsilon$, $\|c_{2n}^\epsilon\|^2 = \left\| \langle c_{2n}^\epsilon, c_{2n}^\epsilon \rangle \right\|_{\Omega(\mathbb{R})} \le \epsilon$. Also, we can show that b is regular element of X similar to the above proof. By Lemma 1.2, we know $-b_n \to -b$ when $b_n \to b$ and $b - b_n \to b - b$. So, for any $\epsilon > 0$ there exists an $n_0 \in \mathbb{N}$ such that the following condition holds for any $n > n_0$:

$$b_n - b_n \le b - b + c_{1n}^\epsilon, \quad b - b \le b_n - b_n + c_{2n}^\epsilon, \quad \|c_{in}^\epsilon\| \le \epsilon.$$

From here, we have $0 = b - b$ since X is an IPQLS. This shows that b is a regular element of X.   □

LEMMA 2.1   Let X be an IPQLS. A floor of any element of IPQLS X may not subspace of X. But, the orthogonal complement of floor of any element of IPQLS X is subspace of X.

Proof   Let $a, b \in F_x$. Definition of floor of an element $a \le x$ and $b \le x$ for a $x \in X$. Since X is an IPQLS, we have

$$\alpha \cdot a + \beta \cdot b \le \alpha \cdot x + \beta \cdot x$$

for every $\alpha, \beta \in \mathbb{R}$. From here, we obtain $\alpha \cdot a + \beta \cdot b \notin F_x$ since $\alpha \cdot x + \beta \cdot x$ may not equal to x for all $\alpha, \beta \in \mathbb{R}$. So, F_x is not a subspace of X. Now, let $z \in F_x$ and $c, d \in F_x^\perp$ for a $x \in X$. From (15), (20) and (21) we have

$$\left\| \langle z, \alpha \cdot c + \beta \cdot d \rangle \right\|_{\Omega_c(\mathbb{R})} \le \left\| \langle z, \alpha \cdot c \rangle \right\|_{\Omega_c(\mathbb{R})} + \left\| \langle z, \beta \cdot d \rangle \right\|_{\Omega_c(\mathbb{R})}$$

$$= \alpha \left\| \langle z, c \rangle \right\|_{\Omega_c(\mathbb{R})} + \beta \left\| \langle z, d \rangle \right\|_{\Omega_c(\mathbb{R})}$$

$$= 0$$

So, we get $\alpha \cdot c + \beta \cdot d \in F_x^{\perp}$ for all $\alpha, \beta \in \mathbb{R}$. □

Remark 2.2 The floor of an subset of $(\Omega_c(\mathbb{R}))_d$ is equal to the largest element according to the order relation of the $\Omega_c(\mathbb{R})$.

Example 2.1 Let $RZ = \{ [n, 0] : n \in \mathbb{R}^- \}$, the right-zero subset of $\Omega_c(\mathbb{R})$. By the definition of floor, we get

$$F_{RZ} = \bigcup_{rz \in RZ} F_{rz}$$

$$= \bigcup_{rz \in RZ} \{ a \in (\Omega_c(\mathbb{R}))_r : a \subseteq rz \}$$

$$= \{ \{a\} : a \in \mathbb{R}^- \} \bigcup \{0\}.$$

Similarly, if we say $LZ = \{ [0, n] : n \in \mathbb{R}^+ \}$, the left-zero subset of $\Omega_c(\mathbb{R})$, we find

$$F_{LZ} = \bigcup_{lz \in LZ} F_{lz}$$

$$= \bigcup_{lz \in LZ} \{ a \in (\Omega_c(\mathbb{R}))_r : a \subseteq lz \}$$

$$= \{ \{a\} : a \in \mathbb{R}^+ \} \bigcup \{0\}.$$

From here, we have

THEOREM 2.2 *Suppose that X is an IPQLS and $A, B \subseteq X$. If $A \bigcup B = X$, then $F_A \bigcup F_B = X_r$.*

$$F_{RZ} \bigcup F_{LZ} = \{ \{a\} : a \in \mathbb{R}^- \} \bigcup \{0\} \bigcup \{ \{a\} : a \in \mathbb{R}^+ \}$$

$$= \{ \{c\} : c \in \mathbb{R} \}.$$

Proof It is easy to see that $x \in X_r$ for every $x \in F_A \bigcup F_B$. Let us consider $x \in X_r$. From here, we know that $x \in X$. Since $A \bigcup B = X, x$ is an element either A or B.

If x is an element of A, $x \in F_A$ since $x \in X_r$.
If x is an element of B, $x \in F_B$ since $x \in X_r$. This implies $x \in F_A \bigcup F_B$. □

Remark 2.3 Although, in an IPQLS X, $F_A \bigcup F_B = X_r$ for all $A, B \subseteq X$, the combination of A and B may not be equal to X.

Example 2.2 Let us consider the IPQLS $X = \Omega_c(\mathbb{R})$ and the subspaces $A = X_s$ and $B = X_r$. Clearly, $A \bigcup B = X$ and $F_A \bigcup F_B = X_r$. If we take $C = LZ$ and $D = RZ$ (RZ and LZ are subset of $\Omega_c(\mathbb{R})$ which is given in Example 2.1), we get $F_{RZ} \bigcup F_{LZ} = X_r$. But $C \bigcup D = RZ \bigcup LZ \ne X$.

Acknowledgements
The authors would like to record their gratitude to the reviewer for his careful reading and making some useful corrections which improved the presentation of the paper.

Funding
The authors received no direct funding for this research.

Author details
Hacer Bozkurt[1]
E-mail: hacer.bozkurt@batman.edu.tr

Yılmaz Yılmaz[2]
E-mail: yyilmaz44@gmail.com
[1] Department of Mathematics, Batman University, 72100 Batman, Turkey.
[2] Department of Mathematics, İnönü University, 44280 Malatya, Turkey.

References

Aseev, S. M. (1986). Quasilinear operators and their application in the theory of multivalued mappings. *Proceedings Steklov Institute Mathematics, 2,* 23–52.

Çakan, S. (2016). *Some new results related to theory of normed quasilinear spaces.* Malatya: University of İnönü.

Lakshmikantham, V., Gnana Bhaskar, T., & Vasundhara Devi, J. (2006). *Theory of set differential equations in metric spaces.* Cambridge: Cambridge Scientific Publishers.

Rojas-Medar, M. A., Jiménez-Gamerob, M. D., Chalco-Canoa, Y., & Viera-Brandão, A. J. (2005). Fuzzy quasilinear spaces and applications. *Fuzzy Sets and Systems, 152,* 173–190.

Yılmaz, Y., Çakan, S., & Aytekin, Ş. (2012). Topological quasilinear spaces. *Abstract and Applied Analysis.* doi:10.1155/2012/951374

8

On some properties of Toeplitz matrices

Dan Kucerovsky[1], Kaveh Mousavand[2] and Aydin Sarraf[3]*

*Corresponding author: Aydin Sarraf, Department of Computer Science, University of New Brunswick, Fredericton, Canada, NB
E-mail: Aydin.Sarraf@unb.ca

Reviewing editor: Nikos Katzourakis, University of Reading, UK

Abstract: In this paper, we investigate some properties of Toeplitz matrices with respect to different matrix products. We also give some results regarding circulant matrices, skew-circulant matrices and approximation by Toeplitz matrices over the field of complex numbers.

Subjects: Advanced Mathematics; Algebra; Analysis - Mathematics; Functional Analysis; Linear & Multilinear Algebra; Mathematics & Statistics; Operator Theory; Science

Keywords: Toeplitz matrices; circulant matrices

1. Introduction

Toeplitz matrices are important both in theory and application. For example, it is known that a large class of matrices are similar to Toeplitz matrices (Heinig, 2001; Mackey, Mackey, & Petrovic, 1999). Moreover, it is shown that every matrix is a product of Toeplitz matrices (Lim & Ye, 2013).

The notation of this paper is as follows. An $n \times n$ Toeplitz matrix T is denoted by $T = [t_{ij}] = [t_{i-j}]$, for $1 \leq i, j \leq n$; which implies all the entries along each of the $2n - 1$ diagonals are the same. A circulant matrix C is a Toeplitz matrix where each row is obtained by applying the permutation $(1 \ 2 \ .. \ n)$ to the previous row. A skew-circulant matrix S differs from a circulant matrix C only by a change in the sign in the subdiagonal entries, i.e. $s_{ij} = c_{ij}$ when $j \geq i$ and $s_{ij} = -c_{ij}$ when $j < i$. A matrix T is involutory if $T^2 = I$, and is called nilpotent of degree k if $T^k = 0$, but $T^{k-1} \neq 0$.

This paper is organized as follows. In Section 2, we characterize those Toeplitz matrices over a commutative ring (of any characteristic) having the property that their product is a Toeplitz matrix.

ABOUT THE AUTHORS

Dan Kucerovsky is a full professor in the Department of Mathematics and Statistics at the University of New Brunswick in Canada. He studied in Oxford and in Paris. He is a former director of the Centre for Noncommutative Geometry and Topology. He has over 50 published papers in several areas of mathematics and statistics.

Kaveh Mousavand is a graduate student in the Department of Mathematics at L'Université du Québec à Montréal (UQAM) in Canada. His research interests are representation theory of finite dimensional algebras and combinatorial algebra.

Aydin Sarraf received his PhD in Mathematics from University of New Brunswick in Canada. His research interests in mathematics include operator theory, C∗-algebras and K-theory, and in computer science include computer vision, computational fluid dynamics and computational finance.

PUBLIC INTEREST STATEMENT

Toeplitz matrices arise in a variety of problems in applied mathematics and engineering such as queuing theory, signal processing, time series analysis, integral equations, etc. Despite the fact that Toeplitz matrices have been around for a long time, there are still a number of open problems regarding Toeplitz matrices and Toeplitz operators. One of them is counting the number of involutory and nilpotent Toeplitz matrices over a finite field. In this paper, we provide some solutions to the problem only in very specific cases. We also give several optimal approximation results for Toeplitz, circulant and skew-circulant matrices over the field of complex numbers. Furthermore, we give a characterization for pairs of Toeplitz matrices over a commutative ring whose product with respect to the usual multiplication or the Kronecker product is Toeplitz.

We consider also Kronecker products of Toeplitz matrices. Furthermore, we also give some combinatorial results over the finite prime field $\mathbb{Z}_p$.

In Section 3, we count the number of involutory and degree two nilpotent Toeplitz matrices in $M_n(\mathbb{Z}_p)$ for particular values of n or p.

In Section 4, Toeplitz matrices over the field of complex numbers are studied. We show that for any matrix, there exists a closest Topelitz matrix (with respect to the Frobenius norm) that approximates it. We describe the equivalence classes of Toeplitz matrices and give several results regarding circulant and skew-circulant matrices.

2. Product of Toeplitz matrices over a commutative ring

We consider three different binary operations, the usual matrix multiplication, Kronecker product and Schur product, denoted by AB, $A \otimes B$ and $A * B$, respectively. With the exception of Schur product, the set of Toeplitz matrices is neither closed under the usual matrix multiplication nor under the Kronecker product. We characterize those pairs of Toeplitz matrices whose product with respect to the usual multiplication or the Kronecker product is Toeplitz again. The following theorem gives a characterization in terms of the usual matrix multiplication:

THEOREM 2.1 *If $A = \{\alpha_{i-j}\}$ and $B = \{\beta_{i-j}\}$ are two $n \times n$ Toeplitz matrices over a commutative ring, then AB is a Toeplitz matrix if and only if the following system of equations, of $(n-1)^2$ equations with $4(n-1)$ variables, holds:*

$$\alpha_i\beta_{-j} - \alpha_{i-n}\beta_{n-j} = 0 \text{ where } 1 \le i,j \le n-1$$

Proof   If we set $C = AB$, by the product formula we have:

$$c_{i+1j+1} = \sum_{k=1}^{n}\alpha_{(i+1)-k}\beta_{k-(j+1)} = \alpha_i\beta_{-j} + \sum_{k=2}^{n}\alpha_{(i+1)-k}\beta_{k-(j+1)}$$

$$= \alpha_i\beta_{-j} + \sum_{k=1}^{n-1}\alpha_{i-k}\beta_{k-j} = \alpha_i\beta_{-j} - \alpha_{i-n}\beta_{n-j} + \sum_{k=1}^{n}\alpha_{i-k}\beta_{k-j} \quad = \alpha_i\beta_{-j} - \alpha_{i-n}\beta_{n-j} + c_{ij}$$

By definition, C is a Toeplitz matrix if and only if $c_{i+1j+1} = c_{ij}$. Therefore, C is Toeplitz if and only if $\alpha_i\beta_{-j} - \alpha_{i-n}\beta_{n-j} = 0$ where $1 \le i,j \le n-1$. □

It is clear that the inverse of an arbitrary nonsingular Toeplitz matrix is not necessarily Toeplitz. However, the following proposition shows how the stability under inversion, with respect to being Toeplitz, is related to stability under multiplication. For an arbitrary $n \times n$ matrix A over a commutative ring, we can compute the coefficients of the characteristic polynomial $p_A(x)$ in many different ways, including the eigenvalues of the matrix, or the entries of the matrix (Brooks, 2006). In particular, we wish to use the following formula to compute the coefficients:

$$p_A(x) = \sum_{m=0}^{n} x^{n-m}(-1)^m tr(\Lambda^m A)$$

where $\Lambda^m A$ is the mth exterior power of A. Furthermore, if the characteristic of the commutative ring is zero then it is known (Winitzki, 2010, 3.9) that $tr(\Lambda^m A) = \frac{1}{m!}\det(C)$ where

$$C = \begin{pmatrix} tr(A) & m-1 & 0 & \cdots & \cdots & \cdots & 0 \\ tr(A^2) & tr(A) & m-2 & 0 & \cdots & \cdots & 0 \\ tr(A^3) & tr(A^2) & tr(A) & m-3 & 0 & \cdots & 0 \\ \vdots & \ddots & \ddots & \ddots & \ddots & \ddots & \vdots \\ \vdots & \ddots & \ddots & \ddots & \ddots & \ddots & 0 \\ tr(A^{m-1}) & tr(A^{m-2}) & \cdots & \ddots & \ddots & \ddots & 1 \\ tr(A^m) & tr(A^{m-1}) & \cdots & \cdots & tr(A^3) & tr(A^2) & tr(A) \end{pmatrix}$$

Proposition 2.2 For an $n \times n$ invertible Toeplitz matrix $A = \{\alpha_i\}$ over a commutative ring, if every A^m, for $m = 2, \ldots, n - 1$, is Toeplitz, then A^{-1} is a Toeplitz matrix. Furthermore, if the Toeplitz matrix is considered over a ring of characteristic zero, we have $tr(A^m) = n(A^m)_0$ where

$$(A^m)_0 = \sum_{k_{m-1}=0}^{n-1} \alpha_{-k_{m-1}} \sum_{k_{m-2}=0}^{n-1} \alpha_{k_{m-1}-k_{m-2}} \sum_{k_{m-3}=0}^{n-1} \alpha_{k_{m-2}-k_{m-3}} \cdots,$$

$k_0 = 0$ and the symbol $\displaystyle\sum_{k_0=0}^{n-1}$ disappears.

Proof By the Cayley–Hamilton Theorem, $A^{-1} = \frac{(-1)^{n-1}}{\det(A)}(A^{n-1} + c_{n-1}A^{n-2} \ldots + c_1 I)$, where each coefficient c_{n-m} of the characteristic polynomial could be explicitly computed as the sum of all principal minors of A of size m. Furthermore, if the matrix is considered over a commutative ring of characteristic zero, then we have $c_{n-m} = \frac{(-1)^m}{m!} \det(C)$. The trace of a Toeplitz matrix is simply n times its diagonal entry, but the diagonal entry of the mth power of A, denoted by $(A^m)_0$ here, can be calculated recursively. □

Remark 2.3 If an upper (lower) triangular Toeplitz matrix is invertible, then its inverse is Toeplitz, because the product of two upper (lower) triangular Toeplitz matrices is again an upper (lower) triangular Toeplitz matrix. Since an upper (lower) unitriangular matrix is always invertible and its inverse is an upper (lower) unitriangular matrix, the inverse of any upper (lower) unitriangular Toeplitz matrix is also an upper (lower) unitriangular Toeplitz matrix.

THEOREM 2.4 *Let $A = \{\alpha_i\}$ be an $n \times n$ Toeplitz matrix and $B = \{\beta_j\}$ be an $m \times m$ Toeplitz matrix over a commutative ring. The $mn \times mn$ matrix $A \otimes B$ is Toeplitz if and only if the following system of equations, of $2(m-1)(n-1)$ equations with $2(n+m) - 3$ variables, holds:*

$$\alpha_i \beta_j = \begin{cases} \alpha_{i-1}\beta_{m+j} & j < 0 \text{ and } i \neq 1 - n \\ \alpha_{i+1}\beta_{j-m} & j > 0 \text{ and } i \neq n - 1 \end{cases}$$

where $i \in \{1-n, 2-n, \ldots, n-2, n-1\}$ and $j \in \{1-m, 2-m, \ldots, m-2, m-1\} - \{0\}$.

Proof With regard to the Kronecker multiplication of two matrices, $A \otimes B$ consists of n^2 blocks of Toeplitz matrices, and it is Toeplitz if and only if each of the $(2nm - 1)$ diagonals has the same entries, i.e. if and only if for each diagonal, which is formed by concatenation of diagonals of the adjacent blocks, the entries agree with each other.

Clearly, for the top right corner block in $A \otimes B$, there is no concatenation, and therefore, the top $m - 1$ diagonals always satisfy the property, and the same fact holds for the bottom $m - 1$ diagonals of $A \otimes B$. Furthermore, the main diagonal of $A \otimes B$ is just the concatenation of n copies of the main diagonal of B, multiplied by α_0, i.e. it is simply $\alpha_0\beta_0$. We also notice that for the diagonals of $A \otimes B$ that are the result of concatenation of the main diagonals of some internal blocks, we have no extra conditions and each of them always has the same entry since A and B are both Toeplitz.

Regarding the remaining diagonals, however, we certainly have concatenation of different diagonals of adjacent blocks, which should agree. A simple computation shows that the above equations are coming from those concatenations. Thus, out of the $2mn - 1$ diagonals of $A \otimes B$, always $(2n - 1) + 2(m - 1)$ of them satisfy the property, from which $2n - 1$ comes from the fact that holds for the main diagonals, and $2(m - 1)$ from the diagonals, coming from the single blocks on the top-right and bottom-left corners. □

Let $\mathbb{F}_q$ be a finite field with q elements and let

$$(\bar{\alpha}, \bar{\beta}) = (\alpha_{1-n}, \ldots \alpha_{-1}, \alpha_1, \ldots, \alpha_{n-1}, \beta_{1-n}, \ldots, \beta_{-1}, \beta_1, \ldots, \beta_{n-1}).$$

Define the following set

$$V_{\mathbb{P}}(n,q) = \{(\bar{\alpha},\bar{\beta}) \in \mathbb{P}^{4n-5}_{\mathbb{F}_q}: \alpha_i\beta_{-j} - \alpha_{i-n}\beta_{n-j} = 0, 1 \le i,j \le n-1\}$$

where $\mathbb{P}^{4n-5}_{\mathbb{F}_q}$ is the $(4n-5)$-dimensional projective space over $\mathbb{F}_q$. Since the equations of Theorem 2.1 are homogeneous, $V_{\mathbb{P}}(n,q)$ is a projective variety and an upper bound of its cardinality can be given by the Lang–Weil estimate (Ghorpade & Lachaud, 2002; Lang & Weil, 1954). Similarly, we can define an affine variety

$$V_{\mathbb{A}}(n,p) = \{(\bar{\alpha},\bar{\beta}) \in \mathbb{F}^{4(n-1)}_p: \alpha_i\beta_{-j} - \alpha_{i-n}\beta_{n-j} = 0, 1 \le i,j \le n-1\}$$

As an example, one can check that $|V_{\mathbb{A}}(2,p)| = p^3 + p^2 - p$ and $|V_{\mathbb{A}}(3,p)| = p^5 + 3p^4 - 2p^3 - 2p^2 + p$.

3. Involutory and nilpotent Toeplitz matrices over the finite field $\mathbb{Z}_p$

Let $\phi(n,p)$ and $\psi(n,p)$, respectively, denote the number of involutory Toeplitz matrices and nilpotent Toeplitz matrices of degree two in $M_n(\mathbb{Z}_p)$. In the following theorem, we calculate $\phi(n,p)$ for specific values of p or n. Clearly, for a Toeplitz matrix A in $M_n(\mathbb{Z}_p)$, A is involutory if and only if $\frac{A+I}{2}$ is idempotent. In this case, since $\frac{A+I}{2}$ is again a Toeplitz matrix in $M_n(\mathbb{Z}_p)$, for $p \ne 2$, $\phi(n,p)$ also gives us the number of idempotent Toeplitz matrices in $M_n(\mathbb{Z}_p)$. Moreover, for $p = 2$, a Toeplitz matrix A is a nontrivial involutory matrix if and only if the Toeplitz matrix $(A - I)$ is nilpotent of degree two. Therefore, $\phi(n,2) - 1$ is also the number of degree two nilpotent Toeplitz matrices when n is odd.

THEOREM 3.1   *The following hold:*

(i) If n is odd then $\phi(n,2) = 2\phi(n-2,2) + 1$ or equivalently $\phi(n,2) = 2^{\frac{n+1}{2}} - 1$

(ii) $\phi(p,p) = 2$ if $p = 3, 5, \dots$

(iii) $\phi(n,p) = 2^{n-2}(p-1) + 2$ if $n \in \{2,4\}$ and $p \in \{3,5,7,\dots\}$ or if $n = 3$ and $p \in \{5,7,\dots\}$

Proof   (i) Let $n = 2k+1$ where $k \ge 1$. Let $T = \{t_{i-j}\}$ be an involutory Toeplitz matrix. The anti-diagonal entries of $C = T^2$ are of the following form:

$$c_{(n-i)(i+1)} = \sum_{l=1}^{n} t_{(n-i)-l}t_{l-(i+1)} = t_{n-i-1}t_{-i} + t_{n-i-2}t_{1-i} + \dots + t_{n-i-\frac{n+1}{2}}t_{\frac{n+1}{2}-(i+1)} + \dots$$

$$+ t_{(n-i)-(n-1)}t_{(n-1)-(i+1)} + t_{-i}t_{n-i-1} = 2t_{(n-i)-(n-1)}t_{(n-1)-(i+1)} + 2t_{-i}t_{n-i-1} + \dots + t^2_{\frac{n-2i-1}{2}} = t_{\frac{n-2i-1}{2}}$$

where $i = 0,\dots,n-1$. If $C = I$ then $c_{(n-i)(i+1)} = t_{\frac{n-2i-1}{2}} = \delta_{(n-i)(i+1)}$. Hence, $c_{\frac{n+1}{2}\frac{n+1}{2}} = t_0 = 1$ if $i = \frac{n-1}{2}$ and $t_{\frac{n-2i-1}{2}} = 0$ if $i = 0,\dots,n-1$ and $i \ne \frac{n-1}{2}$. Therefore, $2k+1$ entries t_l where $l = -k,\dots,-2,-1,0,1,2,\dots,k$ are known, while $2k$ entries t_l where $l = -2k,\dots,-k-1,k+1,\dots,2k$ are unknown. Therefore, the matrix T must be of the following form:

$$\begin{pmatrix}
1 & 0 & \cdots & 0 & t_{-k-1} & \cdots & \cdots & t_{-2k} \\
0 & 1 & 0\cdots & 0 & 0 & t_{-k-1} & \cdots & t_{-2k+1} \\
\vdots & & & & & \vdots & & \\
\vdots & & & & & \vdots & & \\
t_{2k-2} & \cdots & & & \cdots & 1 & 0 & 0 \\
t_{2k-1} & \cdots & t_{k+1} & 0 & \cdots & \cdots & 0 & 1 & 0 \\
t_{2k} & t_{2k-1} & \cdots & t_{k+1} & 0 & \cdots & \cdots & 0 & 1
\end{pmatrix}$$

It is easy to see that for the matrix T which is illustrated above, the equation $T^2 = I$, in $\mathbb{Z}_2$, gives the following k^2 equations in $\mathbb{Z}_2$:

$$t_{i-n}t_{n-j} = 0 \quad 1 \leq i,j \leq k$$

If $t_{i-n} = 0$ for all i, then t_{n-j} can be either zero or one for any j. Therefore, there are 2^k solutions in this case. Similarly, if $t_{n-j} = 0$ for all j, then t_{i-n} can be either zero or one for any i. Therefore, there are 2^k solutions in this case as well. However, we counted the case that $t_{i-n} = t_{n-j} = 0$ for all $1 \leq i,j \leq k$ twice. Hence, there are $2 \times 2^k - 1$ solutions in total. Since $k = \frac{n-1}{2}$, we conclude that there are $2^{\frac{n+1}{2}} - 1$ solutions in total.

(ii) Since $\text{trace}(T) = pt_0 = 0 \pmod p$, the sum of eigenvalues of T must be zero. This is only possible if all eigenvalues of T are 1 or all are -1.

(iii) For $n = 2$, the matrix equation $T^2 = I$, in $\mathbb{Z}_p$, where p is odd, gives the following equations in $\mathbb{Z}_p$:

$$t_0^2 + t_1 t_{-1} = 1, \; t_0 t_{-1} = t_0 t_1 = 0$$

If $t_0 = 0$ then $t_1 t_{-1} = 1$ which has $p - 1$ solutions in $\mathbb{Z}_p$. If $t_0 \neq 0$, then $t_0^2 = 1$ and $t_1 t_{-1} = t_0 t_{-1} = t_0 t_1 = 0$ which has two solutions, $t_0 = 1, t_1 = t_{-1} = 0$ and $t_0 = p - 1, t_1 = t_{-1} = 0$. Therefore, there are $(p - 1) + 2$ solutions when $n = 2$. For $n = 3$, the matrix equation $T^2 = I$, in $\mathbb{Z}_p$, where p is odd and greater than 3, gives the following equations in $\mathbb{Z}_p$:

$$t_0^2 + 2t_1 t_{-1} = 1, \; 2t_0 t_{-1} + t_1 t_{-2} = t_{-1}^2 + 2t_0 t_{-2} = 2t_0 t_1 + t_{-1}t_2 = t_1^2 + 2t_0 t_2 = 0, \; t_1 t_{-1} = t_2 t_{-2}$$

We note that t_0 cannot be zero because if $t_0 = 0$, then $t_1 = 0$ because of the equation $t_1^2 + 2t_0 t_2 = 0$ and the first equation in the above gives a contradiction. Furthermore, it follows from the above equations and invertibility of t_0 that if $t_1 = 0$, then $t_{-1} = t_2 = t_{-2} = 0, t_0^2 = 1$, and if $t_{-1} = 0$, then $t_1 = t_2 = t_{-2} = 0$, $t_0^2 = 1$. Hence, we have two solutions in this case, namely $t_0 = 1, t_1 = t_{-1} = t_2 = t_{-2} = 0$ and $t_0 = p - 1$, $t_1 = t_{-1} = t_2 = t_{-2} = 0$. Therefore, the remaining case is when t_1 and t_{-1} are both invertible which means that by knowing the values of t_0, t_1 and t_{-1}, we can uniquely determine the values of t_2 and t_{-2} from the equations. Thus, it suffices to count the solutions of the equation $t_0^2 + 2t_1 t_{-1} = 1$. Since the equation $t_0^2 + 2t_1 t_{-1} = 1$ has $2(p - 1)$ solutions in $\mathbb{Z}_p$, where p is odd and greater than 3, there are $2(p - 1) + 2$ solutions in total.

For $n = 4$, the matrix equation $T^2 = I$, in $\mathbb{Z}_p$, where p is odd, gives the following equations in $\mathbb{Z}_p$:

$$t_0^2 + 2t_1 t_{-1} + t_2 t_{-2} = 1, \; t_0 t_{-1} + t_1 t_{-2} = t_0 t_{-3} + t_{-1}t_{-2} = t_0 t_1 + t_2 t_{-1} = t_0 t_3 + t_1 t_2 = 0,$$
$$t_3 t_{-3} = t_1 t_{-1}, \; t_1 t_{-2} = t_2 t_{-3}, \; t_2 t_{-1} = t_3 t_{-2}, \; t_{-1}^2 + t_1 t_{-3} + 2t_0 t_{-2} = t_1^2 + t_{-1}t_3 + 2t_0 t_2 = 0$$

The above equations indicate that if $t_0 = 0$, then either $t_1 = t_{-1} = t_3 = t_{-3} = 0$, and $t_2 t_{-2} = 1$ or $t_2 = t_{-2} = 0$, $t_3 = -t_1^{-1}t_2^2$, $t_{-3} = -t_1^{-1}t_{-1}^2$ and $2t_1 t_{-1} = 1$. Since each of the equations $t_2 t_{-2} = 1$ and $2t_1 t_{-1} = 1$ has $p - 1$ solutions in $\mathbb{Z}_p$, there are $2(p - 1)$ solutions when $t_0 = 0$. If $t_0 \neq 0$, then there are two cases to consider: (1) If either t_1 or t_{-1} is zero, then it follows from the above equations that all variables except t_0 must be zero. Therefore, this case leads to the equation $t_0^2 = 1$ which has two solutions. (2) If neither t_1 nor t_{-1} is zero, then the above equations can be simplified as follows:

$$t_{-2} = -t_1^{-1}t_0 t_{-1}, \; t_2 = -t_{-1}^{-1}t_0 t_1, \; t_{-3} = t_{-1}^2 t_1^{-1}, \; t_3 = t_1^2 t_{-1}^{-1}, \; t_1 t_{-1} = t_2 t_{-2} = t_0^2, \; 4t_0^2 = 1$$

The equations $4t_0^2 = 1$ and $t_1 t_{-1} = t_0^2$ lead to equations $t_1 t_{-1} = \frac{p+1}{2}$ and $t_1 t_{-1} = \frac{p^2-1}{2}$ each of which has $2(p - 1)$ solutions. Therefore, if $t_0 \neq 0$, then there are $2(p - 1) + 2$ solutions, and if $t_0 = 0$, then there are $2(p - 1)$ solutions. Hence, there are $4(p - 1) + 2$ solutions in total. $\qquad \square$

CONJECTURE 3.2   *If n is even then $\phi(n, 2) = 2\phi(n - 2, 2) + 2$ or equivalently $\phi(n, 2) = 3 \times 2^{\frac{n}{2}} - 2$.*

THEOREM 3.3   *The following hold:*

(i)

$$\psi(2, p) = \begin{cases} 2p - 2 & \text{if } p \neq 2 \\ 3 & \text{if } p = 2 \end{cases}$$

(ii)

$$\psi(3, p) = \begin{cases} 2p - 2 & \text{if } p \neq 3 \\ 8 & \text{if } p = 3 \end{cases}$$

Proof   (i) The matrix equation $T^2 = 0$ results in the following equations:

$$t_0^2 + t_{-1}t_1 = 0, \ 2t_0t_1 = 0, \ 2t_0t_{-1} = 0$$

If $p = 2$, then there is just one equation $t_0^2 + t_{-1}t_1 = 0$ which has three solutions. Assume $p \neq 2$, if t_1 and t_{-1} are both nonzero, then the equation $t_1 t_{-1} = 0$ is a contradiction. Therefore, either t_1 and t_{-1} are both zero which we do not count or one of them is nonzero which gives $2(p - 1)$ solutions. Hence, there are $2(p - 1)$ solutions in total.

(ii) The matrix equation $T^2 = 0$ results in the following equations:

$$t_0^2 = -2t_{-1}t_1 = -t_2t_{-2} - t_1t_{-1}, \ t_1^2 = -2t_0t_2, \ t_{-1}^2 = -2t_0t_{-2}, \ t_2t_{-1} = -2t_0t_1, \ t_1t_{-2} = -2t_0t_{-1}$$

It follows from the above equations that the cases in which $t_1 = 0$ and $t_{-1} \neq 0$ or $t_{-1} = 0$ and $t_1 \neq 0$ result in contradictions. If $t_1 = t_{-1} = 0$, then $t_0 = 0$ and the equation $t_2t_{-2} = 0$ has $2p - 2$ solutions. If $t_1 \neq 0, t_{-1} \neq 0$ and $p \neq 3$, then the equation $-9t_1t_{-1} = 0$ which follows from the above equations is a contradiction. Therefore, the case $t_1 \neq 0$ and $t_{-1} \neq 0$ is not possible when $p \neq 3$ and there are $2p - 2$ solutions in total. If $t_1 \neq 0, t_{-1} \neq 0$ and $p = 3$, then the equation $-9t_1t_{-1} = 0$ is no longer a contradiction and there are four solutions for this case and $2 \times 3 - 2$ solutions for the previous case which amounts to eight solutions when $p = 3$.   □

Remark 3.4   For matrices in $M_n(\mathbb{Z}_p)$, it is known that the number of invertible, involutory (Hodges, 1958) and nilpotent (Fine & Herstein, 1958) matrices are as follows:

$$g(n, p) = \prod_{i=0}^{n-1}(p^n - p^i)$$

$$I(n, p) = \begin{cases} g(n, p) \sum_{i=0}^{n} \frac{1}{g(i,p)g(n-i,p)} & \text{if } p \text{ is odd} \\ g(n, p) \sum_{0 \leq 2i \leq n} \frac{p^{-(2n-3i)}}{g(i,p)g(n-2i,p)} & \text{if } p \text{ is even} \end{cases}$$

$$l(n, p) = p^{n^2 - n}$$

Therefore, the probabilities of being invertible, involutory and nilpotent are $\frac{g(n,p)}{p^{n^2}}$, $\frac{I(n,p)}{p^{n^2}}$ and $\frac{l(n,p)}{p^{n^2}} = \frac{1}{p^n}$, respectively. Since the number of invertible Toeplitz matrices is $(p - 1)p^{2n-2}$ (Kaltofen & Lobo, 1996, Theorem 4), the probability that a Toeplitz matrix is invertible is $\frac{(p-1)p^{2n-2}}{p^{2n-1}} = 1 - \frac{1}{p}$ which is independent of n. By Theorem 3.1, the probability that a Toeplitz matrix in $M_n(\mathbb{Z}_2)$, where n is odd, is involutory is $2^{1-2n}(2^{\frac{n+1}{2}} - 1) \leq \frac{1}{2}$.

THEOREM 3.5 Let $G(n,p,m) = \{T \in M_n(\mathbb{Z}_p) : \underbrace{T * T... * T}_{m \text{ times}} = J \text{ and } T \text{ is Toeplitz }\}$ where $*$ is the Schur product and J is the matrix with all entries equal to 1. The cardinality of the set G(n, p, m) is $gcd(m, p - 1)^{2n-1}$.

Proof The entries of T must satisfy the equation $t_{ij}^m = 1$ (mod p). Therefore, we are counting elements of $\mathbb{Z}_p$ that their order divides m and their order also divides $p - 1$, i.e. their order divides $gcd(m, p - 1)$. Hence,

$$\sum_{d | gcd(m,p-1)} \phi(d) = gcd(m, p - 1)$$

where ϕ is Euler's totient function. Since there are $2n - 1$ distinct elements, there are $gcd(m, p - 1)^{2n-1}$ different matrices T with the aforementioned properties. □

4. Toeplitz matrices over the complex field $\mathbb{C}$

In this section, we consider Toeplitz matrices over the complex field $\mathbb{C}$. We define a metric that measures how far a matrix is from being Toeplitz. We can associate a Toeplitz matrix T_A to any matrix A as follows:

$$t_{1j} = \frac{\sum_{l=0}^{k} a_{1+lj+l}}{k+1} 1 \leq j \leq n - 1, \quad k = n - j$$

$$t_{i1} = \frac{\sum_{l=0}^{k} a_{i+l1+l}}{k+1} 1 \leq i \leq n - 1, \quad k = n - i, \, t_{ij} = t_{i+lj+l} \quad i+l \leq n, j+l \leq n, \, t_{1n} = a_{1n}, \quad t_{n1} = a_{n1}$$

For example, the associated Toeplitz matrix to $A = \begin{pmatrix} 3 & -1 \\ 1 & 5 \end{pmatrix}$, is $T_A = \begin{pmatrix} 4 & -1 \\ 1 & 4 \end{pmatrix}$ and we observe that $d(A, T_A) = \|A - T_A\|_{op} = 1$ where $\|\ \|_{op}$ is the operator norm. If A is a Toeplitz matrix, then $d(A, T_A)$ is zero.

THEOREM 4.1   If T_A is the associated Toeplitz matrix of A such that $\|T_A - A\|_{op} < \epsilon$ where $\epsilon > 0$, then

(i) $\|[T_A, A]\|_{op} < 2\epsilon \min\{\|T_A\|_{op}, \|A\|_{op}\}$,

(ii) if A is self-adjoint then T_A is self-adjoint and furthermore if $\|A\|_{op} \leq 1$ and $\|T_A\|_{op} \leq 1$, then A and T_A are approximately jointly diagonalizable in the sense of Bronstein and Glashoff (2013).

Proof

(i) We have the following inequalities:

$$\|T_A A - A T_A - T_A^2 + T_A^2\|_{op} < 2\epsilon\|T_A\|_{op} \quad \|T_A A - A T_A - A^2 + A^2\|_{op} < 2\epsilon\|A\|_{op}$$

Therefore, $\|[T_A, A]\|_{op} < 2\epsilon \min\{\|T_A\|_{op}, \|A\|_{op}\}$.

(ii) T_A is self-adjoint because

$$t_{ij} = \frac{\sum_{l=0}^{k} a_{i+lj+l}}{k+1} = \frac{\sum_{l=0}^{k} \bar{a}_{j+li+l}}{k+1} = \bar{t}_{ji}$$

and $t_{1n} = a_{1n}, t_{n1} = a_{n1}$. It follows from Bronstein and Glashoff (2013, [Theorem 4.1]) that if $\|A\|_{op} \leq 1$ and $\|T_A\|_{op} \leq 1$, then A and T_A are approximately jointly diagonalizable. □

We now consider involutory Toeplitz matrices over the complex field $\mathbb{C}$. The following simple Proposition shows that if there exists one such matrix, there exist uncountably many.

Proposition 4.2   If

$$\begin{pmatrix} a_n & a_{n+1} & \cdots & & & \\ a_{n-1} & a_n & a_{n+1} & & & \\ \vdots & a_{n-1} & a_n & a_{n+1} & & \\ & & \ddots & \ddots & \ddots & \\ & & & & & a_{n+1} \\ & & & a_{n-1} & a_n \end{pmatrix}$$

is an involutory Toeplitz matrix over $\mathbb{C}$, and v is a nonzero complex scalar, then

$$\begin{pmatrix} a_n & va_{n-1} & \cdots & & & \\ v^{-1}a_{n-1} & a_n & va_{n-1} & & & \\ \vdots & & v^{-1}a_{n-1} & a_n & va_{n-1} & \\ & & & \ddots & \ddots & \ddots \\ & & & & & va_{n-1} \\ v^{-n+1}a_1 & & & & v^{-1}a_{n-1} & a_n \end{pmatrix}$$

is also an involutory Toeplitz matrix.

Proof   The two matrices are related by a similarity transformation by a diagonal matrix. See for example, Proposition 3.1 in Kucerovsky and Sarraf (2014).   □

In view of the above result, it would appear inevitable that we should consider some equivalence relation or some restriction on the class of involutory Toeplitz matrices considered. We give two such results:

THEOREM 4.3   *There are 2^n equivalence classes of involutory Toeplitz matrices in $M_n(\mathbb{C})$, where the equivalence relation is similarity. Furthermore, every equivalence class contains at least one (involutory) Hermitian circulant matrix.*

Proof   Let T be an involutory Toeplitz matrix. Since the polynomial $p(x) := x^2 - 1$ annihilates T, it follows that the minimal polynomial of T divides p. But because p has distinct roots over $\mathbb{C}$, it follows that the minimal polynomial of T has distinct roots over $\mathbb{C}$, and thus that T is diagonalizable. Thus, T is equivalent under similarity to a diagonal matrix Λ. Let F denote the usual Vandermonde matrix representation of the finite Fourier transform (over $\mathbb{C}$). Then, $F^{-1}\Lambda F$ is equivalent to T, is a Hermitian matrix and is a circulant (Driessel, 2011). Clearly, thus, the number of equivalence classes is given by the number of possible involutory diagonal matrices in $M_n(\mathbb{C})$, which is 2^n.   □

Note: the above proof will work in more general fields: we just need sufficiently many roots of unity and the existence of $1/n$.

We can also look at restricting the class of Toeplitz matrices considered:

THEOREM 4.4   *In $M_n(\mathbb{R})$, the involutory symmetric Toeplitz matrices are all either symmetric real circulants or are symmetric real skew-circulants. If n is even and greater than 2, there are a total of $3 \cdot 2^{\frac{n}{2}} - 2$ such matrices. If n is odd and greater than 1, there is a total of $2^{\frac{k+3}{2}} - 2$ such matrices.*

Proof   We observe that if a symmetric real matrix satisfies $T^2 = \mathrm{Id}$, it is in fact an orthogonal matrix. Orthogonal and symmetric real Toeplitz matrices are either circulants or skew-circulants (Böttcher, 2008, [section 3]). From Theorem 4.1 and Theorem 4.2 in Böttcher (2008), if n is odd and greater than

1, there is a total of $2^{\frac{k+3}{2}} - 2$ such matrices, and if n is even and greater than 2, there are a total of $3 \cdot 2^{\frac{n}{2}} - 2$ such matrices.   □

THEOREM 4.5   *Consider the Frobenius norm $\|\cdot\|$ on the complex n-by-n matrices, $M_n(\mathbb{C})$. Given an arbitrary $A \in M_n(\mathbb{C})$, there exist a unique complex Toeplitz matrix minimizing $\|A - T\|$ over all Toeplitz matrices T. This unique minimizing matrix is the Toeplitz matrix given in each diagonal by the average of the elements of the corresponding diagonal of A. The same statement holds in the real case.*

Proof   Under this norm, $M_n(\mathbb{C})$ is a complex Hilbert space, with the complex Toeplitz matrices forming a subspace $\mathscr{T}$. Let $e_1, \cdots, e_{2n-1}$ denote the Toeplitz matrices that have zero in all diagonals except one, and in that k^{th} diagonal, has elements equal to 1. At the level of Hilbert spaces, this is a finite orthogonal set spanning the subspace $\mathscr{T}$. Thus, we can explicitly construct the projection from $M_n(\mathbb{C})$ onto $\mathscr{T}$ in the form:

$$P(A) := \sum_{k=1}^{2n-1} \frac{<A, e_k> e_k}{\|e_k\|}.$$

It is well known that orthogonal projection onto a closed subspace, in Hilbert space, minimizes the norm. In this case, the subspace $\mathscr{T}$ is closed as a consequence of its finite-dimensionality. It is clear that $P(A)$ is given in each diagonal by the average of the elements of the corresponding diagonal of A. In the real case, exactly the same proof works, but with real Hilbert spaces instead of complex Hilbert spaces.   □

It seems interesting to consider generalizations of the above Theorem to the C*-norm on $M_n(\mathbb{C})$. However, projection onto linear subspaces of C*-algebras has poor properties in general. It is not even true that the operation of projection onto a subspace will not increase the norm distance between elements (Silverman, 1969). The situation improves when we consider projection onto subalgebras. Tomiyama (Tomiyama) showed that each projection of norm one of a C*-algebra onto a C*-subalgebra is a conditional expectation, and Takesaki (1972) showed that in a finite von Neumann algebra, there exist faithful conditional expectations onto von Neumann subalgebras. We will make use of this result, and will explicitly construct faithful conditional expectations onto the circulant algebra and the skew-circulant algebra. We will show these expectations are basically unique.

For information on circulants and skew-circulants, see Driessel, 2011. The circulants of a given order form an abelian subalgebra of $M_n(\mathbb{C})$, as do the skew-circulants. The circulants can be simultaneously diagonalized by (a scalar multiple of) the DFT matrix $F := [\omega^{(i-1)(j-1)}]$ where ω is a primitive (complex) root of unity. The skew-circulants can be simultaneously diagonalized by $F\Omega^{1/2}$ where Ω is a suitable diagonal unitary matrix, see Proposition 31 in Driessel (2011). We will write $W = H^*\Lambda H$, where W is a skew-circulant and Λ is a diagonal matrix, or $C = F^*\Lambda F$, for a circulant C.

THEOREM 4.6   *A faithful, trace-preserving, idempotent, conditional expectation of operator norm 1 of $M_n(\mathbb{C})$ onto the circulants is given by*

$\phi(M) = F^*D(FMF^*)F,$

where D takes all elements that are not on the main diagonal to zero, and preserves the elements of the main diagonal and F is the discrete Fourier transform matrix.

Proof   Since the map $m \mapsto FMF^*$ is a unitary transformation diagonalizing the circulants, we must show that the map D is a faithful, idempotent, trace-preserving, conditional expectation of norm 1 of $M_n(\mathbb{C})$ onto the subalgebra of diagonal matrices. To show that D is a conditional expectation, we verify that the defining property $D(dmd^*) = d\phi(m)d^*$ holds, where d is a diagonal matrix, and m is in $M_n(\mathbb{C})$. It is apparent that $D:M_n(\mathbb{C}) \to M_n(\mathbb{C})$ preserves the trace. Since the diagonal elements of a positive definite matrix are real and nonnegative, it follows that D maps a positive definite matrix to a positive definite (diagonal) matrix. If D were not faithful, there would be some nonzero positive definite matrix that maps to zero under D. But a positive definite matrix with principal diagonal zero is in fact the zero matrix (Horn & Johnson, 1990, p. 398). Conditional expectations are completely positive maps, and a completely positive map of unital algebras has norm one if and only if it takes the unit to the unit. Evidently, the map D is unital, and thus has norm 1. It is clear that D is idempotent.   □

COROLLARY 4.7   There is only one idempotent linear mapping of $M_n(\mathbb{C})$ onto the circulant subalgebra in $M_n(\mathbb{C})$. It is given by the map in the above Theorem.

Proof   By Theorem 1 in Tomiyama (1957), any given idempotent linear mapping L is a conditional expectation of a finite-dimensional von Neumann algebra onto a finite-dimensional von Neumann algebra. By finite-dimensionality, the mapping is necessarily normal. By Proposition 1.2 in Tomiyama (1972), the mapping L is necessarily faithful. By Theorem 6.2.2 (and Theorem 6.2.3) in Arveson (1967), the mapping L is unique. It therefore follows that any two maps satisfying the hypotheses of the Corollary are in fact the same map, and clearly, the map provided by Theorem 4.6 satisfies the hypotheses.   □

From the uniqueness, we have the following further Corollary:

COROLLARY 4.8   The map of Theorem 4.6 minimizes the error in (operator) norm among all idempotent projections of $M_n(\mathbb{C})$ onto the circulant subalgebra in $M_n(\mathbb{C})$.

Replacing circulants by skew-circulants in the proofs of the last three results, we obtain analoguous results for skew-circulants. We summarize as follows:

THEOREM 4.9   *A faithful, trace-preserving, idempotent, conditional expectation of operator norm 1 of $M_n(\mathbb{C})$ onto the skew-circulants is given by*

$$\psi(M) = H^*D(HMH^*)H,$$

where D takes all elements that are not on the main diagonal to zero, and preserves the elements of the main diagonal. This map minimizes the error in (operator) norm among all idempotent projections of $M_n(\mathbb{C})$ onto the skew-circulant subalgebra in $M_n(\mathbb{C})$. H is the matrix provided by Driessel (2011, [Proposition 31]).

Since every Toeplitz matrix can be uniquely decomposed as the sum of a circulant and a skew-circulant, we have the following Remark:

Remark 4.10   For each matrix m, the sum of the maps from Theorems 4.6 and 4.9, $m \mapsto \phi(m) + \psi(m)$, gives a Toeplitz matrix.

We next consider how the usual product and Schur product interact with the circulant property. The finite Fourier transform over the complex numbers can be described most briefly as the unitary transformation that is (up to a scalar multiple) represented by the N-by-N matrix $[a_{ij}]$ with $a_{ij} = \omega^{(i-1)(j-1)}$. The complex number ω is a principal N^{th} root of unity. This definition can be made in a finite field. If ω is a principal N^{th} root of unity in a finite field, it follows that $\omega^N = 1$ and $\sum_{j=0}^{n-1} \omega^{ji} = 0$ for $1 \leq i < N$, and these are the main properties used in establishing the well-known properties of the (complex) finite Fourier transform with respect to convolution. In order to be able to invert the generalized finite Fourier transform, it is necessary that N be not divisible by the characteristic of the field. See Chapter 11 of Elliott and Rao (1982) for more information. If these conditions are met, then the generalized finite Fourier transform takes the convolution operation defined by multiplication by a circulant matrix to multiplication by a diagonal matrix, see section 11.2 and 11.3 of Elliott and Rao (1982). Thus, we are able to diagonalize circulant matrices, even over a finite field. The diagonal matrix obtained will be referred to as the *eigenvalue matrix* of the circulant, and denoted $\Lambda(A)$ where

A is the given circulant. Let us say that a circular convolution of two sequences is the first row of the matrix product of the circulants having the given sequences as their first row. Thus, if $Circ(a_1, \cdots, a_n)Circ(b_1, \cdots, b_n) = Circ(c_1, \cdots, c_n)$, then the sequence $(c_1, \cdots, c_n)$ is the circular convolution of $(a_1, \cdots, a_n)$ and $(b_1, \cdots, b_n)$.

Proposition 4.11   Let A and B be two circulants, over a finite field with characteristic relatively prime to the matrix dimension N of the circulant matrix. The eigenvalues of AB and $A * B$ can be determined in terms of the eigenvalues of A and B. The eigenvalues of AB are given by $\Lambda(A) * \Lambda(B)$, and the eigenvalues of $A * B$ are given by $\frac{1}{N}C$ where C is the circulant matrix whose eigenvalue sequence is the circular convolution of the eigenvalue sequences of A and B.

Proof   Since the generalized Fourier transform is a unitary transformation, W, it is clear that $WABW^{-1} = WAW^{-1}WBW^{-1}$. Thus, the eigenvalues of AB are the elementwise product of the eigenvalues of A and B. The main issue is the behaviour of the Schur product.

To reduce notation, let us consider first the case where both A and B have the property that all the eigenvalues except one are zero, and that eigenvalue is 1. Then, $A = W^{-1}\Lambda(A)W$ is an outer product of a row of W and a column of W^{-1}. Since $W = [\omega^{(i-1)(j-1)}]$ and $W^{-1} = \frac{1}{N}[\omega^{-(i-1)(j-1)}]$, we thus have that $A = \frac{1}{N}[\omega^{(j-i)(k-1)}]$ where k is the index of the nonzero eigenvalue of A. Similarly, $B = \frac{1}{N}[\omega^{(j-i)(\ell-1)}]$ where ℓ is the index of the nonzero eigenvalue of B. Taking the Schur product, we have $A * B = \frac{1}{N^2}[\omega^{(j-i)(\ell+k-2)}]$. We can thus write $A * B = \frac{1}{N}C$ where C is the circulant matrix with all eigenvalues zero except for a 1 in the $(\ell + k - 1)^{th}$ place. Since $\omega^N = 1$, the values of the exponent in $\omega^{(j-i)(\ell+k-2)}$ may be taken as being modulo N. Thus, we see that the eigenvalues of $A * B$ are given by $\frac{1}{N}C$, where the eigenvalue sequence of the circulant matrix C are given by the circular convolution of the eigenvalue sequence of A and the eigenvalue sequence of B. The case of general circulants A and B follows by taking linear combinations.   □

The above result can be rephrased as follows: if we diagonalize a circulant using the generalized discrete Fourier transform, and then put the eigenvalue sequence into the first row of a circulant matrix, then up to a scalar multiple, Schur products are transformed into matrix products. This transformation is invertible.

Funding
This work was partially supported by NSERC [grant number 1-108036-40-01].

Author details
Dan Kucerovsky[1]
E-mail: dan@math.unb.ca
Kaveh Mousavand[2]
E-mail: k.mousavand@lacim.ca
Aydin Sarraf[3]
E-mail: Aydin.Sarraf@unb.ca

[1] Department of Mathematics, University of New Brunswick, Fredericton, Canada, NB, E3B 5A3.
[2] Département de mathématiques, L'Université du Québec à Montréal, Montréal, Canada, H3C 3P8.
[3] Department of Computer Science, University of New Brunswick, Fredericton, Canada, NB, E3B 5A3.

References
Arveson, W. B. (1967). Analyticity in operator algebras. *American Journal of Mathematics, 89,* 578–642.
Böttcher, A. (2008). Orthogonal symmetric Toeplitz matrices. *Complex Analysis and Operator Theory, 2,* 285–298.
Brooks, B. P. (2006). The coefficients of the characteristic polynomial in terms of the eigenvalues and the elements of an $n \times n$ matrix. *Applied Mathematics Letters, 19,* 511–515.
Bronstein, M. M., & Glashoff, K. (2013). *Almost-commuting matrices are almost jointly diagonalizable*, arXiv:1305.2135.
Driessel, K. R. (2011). *A relation between some special centro-skew, near-Toeplitz, tridiagonal matrices and circulant matrices.* arXiv:1102.1953v2.
Elliott, D. F., & Rao, K. R. (1982). *Fast transforms, algorithms, analyses, applications.* New York, NY: Academic Press.
Fine, N. J., & Herstein, I. N. (1958). The probability that a matrix be nilpotent. *Illinois Journal of Mathematics, 2,* 499–504.
Ghorpade, S. R., & Lachaud, G. (2002). Number of solutions of equations over finite fields and a conjecture of Lang and Weil. *Number theory and discrete mathematics* (pp. 269–291). Basel: Springer.
Heinig, G. (2001). Not every matrix is similar to a Toeplitz matrix. *Linear Algebra and its Applications, 332,* 519–531.
Hodges, J. H. (1958). The matrix equation $X^2 - I = 0$ over a finite field. *The American Mathematical Monthly, 65,* 518–520.
Horn, R. A., & Johnson, C. R. (1990). *Matrix analysis.* Cambridge: Cambridge University Press.
Kaltofen, E., & Lobo, A. (1996). On rank properties of Toeplitz matrices over finite fields. In *Proceedings of the 1996 international symposium on Symbolic and algebraic computation*, ACM (pp. 241–249), New York.
Kucerovsky, D., & Sarraf, A. (2014). Schur multipliers and matrix products. *Houston Journal of Mathematics, 40,* 837–850.

Lang, S., & Weil, A. (1954). Number of points of varieties in finite fields. *American Journal of Mathematics, 76,* 819–827.

Lim, L. H., & Ye, K. (2013). *Every matrix is a product of Toeplitz matrices.* arXiv:1307.5132.

Mackey, D. S., Mackey, N., & Petrovic, S. (1999). Is every matrix similar to a Toeplitz matrix? *Linear Algebra and its Applications, 297,* 87–105.

Silverman, E. (1969). A weak projection of C onto a Euclidean subspace. *Transactions of the AMS, 136,* 381–390.

Takesaki, M. (1972). Conditional expectations in von Neumann algebras. *Journal of Functional Analysis, 9,* 306–321.

Tomiyama, J. (1957). On the projection of norm one in W*-algebras. *Proceedings of the Japan Academy, 33,* 608–612.

Tomiyama, J. (1972). On some types of maximal Abelian subalgebras. *Journal of Functional Analysis, 10,* 373–386.

Winitzki, S. (2010). *Linear algebra via exterior products.* San Bernardino: Lulu.

9

Taylor expansions for the generating function of Catalan-like numbers

Lily Li Liu[1]* and Xiaoli Li[1]

*Corresponding author: Lily Li Liu, School of Mathematical Science, Qufu Normal University, Qufu 273165, P.R. China
E-mail: liulily@mail.qfnu.edu.cn

Reviewing editor: Hari M. Srivastava, University of Victoria, Canada

Abstract: In 2002, Eu, Liu and Yeh introduced new Taylor expansions of the generating function of Catalan and Motzkin numbers. And they presented that this Taylor style expansion can be applied to more generating functions satisfying some relations (*Advances in Applied Mathematics*, 29 (2002) 345–357). In this paper, we focus on this Taylor expansion of the generating function of Catalan-like numbers, which are common generalizations of many classic counting coefficients, such as the Catalan numbers, the Motzkin numbers and the Schröder numbers. We present the recurrence relations of the coefficients and bivariate generating functions of the remainders of the new Taylor expansion of the generating function of Catalan-like numbers.

Subjects: Advanced Mathematics; Analysis - Mathematics; Mathematics & Statistics; Pure Mathematics; Science

Keywords: Catalan-like numbers; Catalan numbers; Schröder numbers; Motzkin numbers; Taylor expansion

AMS subject classifications: 05A15; 05A19; 05A20; 11B83; 15A45

ABOUT THE AUTHORS

Lily Li Liu is working as an associate professor in School of Mathematical Science at Qufu Normal University in China. She is pursuing her PhD degree in Mathematics from Dalian University of Technology in 2009. She has published several research papers in national and international journals. Her area of interest is inequalities, the unimodality property and the properties of combinatorial sequences.

Xiaoli Li is a graduate student of the first author in School of Mathematical Science at Qufu Normal University in China. Her research interests are the properties of combinatorial sequences.

PUBLIC INTEREST STATEMENT

Eu, Liu and Yeh introduced new Taylor expansions, whose remainders involve the function itself, of the generating function of Catalan and Motzkin numbers. In this paper, we focus on this Taylor expansion of the generating function of Catalan-like numbers, which are the 0th column of special cases of Riordan arrays. Riordan arrays play an important unifying role in enumerative combinatorics. There have been quite a few papers concerned with combinatorics of Riordan arrays. Our concern is one class of special interesting Riordan arrays—the recursive matrix introduced by Aigner. The 0th column of such recursive matrix includes many classical combinatorial sequences, such as the Catalan numbers, the Motzkin numbers, the large and little Schröder numbers. We present the recurrence relations of the coefficients and bivariate generating functions of the remainders of the new Taylor expansion of the generating function of Catalan-like numbers.

1. Introduction

The Taylor expansion of a real function $f(x)$, which is infinitely differential, at the real number 0 is

$$f(x) = \sum_{i=0}^{n-1} \frac{f^{(i)}(0)}{i!} x^i + R_n(x), \tag{1}$$

where $R_n(x)$ is the nth remainder, which is the error incurred in approximating the function $f(x)$ by its $(n-1)$th Taylor polynomial $\sum_{i=0}^{n-1} \frac{f^{(i)}(0)}{i!} x^i$.

Traditionally, the remainders of Taylor expansions play a central role in the theory of functions, numberical approximations, asymptotic expansions, etc. They are used mainly for quantitative or numerical purposes (Eu, Liu, & Yeh, 2002). Unlike the usual Taylor expansions, the remainders in Taylor expansions considered in this paper involve the generating function itself. Such expansions are quite different from the usual binomial expansions or continued fraction expansions but are not exceptions for combinatorial structures. Eu et al. (2002) introduced these new Taylor expansions for the Catalan numbers C_i and the Motzkin numbers M_i. They showed that

$$C = \sum_{i \geq 0} C_i x^i = \sum_{i=0}^{n-1} C_i x^i + x^n f_n(C),$$

and

$$M = \sum_{i \geq 0} M_i x^i = \sum_{i=0}^{n-1} M_i x^i + x^n g_n(M) + x^{n+1} h_n(M),$$

where the f_n, g_n and h_n are recursively defined polynomials. This new Taylor expansions can be generalized to the Catalan-like numbers (Eu et al., 2002).

The Catalan-like numbers considered in this paper are the 0th columns of special cases of Riordan arrays. Riordan arrays play an important unifying role in enumerative combinatorics (Shapiro, Getu, Woan, & Woodson, 1991). A *(proper) Riordan array*, denoted by $(g(x), f(x))$, is an infinite lower triangular matrix whose generating function of the kth column is $x^k f^k(x) g(x)$ for $k = 0, 1, 2, \ldots$, where $g(0) = 1$ and $f(0) \neq 0$. A Riordan array $R = [d_{n,k}]_{n,k \geq 0}$ can also be characterized by two sequences $(a_n)_{n \geq 0}$ and $(z_n)_{n \geq 0}$ such that

$$d_{0,0} = 1, \quad d_{n+1,0} = \sum_{j \geq 0} z_j d_{n,j}, \quad d_{n+1,k+1} = \sum_{j \geq 0} a_j d_{n,k+j} \tag{2}$$

for $n, k \geq 0$ (see Cheon, Kim, & Shapiro, 2012; He & Sprugnoli, 2009 for instance). Call $(a_n)_{n \geq 0}$ and $(z_n)_{n \geq 0}$ the *A-* and *Z-sequences* of R, respectively. The 0th column of such a Riordan array includes many classical combinatorial sequences, such as the Catalan numbers, the Motzkin numbers, the large and the little Schröder numbers. There have been quite a few papers concerned with combinatorics of Riordan arrays (see Cheon et al., 2012; Ehrenfeucht, Harju, ten Pas, & Rozenberg, 1998; He & Sprugnoli, 2009; Shapiro et al., 1991 for instance). Our concern in the present paper is one class of special interesting Riordan arrays—the recursive matrix introduced by Aigner (1999, 2001). Let p, s, t be three nonnegative numbers. Denote by $R(p; s, t) = [d_{n,k}]_{n,k \geq 0}$, the Riordan array with $Z = (p, t, 0, 0, \ldots)$ and $A = (1, s, t, 0, 0, \ldots)$. More precisely,

$$d_{0,0} = 1, \quad d_{0,k} = 0 \quad (k > 0);$$
$$d_{n,0} = p d_{n-1,0} + t d_{n-1,1} \quad (n \geq 1); \tag{3}$$
$$d_{n,k} = d_{n-1,k-1} + s d_{n-1,k} + t d_{n-1,k+1} \quad (n, k \geq 1).$$

Following Aigner (1999, 2001), the matrix $R(p; s, t)$ is called *the recursive matrix*, and the numbers $C(p; s, t) = d_{n,0}$ are called *the Catalan-like numbers* corresponding to (σ, τ), where $\sigma = (p, s, s, \cdots), \tau = (t, t, t, \cdots)$.

The Catalan-like numbers unify many famous counting coefficients. For example, the numbers $d_{n,0}$ are:

(1) the Catalan numbers $C_n = C(1; 2, 1)$ corresponding to $\sigma = (1, 2, 2, \cdots), \tau = (1, 1, 1, \cdots)$;

(2) the Motzkin numbers $M_n = C(1; 1, 1)$ corresponding to $\sigma = (1, 1, 1, \cdots), \tau = (1, 1, 1, \cdots)$;

(3) the large Schröder numbers $R_n = C(2; 3, 2)$ corresponding to $\sigma = (2, 3, 3, \cdots), \tau = (2, 2, 2, \cdots)$;

(4) the little Schröder numbers $S_n = C(1; 3, 2)$ corresponding to $\sigma = (1, 3, 3, \cdots), \tau = (2, 2, 2, \cdots)$;

(5) the restricted hexagonal numbers $r_n = C(3; 3, 1)$ corresponding to $\sigma = (3, 3, 3, \cdots), \tau = (1, 1, 1, \cdots)$.

In this paper, we discuss the remainders of new Taylor expansions for the generating function of Catalan-like numbers. More precisely, we obtain the recurrence relations of the coefficients and bi-variate generating functions of the remainders.

2. The Catalan-like numbers' Taylor expansion

Let $d_{n,0} = c_n$ be the nth Catalan-like numbers. Denote by $C(x) = \sum_{n \geq 0} c_n x^n$ the generating function of Catalan-like numbers. Then by the theory of Riordan array, we have

$$C(x) = \cfrac{2}{1 + (s - 2p)x + \sqrt{1 - 2sx + (s^2 - 4rt)x^2}}$$

$$= \cfrac{1 + (s - 2p)x - \sqrt{1 - 2sx + (s^2 - 4rt)x^2}}{2(s - p)x + 2(p^2 - ps + rt)x^2} \tag{4}$$

Let $a = (s - p)x + (p^2 - ps + rt)x^2, b = -1 - (s - 2p)x, c = 1$. So $C(x)$ satisfies the following recurrence relation

$$\left[(s - p)x + (p^2 - ps + rt)x^2\right]C^2(x) - \left[1 + (s - 2p)x\right]C(x) + 1 = 0.$$

Also let $a_1 = s - p, a_2 = p^2 - ps + rt, b_1 = -(s - 2p)$.

Then we have

$$C = 1 + b_1 xC + \left(a_1 x + a_2 x^2\right)C^2$$

$$= 1 + x\left(b_1 C + a_1 C^2\right) + x^2\left(a_2 C^2\right);$$

$$C^2 = C + x\left(b_1 C^2 + a_1 C^3\right) + x^2\left(a_2 C^3\right)$$

$$= 1 + x\left(b_1 C + b_1 C^2 + a_1 C^2 + a_1 C^3\right) + x^2\left(a_2 C^2 + a_2 C^3\right);$$

$$C^3 = C + x\left(b_1 C^2 + b_1 C^3 + a_1 C^3 + a_1 C^4\right) + x^2\left(a_2 C^3 + a_2 C^4\right)$$

$$= 1 + x\left(b_1 C + b_1 C^2 + b_1 C^3 + a_1 C^2 + a_1 C^3 + a_1 C^4\right) + x^2\left(a_2 C^2 + a_2 C^3 + a_2 C^4\right).$$

After arrangement successively, we have

$$C^k = 1 + x\left(\sum_{i=1}^{k} b_1 C^i + \sum_{i=1}^{k} a_1 C^{i+1}\right) + x^2 \sum_{i=1}^{k} a_2 C^{i+1}$$

$$= 1 + x\sum_{i=1}^{k} (b_1 + a_1 C)C^i + x^2 \sum_{i=1}^{k} a_2 C^{i+1}. \tag{5}$$

Then following Eu et al. (2002), Taylor expansions for the generating function of Catalan-like numbers are

$$C = \sum_{k=0}^{n-1} c_k x^k + x^n g_n(C) + x^{n+1} h_n(C). \tag{6}$$

where g_n and h_n are polynomials, and $x^n g_n(C) + x^{n+1} h_n(C)$ is the nth remainder.

So in order to get the uniqueness of g_n and h_n, it suffices to prove that the equation $a(C) + xb(C) = 0$ only has the trivial solution for polynomials $a(y)$ and $b(y)$. Note that

$$C(x) = \frac{1 + (s - 2p)x - \sqrt{1 - 2sx + (s^2 - 4rt)x^2}}{2(s - p)x + 2(p^2 - ps + rt)x^2}$$

is a continuous function. For any y, there exist different α, β, so that $C(\alpha) = C(\beta) = y$. By inserting $x = \alpha, \beta$ into $a(C) + xb(C) = 0$, we can get $a(y) = b(y) = 0$. As to the polynomial, a and b must be the constant zero.

Since

$$C = 1 + x(b_1 C + a_1 C^2) + x^2(a_2 C^2), \tag{7}$$

we set

$$C = 1 + x g_1(C) + x^2 h_1(C)$$
$$= 1 + x\left(g_{1,1} C + g_{1,2} C^2\right) + x^2 h_{1,1} C^2,$$

where

$$g_{1,1} = b_1, \quad g_{1,2} = a_1, \quad h_{1,1} = a_2.$$

Now, we replace C with $1 + x(b_1 C + a_1 C^2) + x^2(a_2 C^2)$. So we have

$$C = 1 + x(b_1 C + a_1 C^2) + x^2(a_2 C^2)$$
$$= 1 + x(b_1 + a_1 C) + x^2\left[b_1^2 C + (2b_1 a_1 + a_2)C^2 + a_1^2 C^3\right] + x^3(b_1 a_2 C^2 + a_1 a_2 C^3)$$
$$= 1 + b_1 x + a_1 x\left[1 + x(b_1 C + a_1 C^2) + x^2(a_2 C^2)\right] + x^2\left[b_1^2 C + (2b_1 a_1 + a_2)C^2 + a_1^2 C^3\right]$$
$$+ x^3(b_1 a_2 C^2 + a_1 a_2 C^3)$$
$$= 1 + x(a_1 + b_1) + x^2\left[(b_1^2 + a_1 b_1)C + (2b_1 a_1 + a_2 + a_1^2)C^2 + a_1^2 C^3\right]$$
$$+ x^3\left[(b_1 a_2 + a_1 a_2)C^2 + a_1 a_2 C^3\right].$$

Then we set

$$C = 1 + c_1 x + x^2 g_2(C) + x^3 h_2(C)$$
$$= 1 + c_1 x + x^2(g_{2,1} C + g_{2,2} C^2 + g_{2,3} C^3) + x^3(h_{2,1} C^2 + h_{2,2} C^3), \tag{8}$$

where

$$g_{2,1} = b_1^2 + a_1 b_1 = b_1(g_{1,1} + g_{1,2});$$
$$g_{2,2} = 2b_1 a_1 + a_2 + a_1^2 = b_1 g_{1,2} + a_1(g_{1,1} + g_{1,2}) + h_{1,1};$$
$$g_{2,3} = a_1^2 = a_1 g_{1,2};$$
$$h_{2,1} = b_1 a_2 + a_1 a_2 = a_2(g_{1,1} + g_{1,2});$$
$$h_{2,2} = a_1 a_2 = a_2 g_{1,2}.$$

We replace C with $1 + x(b_1 C + a_1 C^2) + x^2(a_2 C^2)$. And we have

$$
\begin{aligned}
g_2(C) &= g_{2,1}C + g_{2,2}C^2 + g_{2,3}C^3 \\
&= g_{2,1}C + g_{2,2}C^2 + g_{2,3}\left[C^2 + x(b_1 C^3 + a_1 C^4) + x^2(a_2 C^4)\right] \\
&= g_{2,1}C + (g_{2,2} + g_{2,3})C^2 + x(b_1 g_{2,3}C^3 + a_1 g_{2,3}C^4) + x^2(a_2 g_{2,3}C^4) \\
&= g_{2,1}C + (g_{2,2} + g_{2,3})\left[C + x(b_1 C^2 + a_1 C^3) + x^2(a_2 C^3)\right] + x(b_1 g_{2,3}C^3 + a_1 g_{2,3}C^4) \\
&\quad + x^2(a_2 g_{2,3}C^4) \\
&= (g_{2,1} + g_{2,2} + g_{2,3})C + x\left\{b_1(g_{2,2} + g_{2,3})C^2 + [a_1(g_{2,2} + g_{2,3}) + b_1 g_{2,3}]C^3 + a_1 g_{2,3}C^4\right\} \\
&\quad + x^2\left[a_2(g_{2,2} + g_{2,3})C^3 + a_2 g_{2,3}C^4\right] \\
&= (g_{2,1} + g_{2,2} + g_{2,3})\left[1 + x(b_1 C + a_1 C^2) + x^2(a_2 C^2)\right] + x^2\left[a_2(g_{2,2} + g_{2,3})C^3 + a_2 g_{2,3}C^4\right] \\
&\quad + x\left\{b_1(g_{2,2} + g_{2,3})C^2 + [a_1(g_{2,2} + g_{2,3}) + b_1 g_{2,3}]C^3 + a_1 g_{2,3}C^4\right\}.
\end{aligned}
\tag{9}
$$

So inserting (Equation (9)) into (Equation (8)), we get

$$
\begin{aligned}
C &= 1 + c_1 x + x^2 g_2(C) + x^3 h_2(C) \\
&- 1 + c_1 x + x^2(g_{2,1} + g_{2,2} + g_{2,3}) + x^4\left[a_2(g_{2,1} + g_{2,2} + g_{2,3})C^2 + a_2(g_{2,2} + g_{2,3})C^3 + a_2 g_{2,3}C^4\right] \\
&\quad + x^3\left\{b_1(g_{2,1} + g_{2,2} + g_{2,3})C + [a_1(g_{2,1} + g_{2,2} + g_{2,3}) + b_1(g_{2,2} + g_{2,3})]C^2\right\} \\
&\quad + x^3\left\{[a_1(g_{2,2} + g_{2,3}) + b_1 g_{2,3}]C^3 + a_1 g_{2,3}C^4 + x^3(h_{2,1}C^2 + h_{2,2}C^3)\right\}.
\end{aligned}
$$

Then we set

$$
\begin{aligned}
C &= 1 + c_1 x + c_2 x^2 + x^3 g_3 C + x^4 h_3 C \\
&= 1 + c_1 x + c_2 x^2 + x^3\left(g_{3,1}C + g_{3,2}C^2 + g_{3,3}C^3 + g_{3,4}C^4\right) + x^4\left(h_{3,1}C^2 + h_{3,2}C^3 + h_{3,3}C^4\right),
\end{aligned}
$$

where

$$g_{3,1} = b_1(g_{2,1} + g_{2,2} + g_{2,3});$$
$$g_{3,2} = a_1(g_{2,1} + g_{2,2} + g_{2,3}) + b_1(g_{2,2} + g_{2,3}) + h_{2,1};$$
$$g_{3,3} = a_1(g_{2,2} + g_{2,3}) + b_1 g_{2,3} + h_{2,2};$$
$$g_{3,4} = a_1 g_{2,3};$$
$$h_{3,1} = a_2(g_{2,1} + g_{2,2} + g_{2,3});$$
$$h_{3,2} = a_2(g_{2,2} + g_{2,3});$$
$$h_{3,3} = a_2 g_{2,3}.$$

After arrangement successively, we can get the nth Taylor expansion for the generating function of Catalan-like numbers as follows (Eu et al., 2002).

$$C = \sum_{k=0}^{n-1} c_k x^k + x^n \sum_{k=1}^{n+1} g_{n,k} C^k + x^{n+1} \sum_{k=1}^{n} h_{n,k} C^{k+1},$$

where

$$g_n(C) = \sum_{k=1}^{n+1} g_{n,k} C^k, \qquad h_n(C) = \sum_{k=1}^{n} h_{n,k} C^{k+1},$$

and $g_{n,i} = h_{n,j} = 0$, for $i > n + 1$ and $j > n$. Now we present the main result of this paper based on the method proposed by Eu et al. (2002).

THEOREM 1   Let

$$C = \sum_{k=0}^{n-1} c_k x^k + x^n \sum_{k=1}^{n+1} g_{n,k} C^k + x^{n+1} \sum_{k=1}^{n} h_{n,k} C^{k+1}$$

be the nth Taylor expansion for the generating fuction of Catalan-like numbers defined by Equation (3). Then we have the following.

(i) $g_{n,1} = b_1 c_{n-1}$, $\quad h_{n,1} = a_2 c_{n-1}$, $\qquad$ for $n \geq 1$;

(ii) $g_{n,2} = (b_1 + a_1)c_{n-1} - \left(b_1^2 - a_2\right)c_{n-2}$ and $h_{n,2} = a_2(c_{n-1} - b_1 c_{n-2})$, $\qquad$ for $n \geq 2$;

(iii) $g_{n,k} = \frac{b_1}{a_2} h_{n,k} + \frac{a_1}{a_2} h_{n,k-1} + h_{n-1,k-1}$, $\qquad$ for $n, k \geq 2$, where $h_{n-1,0} = 0$;

(iv) $g_{n,k}$ and $h_{n,k}$ satisfy the following recurrence relations.

$$g_{n,k} = g_{n,k+1} + b_1 g_{n-1,k} + a_1 g_{n-1,k-1} + a_2 g_{n-2,k-1}, \quad \text{for } n \geq 3 \text{ and } 3 \leq k \leq n;$$
$$h_{n,k} = h_{n,k+1} + b_1 h_{n-1,k} + a_1 h_{n-1,k-1} + a_2 h_{n-2,k-1}, \quad \text{for } n \geq 3 \text{ and } 2 \leq k \leq n-1;$$

where $g_{1,1} = b_1, g_{1,2} = a_1, h_{1,1} = a_2$.

Proof   It is trivial for $n = 1$. Let $n \geq 2$. Replacing each C^k in $(n-1)$th Taylor expansion of Equation (6) with $1 + x(b_1 \sum_{i=1}^{k} C^i + a_1 \sum_{i=1}^{k} C^{i+1}) + x^2 \sum_{i=1}^{k} a_2 C^{i+1}$, we get

$$C = \sum_{k=0}^{n-2} c_k x^k + x^{n-1} \sum_{k=1}^{n} g_{n-1,k} C^k + x^n \sum_{k=1}^{n-1} h_{n-1,k} C^{k+1}$$

$$= \sum_{k=0}^{n-1} c_k x^k + x^n \left(\sum_{k=1}^{n-1} h_{n-1,k} C^{k+1} + \sum_{i=1}^{n} b_1 \sum_{k=i}^{n} g_{n-1,k} C^i + \sum_{i=1}^{n} a_1 \sum_{k=i}^{n} g_{n-1,k} C^{i+1} \right)$$

$$+ x^{n+1} \sum_{i=1}^{n} a_2 \sum_{k=i}^{n} g_{n-1,k} C^{i+1}.$$

Note that

$$C = \sum_{k=0}^{n-1} c_k x^k + x^n \sum_{k=1}^{n+1} g_{n,k} C^k + x^{n+1} \sum_{k=1}^{n} h_{n,k} C^{k+1}.$$

Hence comparing coefficients of each term of the remainder on the above two equations, we can get

$$g_{n,1} = b_1 \sum_{k=1}^{n} g_{n-1,k};$$

$$g_{n,i} = b_1 \sum_{k=i}^{n} g_{n-1,k} + a_1 \sum_{k=i-1}^{n} g_{n-1,k} + h_{n-1,i-1} \quad (i \geq 2, n \geq 2);$$

$$g_{n,n+1} = a_1 g_{n-1,n};$$

$$h_{n,i} = a_2 \sum_{k=i}^{n} g_{n-1,k} \quad (i \geq 1);$$

$$(10)$$

where $h_{n-1,0} = 0, h_{n-1,n} = 0, g_{n-1,n+1} = 0.$

(i) If $k = 1$, then we have

$$g_{n,1} = b_1 \sum_{k=1}^{n} g_{n-1,k} = b_1 c_{n-1};$$

$$h_{n,1} = a_2 \sum_{k=1}^{n} g_{n-1,k} = a_2 c_{n-1}.$$

(ii) If $k = 2$, then we have

$$g_{n,2} = b_1 \sum_{k=2}^{n} g_{n-1,k} + a_1 \sum_{k=1}^{n} g_{n-1,k} + h_{n-1,1}$$

$$= b_1 (\sum_{k=1}^{n} g_{n-1,k} - g_{n-1,1}) + a_1 \sum_{k=1}^{n} g_{n-1,k} + h_{n-1,1}$$

$$= (b_1 + a_1) \sum_{k=1}^{n} g_{n-1,k} - b_1 g_{n-1,1} + h_{n-1,1}$$

$$= (b_1 + a_1) c_{n-1} - b_1^2 c_{n-2} + a_2 c_{n-2}$$

$$= (b_1 + a_1) c_{n-1} - \left(b_1^2 - a_2 \right) c_{n-2};$$

$$h_{n,2} = a_2 \sum_{k=2}^{n} g_{n-1,k} = a_2 (\sum_{k=1}^{n} g_{n-1,k} - g_{n-1,1}) = a_2 (c_{n-1} - b_1 c_{n-2}).$$

(iii) If $n, i \geq 2$, then we have

$$a_2 g_{n,i} = a_2 (b_1 \sum_{k=i}^{n} g_{n-1,k} + a_1 \sum_{k=i-1}^{n} g_{n-1,k} + h_{n-1,i-1})$$

$$= a_2 b_1 \sum_{k=i}^{n-1} g_{n-1,k} + a_2 a_1 \sum_{k=i-1}^{n-1} g_{n-1,k} + a_2 h_{n-1,i-1}$$

$$= b_1 h_{n,i} + a_1 h_{n,i-1} + a_2 h_{n-1,i-1}.$$

Hence, we can get

$$g_{n,i} = \frac{b_1}{a_2} h_{n,i} + \frac{a_1}{a_2} h_{n,i-1} + h_{n-1,i-1}.$$

(iv) If $n \geq 3$, $3 \leq i \leq n$, then we have

$$g_{n,i} - g_{n,i+1} = b_1 \sum_{k=i}^{n} g_{n-1,k} + a_1 \sum_{k=i-1}^{n} g_{n-1,k} + h_{n-1,i-1} - (b_1 \sum_{k=i+1}^{n} g_{n-1,k} + a_1 \sum_{k=i}^{n} g_{n-1,k} + h_{n-1,i})$$

$$= b_1 g_{n-1,i} + a_1 g_{n-1,i-1} + h_{n-1,i-1} - h_{n-1,i}$$

$$= b_1 g_{n-1,i} + a_1 g_{n-1,i-1} + a_2 \sum_{k=i-1}^{n-1} g_{n-2,k} - a_2 \sum_{k=i}^{n-1} g_{n-2,k}$$

$$= b_1 g_{n-1,i} + a_1 g_{n-1,i-1} + a_2 g_{n-2,i-1}.$$

If $n \geq 3$, $2 \leq i \leq n - 1$, then we have

$$h_{n,i} - h_{n,i+1} = a_2 \sum_{k=i}^{n} g_{n-1,k} - a_2 \sum_{k=i+1}^{n} g_{n-1,k}$$

$$= a_2 g_{n-1,i}$$

$$= a_2 \left(b_1 \sum_{k=i}^{n-1} g_{n-2,k} + a_1 \sum_{k=i-1}^{n-1} g_{n-2,k} + h_{n-2,i-1} \right)$$

$$= a_2 b_1 \sum_{k=i}^{n-1} g_{n-2,k} + a_2 a_1 \sum_{k=i-1}^{n-1} g_{n-2,k} + a_2 h_{n-2,i-1}$$

$$= b_1 h_{n-1,i} + a_1 h_{n-1,i-1} + a_2 h_{n-2,i-1}.$$

$\square$

In the proof of Theorem (1), we can get recurrence relations (Equation (10)) of $g_{n,k}$ and $h_{n,k}$ with $g_{1,1} = b_1$ and $h_{1,1} = a_2$. So we can obtain a relation between $g_n(C)$ and $h_n(C)$ according to (iii).

COROLLARY 2   The two functions g_n and h_n satisfy

$$g_n(C) = \frac{(b_1 + a_1 C) h_n(C)}{a_2 C} + h_{n-1}(C),$$

for $n \geq 1$ with $h_0(C) = 0$.

Proof   Since

$$g_n(C) = \sum_{k=1}^{n+1} g_{n,k} C^k \quad \text{and} \quad h_n(C) = \sum_{k=1}^{n} h_{n,k} C^{k+1},$$

and

$$a_2 g_{n,i} = b_1 h_{n,i} + a_1 h_{n,i-1} + a_2 h_{n-1,i-1}, \quad (n, i \geq 2),$$

$\square$

we have

$$g_n(C) = \sum_{k=1}^{n+1} g_{n,k} C^k$$

$$= g_{n,1} C + \frac{1}{a_2} \sum_{k=2}^{n+1} a_2 g_{n,k} C^k$$

$$= g_{n,1} C + \frac{1}{a_2} \sum_{k=2}^{n+1} (b_1 h_{n,k} + a_1 h_{n,k-1} + a_2 h_{n-1,k-1}) C^k$$

$$= b_1 c_{n-1} C + \frac{b_1}{a_2} \sum_{k=2}^{n+1} h_{n,k} C^k + \frac{a_1}{a_2} \sum_{k=2}^{n+1} h_{n,k-1} C^k + \sum_{k=2}^{n+1} h_{n-1,k-1} C^k$$

$$= b_1 c_{n-1} C + \frac{b_1}{a_2 C} \sum_{k=2}^{n+1} h_{n,k} C^{k+1} + \frac{a_1}{a_2} h_n(C) + \sum_{k=2}^{n+1} h_{n-1,k-1} C^k$$

$$= b_1 c_{n-1} C + \frac{b_1}{a_2 C} \left(\sum_{k=1}^{n+1} h_{n,k} C^{k+1} - h_{n,1} C^2 \right) + \frac{a_1}{a_2} h_n(C) + h_{n-1}(C)$$

$$= b_1 c_{n-1} C + \frac{b_1}{a_2 C} \left(\sum_{k=1}^{n} h_{n,k} C^{k+1} - h_{n,1} C^2 \right) + \frac{a_1}{a_2} h_n(C) + h_{n-1}(C)$$

$$= b_1 c_{n-1} C + \frac{b_1}{a_2 C} h_n(C) - \frac{b_1}{a_2 C} a_2 c_{n-1} C^2 + \frac{a_1}{a_2} h_n(C) + h_{n-1}(C)$$

$$= \frac{b_1}{a_2 C} h_n(C) + \frac{a_1}{a_2} h_n(C) + h_{n-1}(C)$$

$$= \frac{(b_1 + a_1 C) h_n(C)}{a_2 C} + h_{n-1}(C).$$

COROLLARY 3   The generating functions

$$G(x,y) = \sum_{n\geq k\geq 1} g_{n,k}x^{n-1}y^{k-1} + \sum_{n\geq 1} g_{n,n+1}x^{n-1}y^n \quad \text{and} \quad H(x,y) = \sum_{n\geq k\geq 1} h_{n,k}x^{n-1}y^{k-1}$$

have closed forms:

$$G = \left(\frac{b_1 + a_1 y}{a_2} + xy\right)H \quad \text{and} \quad H = \frac{a_2 C - y\sum_{n=1}^{\infty} h_{n,n}x^{n-1}y^{n-1}(1 - a_1 xy)}{1 - y + b_1 xy + a_1 xy^2 + a_2 x^2 y^2}.$$

Proof   Since

$$a_2 g_{n,i} = b_1 h_{n,i} + a_1 h_{n,i-1} + a_2 h_{n-1,i-1}, \quad (n, i \geq 2),$$

and

$$g_{n,1} = b_1 c_{n-1}, \quad h_{n,1} = a_2 c_{n-1}, \quad g_{n,n+1} = a_1 g_{n-1,n}, \quad h_{n,n} = a_2 g_{n-1,n},$$

we have

$$\begin{aligned}
G(x,y) &= \sum_{n\geq k\geq 1} g_{n,k}x^{n-1}y^{k-1} + \sum_{n\geq 1} g_{n,n+1}x^{n-1}y^n \\
&= \sum_{n=1}^{\infty}\sum_{k=1}^{n} g_{n,k}x^{n-1}y^{k-1} + \sum_{n=1}^{\infty} g_{n,n+1}x^{n-1}y^n \\
&= \sum_{n=1}^{\infty}\sum_{k=2}^{n} g_{n,k}x^{n-1}y^{k-1} + \sum_{n=1}^{\infty} g_{n,1}x^{n-1} + \sum_{n=1}^{\infty} g_{n,n+1}x^{n-1}y^n \\
&= \frac{1}{a_2}\sum_{n=1}^{m}\sum_{k=2}^{n} a_2 g_{n,k}x^{n-1}y^{k-1} + \sum_{n=1}^{\infty} g_{n,1}x^{n-1} + \sum_{n=1}^{\infty} g_{n,n+1}x^{n-1}y^n \\
&= \frac{1}{a_2}\sum_{n=1}^{\infty}\sum_{k=2}^{n} (b_1 h_{n,k} + a_1 h_{n,k-1} + a_2 h_{n-1,k-1})x^{n-1}y^{k-1} + \sum_{n=1}^{\infty} g_{n,1}x^{n-1} \\
&\quad + \sum_{n=1}^{\infty} g_{n,n+1}x^{n-1}y^n \\
&= \frac{b_1}{a_2}\sum_{n=1}^{\infty}\sum_{k=2}^{n} h_{n,k}x^{n-1}y^{k-1} + \frac{a_1}{a_2}\sum_{n=1}^{\infty}\sum_{k=2}^{n} h_{n,k-1}x^{n-1}y^{k-1} + \sum_{n=1}^{\infty} b_1 c_{n-1}x^{n-1} \\
&\quad + \sum_{n=1}^{\infty}\sum_{k=2}^{n} h_{n-1,k-1}x^{n-1}y^{k-1} + \sum_{n=1}^{\infty} a_1 g_{n-1,n}x^{n-1}y^n,
\end{aligned} \tag{11}$$

where

$$\begin{aligned}
\sum_{n-1}^{\infty}\sum_{k-2}^{n} h_{n,k}x^{n-1}y^{k-1} &= \sum_{n=1}^{\infty}\sum_{k=1}^{n} h_{n,k}x^{n-1}y^{k-1} - \sum_{n-1}^{\infty} h_{n,1}x^{n-1} \\
&= H - \sum_{n=1}^{\infty} h_{n,1}x^{n-1} = H - \sum_{n=1}^{\infty} a_2 c_{n-1}x^{n-1} \tag{12}\\
&= H - a_2 C,
\end{aligned}$$

$$\begin{aligned}
\sum_{n=1}^{\infty}\sum_{k=2}^{n} h_{n,k-1}x^{n-1}y^{k-1} &= \sum_{n=1}^{\infty}\sum_{v=1}^{n-1} h_{n,v}x^{n-1}y^v \\
&= y\sum_{n=1}^{\infty}\sum_{v=1}^{n-1} h_{n,v}x^{n-1}y^{v-1} \\
&= y(H - \sum_{n=1}^{\infty} h_{n,n}x^{n-1}y^{n-1}) = y(H - \sum_{n=1}^{\infty} a_2 g_{n-1,n}x^{n-1}y^{n-1}), \tag{13}
\end{aligned}$$

and

$$\sum_{n=1}^{\infty}\sum_{k=2}^{n}h_{n-1,k-1}x^{n-1}y^{k-1} = xy\sum_{n=1}^{\infty}\sum_{k=2}^{n}h_{n-1,k-1}x^{n-2}y^{k-2} = xyH.$$

(14)

Now inserting (Equation (12))–(Equation (14)) into (Equation (11)), we can get

$$G(x,y) = \frac{b_1}{a_2}(H - a_2 C) + \frac{a_1}{a_2}y(H - \sum_{n=1}^{\infty}a_2 g_{n-1,n}x^{n-1}y^{n-1}) + xyH + \sum_{n=1}^{\infty}b_1 c_{n-1}x^{n-1}$$

$$+ \sum_{n=1}^{\infty}a_1 g_{n-1,n}x^{n-1}y^n$$

$$= \frac{b_1}{a_2}H - b_1 C + \frac{a_1}{a_2}yH - \sum_{n=1}^{\infty}a_1 g_{n-1,n}x^{n-1}y^n + xyH + b_1 C + \sum_{n=1}^{\infty}a_1 g_{n-1,n}x^{n-1}y^n$$

$$= (\frac{b_1 + a_1 y}{a_2} + xy)H.$$

Also since

$$h_{n,k} = h_{n,k+1} + b_1 h_{n-1,k} + a_1 h_{n-1,k-1} + a_2 h_{n-2,k-1}, \quad n \geq 3 \text{ and } 2 \leq k \leq n-1,$$

and

$$h_{n,1} = a_2 c_{n-1}, \quad h_{n,2} = a_2(c_{n-1} - b_1 c_{n-2}), \quad n \geq 2,$$

we have

$$H(x,y) = \sum_{n \geq k \geq 1}h_{n,k}x^{n-1}y^{k-1} = \sum_{n=1}^{\infty}\sum_{k=1}^{n}h_{n,k}x^{n-1}y^{k-1}$$

$$= \sum_{n=3}^{\infty}\sum_{k=1}^{n}h_{n,k}x^{n-1}y^{k-1} + \sum_{n=1}^{2}\sum_{k=1}^{n}h_{n,k}x^{n-1}y^{k-1}$$

$$= \sum_{n=3}^{\infty}\sum_{k=2}^{n-1}h_{n,k}x^{n-1}y^{k-1} + \sum_{n=3}^{\infty}(h_{n,1}x^{n-1} + h_{n,n}x^{n-1}y^{n-1}) + h_{1,1} + h_{2,1}x + h_{2,2}xy$$

$$= \sum_{n=3}^{\infty}\sum_{k=2}^{n-1}h_{n,k}x^{n-1}y^{k-1} + \sum_{n=1}^{\infty}h_{n,1}x^{n-1} + \sum_{n=2}^{\infty}h_{n,n}x^{n-1}y^{n-1}$$

$$= \sum_{n=3}^{\infty}\sum_{k=2}^{n-1}(h_{n,k+1} + b_1 h_{n-1,k} + a_1 h_{n-1,k-1} + a_2 h_{n-2,k-1})x^{n-1}y^{k-1}$$

$$+ a_2 C + \sum_{n=2}^{\infty}h_{n,n}x^{n-1}y^{n-1},$$

(15)

where

$$\sum_{n=3}^{\infty}\sum_{k=2}^{n-1}h_{n-1,k-1}x^{n-1}y^{k-1} = \sum_{u=2}^{\infty}\sum_{v=1}^{u-1}h_{u,v}x^u y^v = xy\sum_{n=2}^{\infty}\sum_{k=1}^{n-1}h_{n,k}x^{n-1}y^{k-1}$$

$$= xy\left(\sum_{n=1}^{\infty}\sum_{k=1}^{n}h_{n,k}x^{n-1}y^{k-1} - \sum_{n=1}^{1}\sum_{k=1}^{n}h_{n,k}x^{n-1}y^{k-1} - \sum_{n=2}^{\infty}h_{n,n}x^{n-1}y^{n-1}\right)$$

$$= xy(H - \sum_{n=1}^{\infty}h_{n,n}x^{n-1}y^{n-1}),$$

(16)

$$\sum_{n=3}^{\infty}\sum_{k=2}^{n-1}h_{n-2,k-1}x^{n-1}y^{k-1} = x^2y\sum_{n=1}^{\infty}\sum_{k=1}^{n}h_{n,k}x^{n-1}y^{k-1} = x^2yH, \tag{17}$$

$$\sum_{n=3}^{\infty}\sum_{k=2}^{n-1}h_{n,k+1}x^{n-1}y^{k-1} = \sum_{n=3}^{\infty}\sum_{u=3}^{n}h_{n,u}x^{n-1}y^{u-2} = \sum_{n=3}^{\infty}\sum_{k=3}^{n}h_{n,k}x^{n-1}y^{k-2}$$

$$= \sum_{n=1}^{\infty}\sum_{k=1}^{n}h_{n,k}x^{n-1}y^{k-2} - \sum_{n=1}^{2}\sum_{k=1}^{n}h_{n,k}x^{n-1}y^{k-2} - \sum_{n=3}^{\infty}\sum_{k=1}^{2}h_{n,k}x^{n-1}y^{k-2}$$

$$= \frac{1}{y}(H - h_{1,1} - h_{2,1}x - h_{2,2}xy - \sum_{n=3}^{\infty}h_{n,1}x^{n-1} - \sum_{n=3}^{\infty}h_{n,2}x^{n-1}y) \tag{18}$$

$$= \frac{1}{y}\left(H - \sum_{n=1}^{\infty}h_{n,1}x^{n-1} - \sum_{n=2}^{\infty}h_{n,2}x^{n-1}y\right)$$

$$= \frac{1}{y}\left[H - \sum_{n=1}^{\infty}a_2c_{n-1}x^{n-1} - \sum_{n=2}^{\infty}a_2(c_{n-1} - b_1c_{n-2})x^{n-1}y\right]$$

$$= \frac{1}{y}(H - a_2C - a_2yC + a_2y + a_2b_1xyC),$$

and

$$\sum_{n=3}^{\infty}\sum_{k=2}^{n-1}h_{n-1,k}x^{n-1}y^{k-1} = \sum_{u=2}^{\infty}\sum_{k=2}^{u}h_{u,k}x^{u}y^{k-1} = x\sum_{n=2}^{\infty}\sum_{k=2}^{n}h_{n,k}x^{n-1}y^{k-1}$$

$$= x\left(\sum_{n=1}^{\infty}\sum_{k=1}^{n}h_{n,k}x^{n-1}y^{k-1} - \sum_{n=1}^{1}\sum_{k=1}^{n}h_{n,k}x^{n-1}y^{k-1} - \sum_{n=2}^{\infty}\sum_{k=1}^{1}h_{n,k}x^{n-1}y^{k-1}\right)$$

$$= x\left(H - h_{1,1} - \sum_{n=2}^{\infty}h_{n,1}x^{n-1}\right) \tag{19}$$

$$= x(H - \sum_{n=1}^{\infty}h_{n,1}x^{n-1}) = x(H - \sum_{n=1}^{\infty}a_2c_{n-1}x^{n-1})$$

$$= x(H - a_2C).$$

Now inserting (Equation (16))–(Equation (19)) into (Equation (15)), we have

$$H(x,y) = \frac{1}{y}(H - a_2C - a_2yC + a_2y + a_2b_1xyC) + b_1x(H - a_2C)$$

$$+ a_1xy(H - \sum_{n=1}^{\infty}h_{n,n}x^{n-1}y^{n-1}) + a_2x^2yH + a_2C + \sum_{n=2}^{\infty}h_{n,n}x^{n-1}y^{n-1}. \tag{20}$$

Multiplying by y on both sides of (Equation 20) and then after arrangement, we have

$$Hy = H - a_2C + b_1xyH + a_1xy^2H + a_2x^2y^2H + y\sum_{n=1}^{\infty}h_{n,n}x^{n-1}y^{n-1}(1 - a_1xy).$$

So we get

$$H = \frac{a_2C - y\sum_{n=1}^{\infty}h_{n,n}x^{n-1}y^{n-1}(1 - a_1xy)}{1 - y + b_1xy + a_1xy^2 + a_2x^2y^2}. \qquad \square$$

3. Remarks

In the paper, we study the remainders of new Taylor expansions for the generating function of Catalan-like numbers. Since the Catalan-like numbers, such as the Motzkin numbers (Aigner, 1998; Donaghey & Shapiro, 1977; Sulanke, 2001), the Catalan numbers (Aigner, 2001; He, 2013; Mahmoud & Qi, 2016), the large and little Schröder numbers (Ehrenfeucht et al., 1998; Qi, Shi, & Guo, 2016a, 2016b; Stanley, 1997) naturally appear in the combinatorial objects, their Taylor expansions can be interpreted in distinct combinatorial ways.

Acknowledgements

The authors thank the anonymous referees for their constructive comments and helpful suggestions which have greatly improved the original manuscript.

Funding

This work was supported in part by the Domestic Visiting Scholar Program of the Young Teacher of Shandong Province and the Program for Scientific Research Innovation Team in Applied Probability and Statistics of Qufu Normal University [grant number 0230518].

Author details

Lily Li Liu[1]
E-mail: liulily@mail.qfnu.edu.cn
Xiaoli Li[1]
E-mail: 971253130@qq.com
[1] School of Mathematical Science, Qufu Normal University, Qufu 273165, P.R. China.

References

Aigner, M. (1998). Motzkin numbers. *European Journal of Combinatorics, 19*, 663–675.

Aigner, M. (1999). Catalan-like numbers and determinants. *Journal of Combinatorial Theory, Series A, 87*, 33–51.

Aigner, M. (2001). Catalan and other numbers—A recurrent theme. In H. Crapo & D. Senato (Eds.), *Algebraic combinatorics and computer science* (pp. 347–390). Berlin: Springer.

Cheon, G.-S., Kim, H., & Shapiro, L. W. (2012). Combinatorics of Riordan arrays with identical A and Z sequences. *Discrete Mathematics, 312*, 2040–2049.

Donaghey, R., & Shapiro, L. W. (1977). Motzkin numbers. *Journal of Combinatorial Theory, Series, 23*, 291–301.

Ehrenfeucht, A., Harju, T., ten Pas, P., & Rozenberg, G. (1998). Permutations, parenthesis words, and Schröder numbers. *Discrete Mathematics, 190*, 259–264.

Eu, S.-P., Liu, S.-C., & Yeh, Y.-N. (2002). Taylor expansions for Catalan and Motzkin numbers. *Advances in Applied Mathematics, 29*, 345–357.

He, T.-X. (2013). Parametric Catalan numbers and Catalan triangles. *Linear Algebra and its Applications, 438*, 1467–1484.

He, T.-X., & Sprugnoli, R. (2009). Sequence characterization of Riordan arrays. *Discrete Mathematics, 309*, 3962–3974.

Mahmoud, M., & Qi, F. (2016). Three identities of the Catalan-Qi numbers. *Mathematics, 4*, Article 35, 7 p. doi:10.3390/math4020035

Qi, F., Shi, X.-T. & Guo, B.-N. (2016a). *Integral representations of the large and little Schröder numbers.* doi:10.13140/RG.2.1.1988.3288

Qi, F., Shi, X.-T., & Guo, B.-N. (2016b). Two explicit formulas of the Schröder numbers. *Integers, 16*, Paper No. A23, 15 p.

Shapiro, L. W., Getu, S., Woan, W.-J., & Woodson, L. C. (1991). The Riordan group. *Discrete Applied Mathematics, 34*, 229–239.

Stanley, R. P. (1997). Hipparchus, Plutarch, Schröder and Hough. *The American Mathematical Monthly, 104*, 344–350.

Sulanke, R. A. (2001). Bijective recurrences for Motzkin paths. *Advances in Applied Mathematics, 27*, 627–640.

10

Graded fuzzy topological spaces

Ismail Ibedou[1,2]*

*Corresponding author: Ismail Ibedou, Faculty of Science, Department of Mathematics, Benha University, 13518 Benha, Egypt; Faculty of Science, Department of Mathematics, Jazan University, KSA
E-mail: ismail.ibedou@gmail.com
Reviewing editor: Hari M. Srivastava, University of Victoria, Canada

Abstract: In this paper, graded fuzzy topological spaces based on the notion of neighbourhood system of graded fuzzy neighbourhoods at ordinary points are introduced and studied. These graded fuzzy neighbourhoods at ordinary points and usual subsets played the main role in this study.

Subjects: Advanced Mathematics; Foundations & Theorems; Mathematics & Statistics; Science

Keywords: neighbourhood systems; fuzzy filters; fuzzy neighbourhood filters; fuzzy topological spaces; separation axioms

AMS Subject classification: 54A40

1. Introduction
Kubiak (1985) and Šostak (1985) introduced the fundamental concept of a fuzzy topological structure as an extension of both crisp topology and fuzzy topology Chang (1968), in the sense that both objects and axioms are fuzzified and we may say they began the graded fuzzy topology. Bayoumi and Ibedou (2001, 2002, 2002b, 2004) introduced and studied the separation axioms in the fuzzy case in Chang's topology (1968) using the notion of fuzzy filter defined by Gähler (1995a,1995b).

Now, we will try to investigate fuzzy topological spaces in sense of Šostak, not using fuzzy filters but starting from a neighbourhood system of graded fuzzy neighbourhoods at ordinary points and usual sets. From that neighbourhood system, we can build a fuzzy topology in sense of Šostak and moreover, this fuzzy topology is itself the fuzzy topology in sense of Chang associated with the fuzzy neighbourhoterest Satementod filter $\mathcal{N}_x$ (Gähler, 1995b) at ordinary point $x \in X$ defined by Gähler. Interior operator and closure operator are defined using these graded fuzzy neighbourhoods; also

ABOUT THE AUTHOR
My research interests are: Fuzzy Topology, Fuzzy Topological Groups, Fuzzy Sets, Soft Sets, Soft Topological Spaces, and their applications. My research in before was concerning separation axioms in Chang's Fuzzy Topology, and its relations with Fuzzy Compactness, Fuzzy Proximity, Fuzzy Uniformities and other types of Fuzzy separation axioms. Also, a wide research was done for Fuzzy Topological Groups and studying its uniformizability and metrizability. These separation axioms in Fuzzy Bitopological spaces are introduced. My research was mainly done with Professor Fatma Bayoumi, fatma_bayoumi@yahoo.com. In this paper a continuation to my research dealing with the graded fuzzy separation axioms. The start step is defining a fuzzy neighbourhood system of fuzzy graded neighbourhoods at ordinary points and usual subsets.

PUBLIC INTEREST STATEMENT
Separation axioms depend on the concept of neighbourhoods and so, for the fuzzy case, fuzzy neighbourhoods or valued fuzzy neighbourhoods means neighbourhoods with some degree in [0, 1]. These grades to be a fuzzy neighbourhood forced the fuzzy separation axioms to be graded. In the fuzzy case, separation axioms are not sharp concepts. For example, there is no T_0 topological space, but there are $(\alpha, \beta) - T_0$ topological spaces depending on the existence of the fuzzy neighbourhood with grade α at a point or the existence of the fuzzy neighbourhood with grade β at the other distinct point. In this paper, I introduced these graded fuzzy separation axioms. The main section was for defining a fuzzy neighbourhood system of fuzzy graded neighbourhoods at ordinary points and usual subsets.

their associated fuzzy topologies coincide with this fuzzy topology in sense of Chang associated with the fuzzy neighbourhood filter of Gähler. Fuzzy continuous, fuzzy open and fuzzy closed mappings are defined with grades according to these graded fuzzy neighbourhoods.

Separation axioms in the fuzzy case are introduced based on these graded fuzzy neighbourhoods and thus, axioms are graded. These axioms satisfy common results and implications. These graded axioms are a good extension in sense of Lowen (1978). In *Fuzzy neuro systems for machine learning for large data sets* (2009) and DCPE Co-Training for Classification (2012), there are some applications based on fuzzy sets.

2. Preliminaries
Throughout the paper, let $I_0 = (0, 1]$ and $I_1 = [0, 1)$.

A fuzzy topology $\tau{:}I^X \to I$ is defined by Kubiak (1985) and Šostak (1985):

(1) $\tau(\overline{0}) = \tau(\overline{1}) = 1$,

(2) $\tau(f \wedge g) \geq \tau(f) \wedge \tau(g)$ for all $f, g \in I^X$,

(3) $\tau(\bigvee_{j \in J} \mu_j) \geq \bigwedge_{j \in J} \tau(\mu_j)$ for any family of $(\mu_j)_{j \in J} \in I^X$. Let τ_1 and τ_2 be fuzzy topologies on X. Then, τ_1 is finer than τ_2 (τ_2, which is coarser than τ_1), denoted by $\tau_1 \leq \tau_2$, if $\tau_2(\mu) \leq \tau_1(\mu)$ for all $\mu \in I^X$.

For each fuzzy set $f \in I^X$, the weak α cut-off f is given by $w_\alpha f = \{x \in X \mid f(x) \geq \alpha\}$; the strong α cut-off f is the subset of X, $s_\alpha f = \{x \in X \mid f(x) > \alpha\}$.

If T is an ordinary topology on X, then the *induced fuzzy topology* on X is given by $\omega(T) = \{f \in I^X \mid s_\alpha f \in T$ for all $\alpha \in I_1\}$.

fuzzy filters. Let X be a non- empty set. A fuzzy *filter* on X (Eklund, 1992; Gähler, 1995a) is a mapping $\mathcal{M}{:}I^X \to I$ such that the following conditions are fulfilled:

(F1) $\mathcal{M}(\overline{\alpha}) \leq \alpha$ holds for all $\alpha \in I$ and $\mathcal{M}(\overline{1}) = 1$;

(F2) $\mathcal{M}(f \wedge g) = \mathcal{M}(f) \wedge \mathcal{M}(g)$ for all $f, g \in I^X$.

If $\mathcal{M}$ and $\mathcal{N}$ are fuzzy filters on X, the.n $\mathcal{M}$ is said to be *finer than* $\mathcal{N}$, denoted by $\mathcal{M} \leq \mathcal{N}$, provided that $\mathcal{M}(f) \geq \mathcal{N}(f)$ for every $f \in I^X$. By $\mathcal{M} \nleq \mathcal{N}$ we mean that $\mathcal{M}$ is not finer than $\mathcal{N}$. $\mathcal{M} \nleq \mathcal{N} \iff$ there is $f \in I^X$ such that $\mathcal{M}(f) < \mathcal{N}(f)$.

A non-empty subset $\mathcal{F}$ of I^X is called a *prefilter* on X (Lowen, Lowen), provided that the following conditions are fulfilled:

(1) $\overline{0} \notin \mathcal{F}$;

(2) $f, g \in \mathcal{F}$ implies $f \wedge g \in \mathcal{F}$;

(3) $f \in \mathcal{F}$ and $f \leq g$ imply $g \in \mathcal{F}$.

For each fuzzy filter $\mathcal{M}$ on X, the subset

α-pr $\mathcal{M}$ of I^X defined by:

α-pr $\mathcal{M} = \{f \in I^X \mid \mathcal{M}(f) \geq \alpha\}$

is a prefilter on X.

Proposition 2.1 (Gähler, 1995a) There is a one-to-one correspondence between fuzzy filters $\mathcal{M}$ on X and the families $(\mathcal{M}_\alpha)_{\alpha \in I_0}$ of prefilters on X which fulfill the following conditions:

(1) $f \in \mathcal{M}_\alpha$ implies $\alpha \leq \sup f$.

(2) $0 < \alpha \leq \beta$ implies $\mathcal{M}_\alpha \supseteq \mathcal{M}_\beta$.

(3) For each $\alpha \in I_0$ with $\bigvee_{0<\beta<\alpha} \beta = \alpha$, we have $\bigcap_{0<\beta<\alpha} \mathcal{M}_\beta = \mathcal{M}_\alpha$. This correspondence is given by $\mathcal{M}_\alpha = \alpha\text{-pr}\,\mathcal{M}$ for all $\alpha \in I_0$ and $\mathcal{M}(f) = \bigvee_{g \in \mathcal{M}_\alpha, g \leq f} \alpha$ for all $f \in I^X$.

Proposition 2.2   (Eklund, 1992) Let A be a set of fuzzy filters on X. Then, the following are equivalent.

(1) The infimum $\bigwedge_{\mathcal{M} \in A} \mathcal{M}$ of A with respect to the finer relation of fuzzy filters exists,

(2) For each non-empty finite subset $\{\mathcal{M}_1, \ldots, \mathcal{M}_n\}$ of A, we have $\mathcal{M}_1(f_1) \wedge \cdots \wedge \mathcal{M}_n(f_n) \leq \sup(f_1 \wedge \cdots \wedge f_n)$ for all $f_1, \ldots, f_n \in I^X$,

(3) For each $\alpha \in I_0$ and each non-empty finite subset $f_1, \ldots, f_n$ of $\bigcup_{\mathcal{M} \in A} \alpha\text{-pr}\,\mathcal{M}$, we have $\alpha \leq \sup(f_1 \wedge \cdots \wedge f_n)$.

Recall that $(\bigwedge_{\mathcal{M} \in A} \mathcal{M})(f) = \bigvee_{\mathcal{M} \in A} \mathcal{M}(f)$ and $(\bigvee_{\mathcal{M} \in A} \mathcal{M})(f) = \bigwedge_{\mathcal{M} \in A} \mathcal{M}(f)$ for all $f \in I^X$.

Fuzzy neighbourhood filters. For each fuzzy topological space (X, τ) and each $x \in X$, the mapping $\mathcal{N}_x : I^X \to I$ defined by (Gähler, Gahler):

$$\mathcal{N}_x(\lambda) = \text{int}_\tau \lambda(x) \text{ for all } \lambda \in I^X$$

is a fuzzy filter on X, called the fuzzy *neighbourhood filter* of the space (X, τ) at the point x, and for short is called a fuzzy neighbourhood filter at x. The mapping $\dot{x} : I^X \to I$ is defined by $\dot{x}(\lambda) = \lambda(x)$ for all $\lambda \in I^X$. The fuzzy neighbourhood filters fulfil the following conditions:

(1) $\dot{x} \leq \mathcal{N}_x$ holds for all $x \in X$;

(2) $(\mathcal{N}_x)(\text{int}_\tau f) = (\mathcal{N}_x)(f)$ for all $x \in X$ and $f \in I^X$.

A fuzzy filter $\mathcal{M}$ is said to *converge to* $x \in X$, denoted by $\mathcal{M} \xrightarrow{\tau} x$, if $\mathcal{M} \leq \mathcal{N}_x$ (Gähler, 1995b).

The fuzzy neighbourhood filter $\mathcal{N}_F$ at an ordinary subset F of X is the fuzzy filter on X defined in Bayoumi and Ibedou (2002b), by means of $\mathcal{N}_x$, $x \in F$ as:

$$\mathcal{N}_F = \bigvee_{x \in F} \mathcal{N}_x.$$

The fuzzy filter $\dot{F}$ is defined by

$$\dot{F} = \bigvee_{x \in F} \dot{x}.$$

$\dot{F} \leq \mathcal{N}_F$ holds for all $F \subseteq X$. Also, recall that the fuzzy filter $\dot{\lambda}$ and the fuzzy neighbourhood filter $\mathcal{N}_\lambda$ at a fuzzy subset λ of X are defined by

$$\dot{\lambda} = \bigvee_{0 < \lambda(x)} \dot{x} \quad \text{and} \quad \mathcal{N}_\lambda = \bigvee_{0 < \lambda(x)} \mathcal{N}_x, \tag{1}$$

respectively. $\dot{\lambda} \leq \mathcal{N}_\lambda$ holds for all $\lambda \in I^X$ (Bayoumi & Ibedou, 2004).

For each fuzzy topological space (X, τ) the closure operator cl which assigns to each fuzzy filter $\mathcal{M}$ on X, the fuzzy filter cl $\mathcal{M}$ is defined by

$$\text{cl}\,\mathcal{M}(f) = \bigvee_{\text{cl}_\tau g \leq f} \mathcal{M}(g). \tag{2}$$

cl $\mathcal{M}$ is called the *closure* of $\mathcal{M}$. cl is isotone, hull and idempotent operator, that is for all fuzzy filters $\mathcal{M}$ and $\mathcal{N}$ on X, we have (Gähler, 1995b):

$$\mathcal{M} \leq \mathcal{N} \quad \text{implies } \mathrm{cl}\mathcal{M} \leq \mathrm{cl}\mathcal{N}, \tag{3}$$

$$\mathcal{M} \leq \mathrm{cl}\mathcal{M}, \tag{4}$$

3. Neighbourhood systems

Definition 3.1   A family $(\mathcal{N}_x^{\alpha})_{x \in X}$ of fuzzy sets $\mathcal{N}_x^{\alpha}$ is said to be a neighbourhood system with grade $\alpha \in I_0$ on X if it satisfies the following conditions:

(Nb1) *For all* $f \in \mathcal{N}_x^{\alpha}$, *we have* $\alpha \leq f(x)$,

(Nb2) $\overline{1} \in \mathcal{N}_x^{\alpha}$,

(Nb3) $f, g \in \mathcal{N}_x^{\alpha}$ *implies that* $f \wedge g \in \mathcal{N}_x^{\alpha}$,

(Nb4) $f \in \mathcal{N}_x^{\alpha}, f \leq g$ *imply that* $g \in \mathcal{N}_x^{\alpha}$,

(Nb5) *If* $f \in \mathcal{N}_x^{\alpha}$, *then there is* $g \in \mathcal{N}_x^{\alpha}$, *such that for all* $y \in X$ *with* $0 < g(y)$, *we have* $f \in \mathcal{N}_y^{\alpha}$.

LEMMA 3.1   *These families of prefilters* $(\mathcal{N}_x^{\alpha})_{\alpha \in I_0}$ *at* $x \in X$ *satisfy the following conditions:*

(Pr1) $f \in \mathcal{N}_x^{\alpha}$ *implies that* $\alpha \leq \sup f$,

(Pr2) $0 < \beta \leq \alpha$ *implies that* $\mathcal{N}_x^{\alpha} \subseteq \mathcal{N}_x^{\beta}$,

(Pr3) *For every* $\alpha \in I_0$ *with* $\bigvee_{0 < \beta < \alpha} \beta = \alpha$, *we have* $\bigcap_{0 < \beta < \alpha} \mathcal{N}_x^{\beta} = \mathcal{N}_x^{\alpha}$.

Proof   Clear.   □

Remark 3.1   For any subset A of X, let us define $\mathcal{N}_A^{\alpha}$ by $\mathcal{N}_A^{\alpha} = \bigcap_{x \in A} \mathcal{N}_x^{\alpha}$, that is $f \in \mathcal{N}_A^{\alpha}$ iff $\alpha \leq \bigwedge_{x \in A} \mathcal{N}_x(f)$ iff $\alpha \leq \mathcal{N}_A(f)$. $\mathcal{N}_x^{\alpha} = \mathcal{N}_{\{x\}}^{\alpha}, \mathcal{N}_x^{\alpha} \cap \mathcal{N}_x^{\beta} = \mathcal{N}_x^{\alpha \vee \beta}, \mathcal{N}_x^{\alpha} \cup \mathcal{N}_x^{\beta} = \mathcal{N}_x^{\alpha \wedge \beta}, \bigcap_j \mathcal{N}_x^{\alpha_j} = \mathcal{N}_x^{\bigvee_j \alpha_j}, \bigcup_j \mathcal{N}_x^{\alpha_j} = \mathcal{N}_x^{\bigwedge_j \alpha_j}$. For all $\alpha \neq \beta$ in I_0, we have $\mathcal{N}_x^{\alpha} \neq \mathcal{N}_x^{\beta}$.

For any $\alpha, \beta, \gamma \in I_0$, we have $\mathcal{N}_x^{\alpha} \subseteq \mathcal{N}_x^{\alpha}, \mathcal{N}_x^{\alpha} \subseteq \mathcal{N}_x^{\beta}$ and $\mathcal{N}_x^{\beta} \subseteq \mathcal{N}_x^{\alpha}$ implies that $\alpha = \beta, \mathcal{N}_x^{\alpha} \subseteq \mathcal{N}_x^{\beta}$ and $\mathcal{N}_x^{\beta} \subseteq \mathcal{N}_x^{\gamma}$ implies that $\mathcal{N}_x^{\alpha} \subseteq \mathcal{N}_x^{\gamma}$. Also, for all $\alpha \neq \beta \in I_0$, we have either $\mathcal{N}_x^{\alpha} \subseteq \mathcal{N}_x^{\beta}$ or $\mathcal{N}_x^{\beta} \subseteq \mathcal{N}_x^{\alpha}$.

(α) fuzzy open sets, fuzzy open sets.

Let us define an (α) fuzzy *open* set as follows:

$$\alpha \leq \tau(f) \text{ iff for all } x \in X \text{ there is } \alpha \in I_0 \text{ such that } f \in \mathcal{N}_x^{\alpha} \text{ and } f(x) \leq \alpha. \tag{5}$$

An (α) fuzzy *closed* set is the complement of an (α) fuzzy open set.

A set $f \in I^X$ is said to be fuzzy *open* if it is (α) fuzzy open for all $\alpha \in I_0$. In other words, if for all $x \in X$ and for all $\alpha \in I_0$, we have $f \in \mathcal{N}_x^{\alpha}$ and $f(x) \leq \alpha$.

It is called a fuzzy *closed* if it is the complement of a fuzzy open set. These notations are restricted to the usual open and closed sets in fuzzy topology and usual topology.

Starting from a neighbourhood system $(\mathcal{N}_x^{\alpha})_{x \in X}$ with grade $\alpha \in I_0$, we can define an interior operator and a closure operator as follows:

$$\text{int}f(x) = \bigvee_{g \in \mathcal{N}_x^{\alpha}} \bigwedge_{0 < g(y)} f(y), \tag{6}$$

$$\text{cl}f(x) = \bigwedge_{g \in \mathcal{N}_x^{\alpha}} \bigvee_{0 < g(y)} f(y). \tag{7}$$

For every $x \in X$, $\mathcal{N}_x^{\alpha}$ satisfying (Nb1) to (Nb4) is exactly a prefilter on X of all neighbourhoods of $x \in X$ with grade $\alpha \in I_0$. That is, $(\mathcal{N}_x^{\alpha})_{x \in X}$ is a family of prefilters with grade $\alpha \in I_0$ at every $x \in X$ constructing after adding condition (Nb5) a neighbourhood system on X with grade $\alpha \in I_0$. The pair $(X, (\mathcal{N}_x^{\alpha})_{x \in X})$ is called a neighbourhood space with a grade $\alpha \in I_0$.

From Lemma 3.1 and from the correspondence given in Proposition 2.1 between the fuzzy filters and the families satisfying the conditions (Pr1) to (Pr3), we can say this family $(\mathcal{N}_x^{\alpha})_{\alpha \in I_0}$ is a representation of the fuzzy neighbourhood filter $\mathcal{N}_x$ as a family of prefilters. This is given by the following two conditions (Nb) and (Pr):

(Pr) $\mathcal{N}_x(f) = \bigvee_{g \in \mathcal{N}_x^{\alpha}, g \leq f} \alpha$ for all $f \in I^X$.

(Nb) $\mathcal{N}_x^{\alpha} = \{f \in I^X \mid \alpha \leq \mathcal{N}_x(f)\}$. Denote the subset $\mathcal{N}_x^{\alpha} \subseteq I^X$ as the fuzzy *neighbourhoods with grade* $\alpha \in I_0$ of $x \in X$.

Clearly, both the interior operator and closure operator satisfy the common axioms of interior operator and closure operator, respectively. A fuzzy topology on X could be generated by this interior operator given by (2.2) or this closure operator given by (2.3), using the properties of $\mathcal{N}_x^{\alpha}$ stated in (Nb1)—(Nb5). That fuzzy topology is exactly the fuzzy topology τ associated with the fuzzy neighbourhood filters $\mathcal{N}_x$ given by an interior operation as in (2.2) so that

$$\mathcal{N}_x(f) = \text{int}_\tau f(x) \text{ for all } f \in I^X.$$

Also, we can consider

$$\mathcal{N}_x^{\alpha} = \{f \in I^X \mid \alpha \leq \text{int}_\tau f(x)\} \tag{8}$$

and then, (2.1) for an (α) fuzzy open set could be rewritten as

$$\alpha \leq \tau(f) \text{ iff for all } x \in X, \text{ there is } \alpha \in I_0 \text{ so that } \alpha \leq \text{int}_\tau f(x), f(x) \leq \alpha. \tag{9}$$

That is, from a neighbourhood system of graded neighbourhoods, we can deduce interior operation by which it is introduced a graded fuzzy topology and the converse is true.

From (1.2) and (1.4) for all $x \in X$ and all $\alpha \in I_0$, we can define $\text{cl}\mathcal{N}_x^{\alpha}$ by

$$\text{cl}\mathcal{N}_x^{\alpha} = \{f \in I^X \mid \alpha \leq \text{cl}\mathcal{N}_x(f)\}, \tag{10}$$

and equivalently,

$$\text{cl}\mathcal{N}_x^{\alpha} = \{f \in I^X \mid \text{ there is } h \in \mathcal{N}_x^{\alpha}, \text{cl}h \leq f\}. \tag{11}$$

For all $x \in X$ and all $\alpha \in I_0$, we have $\text{cl}\mathcal{N}_x^{\alpha} \subseteq \mathcal{N}_x^{\alpha}$.

Definition 3.2   *Let (X, τ_1) and (Y, τ_2) be fuzzy topological spaces, and $f{:}X \to Y$ a map. Then, for some $\alpha \in I_0$, f is called an (α) fuzzy continuous if for all (α) fuzzy open set μ with respect to τ_2, we have $f^{-1}(\mu)$ is an (α) fuzzy open set with respect to τ_1 for all $\mu \in I^Y$.*

f is called fuzzy continuous if for all fuzzy open set μ with respect to τ_2, we have $f^{-1}(\mu)$ is a fuzzy open set with respect to τ_1 for all $\mu \in I^Y$.

Definition 3.3   *Let (X, τ_1) and (Y, τ_2) be fuzzy topological spaces. Then, the mapping $f:(X, \tau_1) \to (Y, \tau_2)$ is called (α) fuzzy open ((α) fuzzy closed) mapping if the image $f(g)$ of the (α) fuzzy open ((α) fuzzy closed) set g with respect to τ_1 is (α) fuzzy open ((α) fuzzy closed) set with respect to τ_2.*

The mapping $f:(X, \tau_1) \to (Y, \tau_2)$ is called fuzzy open (fuzzy closed) mapping if the image $f(g)$ of the fuzzy open (fuzzy closed) set g with respect to τ_1 is fuzzy open (fuzzy closed) set with respect to τ_2.

Now, we define the continuity locally at a point $x_0 \in X$ between two fuzzy topological spaces using these graded neighbourhoods.

Definition 3.4   *Let (X, τ) and (Y, σ) be two fuzzy topological spaces. Then, the mapping $f : (X, \tau) \to (Y, \sigma)$ is called (α) fuzzy continuous at a point x_0 provided that for all $g \in \mathcal{N}^{\alpha}_{f(x_0)}$,*

there exists $h \in \mathcal{N}^{\alpha}_{x_0}$ such that $h \leq f^{-1}(g)$ for some $\alpha \in I_0$. f is (α) fuzzy continuous if it is (α) fuzzy continuous at every $x \in X$. f is an fuzzy continuous if it is (α) fuzzy continuous for all $\alpha \in I_0$.

This is an equivalent definition with Definition 3.2 for the (α) fuzzy continuous mapping and fuzzy continuous mapping.

4. $(\alpha, \beta)T_0$-spaces and $(\alpha, \beta)T_1$-spaces

This section is devoted to introduce the notions of T_0-spaces and T_1-spaces using the notion of α-neighbourhoods at ordinary points. We will introduce different equivalent definitions, and we show that these notions are good extensions in sense of Lowen (1978]).

Definition 4.1   *A fuzzy topological space (X, τ) is called an $(\alpha, \beta)T_0$-space if for all $x \neq y$ in X, there exists $f \in \mathcal{N}^{\alpha}_x$ such that $f(y) < \alpha$; $\alpha \in I_0$ or there exists $g \in \mathcal{N}^{\beta}_y$ such that $g(x) < \beta$; $\beta \in I_0$.*

Definition 4.2   *A fuzzy topological space (X, τ) is called an $(\alpha, \beta)T_1$-space if for all $x \neq y$ in X there exist $f \in \mathcal{N}^{\alpha}_x$ and $g \in \mathcal{N}^{\beta}_y$ such that $f(y) < \alpha$ and $g(x) < \beta$; $\alpha, \beta \in I_0$.*

Example 4.1   *Let $X = \{x, y\}$, and*

$$\tau(f) = \begin{cases} 1 & \text{at } \overline{0} \text{ or } \overline{1} \\ \frac{1}{3} & \text{at } x_{\frac{1}{2}} \\ 0 & \text{otherwise.} \end{cases}$$

Taking $\alpha = \frac{1}{3}$, we get that there is $f = x_{\frac{1}{2}}$ in $\mathcal{N}^{\alpha}_x$ such that $f(y) < \alpha$. For all $\alpha \in I_0$, we can not find any f in $\mathcal{N}^{\alpha}_y$ such that $f(x) < \alpha$. That is, (X, τ) is an $(\alpha, \beta)T_0$-space.

Example 4.2   *Let $X = \{x, y\}$, and*

$$\tau(f) = \begin{cases} 1 & \text{at } \overline{0} \text{ or } \overline{1} \\ 0 & \text{otherwise.} \end{cases}$$

Only there is $f = \overline{1}$ which is a graded neighbourhood but for both of x, y. Hence, for all $\alpha \in I_0$, $\mathcal{N}^{\alpha}_x = \mathcal{N}^{\alpha}_y$ and therefore, (X, τ) is not $(\alpha, \beta)T_0$-space.

Proposition 4.1   *Every $(\alpha, \beta)T_1$-space is an $(\alpha, \beta)T_0$-space.*

Proof   Clear.   □

Example 4 is an $(\alpha, \beta)T_0$-space but not $(\alpha, \beta)T_1$-space.

Example 4.3   Let $X = \{x, y\}$, and

Taking $\alpha = \frac{1}{3}$ and $\beta = \frac{1}{3}$, we get that there is $f = x_{\frac{1}{2}}$ in $\mathcal{N}_x^\alpha$ and $g = y_{\frac{1}{2}}$ in $\mathcal{N}_y^\beta$ such that $f(y) < \alpha$ and $g(x) < \beta$, for some $\alpha, \beta \in I_0$. Hence, (X, τ) is an $(\alpha, \beta)T_1$-space.

$$
\tau(f) = \begin{cases} 1 & \text{at } \overline{0} \text{ or } \overline{1} \\ \frac{1}{3} & \text{at } x_{\frac{1}{2}} \\ \frac{1}{3} & \text{at } y_{\frac{1}{2}} \\ 0 & \text{otherwise.} \end{cases}
$$

In the following theorems, there will be introduced some equivalent definitions for the $(\alpha, \beta)T_0$-spaces and $(\alpha, \beta)T_1$-spaces.

THEOREM 4.1   *Let (X, τ) be a fuzzy topological space. Then, the following statements are equivalent.*

(1) (X, τ) is $(\alpha, \beta)T_0$.

(2) For all $x \neq y$ in X and for all $\alpha \in I_0$, $\mathcal{N}_x^\alpha \neq \mathcal{N}_y^\alpha$.

(3) For all $x \neq y$ in X, there exists $f \in I^X$ such that $f(y) < \alpha \leq \text{int}_\tau f(x)$; $\alpha \in I_0$ or there exists $g \in I^X$ such that $g(x) < \beta \leq \text{int}_\tau g(y)$; $\beta \in I_0$.

(4) For all $x \neq y$ in X, there exists $f \in I^X$ such that $f(y) > \text{cl}_\tau f(x)$ or there exists $g \in I^X$ such that $g(x) > \text{cl}_\tau g(y)$.

Proof   (1) $\Rightarrow$ (2): From (1), there is $f \in I^X$ such that $\text{int}_\tau f(y) \leq f(y) < \alpha \leq \text{int}_\tau f(x)$; $\alpha \in I_0$ and then, $f \in \mathcal{N}_x^\alpha$ and $f \notin \mathcal{N}_y^\alpha$. Hence, $\mathcal{N}_x^\alpha \neq \mathcal{N}_y^\alpha$; $\alpha \in I_0$ and thus, (2) holds.

(2) $\Rightarrow$ (3): There exists $f \in I^X$ such that $\text{int}_\tau f(y) < \alpha \leq \text{int}_\tau f(x)$; $\alpha \in I_0$ and then, for $g = \text{int}_\tau f$, we can say $g(y) < \alpha \leq \text{int}_\tau g(x)$; $\alpha \in I_0$. The other case is similar and thus, (3) is satisfied.

(3) $\Rightarrow$ (4): From Equation 7, we get that $\text{cl}_\tau f(x) < f(y)$ for all $\text{int}_\tau f(y) \geq \alpha > f(x)$, then (4) holds.

(4) $\Rightarrow$ (1): Since $f(y) < \bigwedge_{h \in \mathcal{N}_x^\alpha} \bigvee_{0 < h(z)} f(z) = \text{cl}_\tau f(x)$ implies that z could not be y with $0 < h(y)$ for all $h \in \mathcal{N}_x^\alpha$; $\alpha \in I_0$, which means that there is $h \in \mathcal{N}_x^\alpha$ such that $h(y) = 0 < \alpha \leq \text{int}_\tau h(x)$; $\alpha \in I_0$. The other case is similar and thus, (1) holds.   □

THEOREM 4.2   *Let (X, τ) be a fuzzy topological space. Then, the following statements are equivalent.*

(1) (X, τ) is $(\alpha, \beta)T_1$.

(2) For all $x \in X$, we have $\text{cl}_\tau x_1 = x_1$.

(3) For all $x \neq y$ in X, there exist $f, g \in I^X$ such that $f(y) < \alpha \leq \text{int}_\tau f(x)$ and $g(x) < \beta \leq \text{int}_\tau g(y)$; $\alpha, \beta \in I_0$.

(4) For all $x \neq y$ in X, there exist $f, g \in I^X$ such that $f(y) > \text{cl}_\tau f(x)$ and $g(x) > \text{cl}_\tau g(y)$.

Proof   (1) $\Rightarrow$ (2): Let $y \neq x$ in X. Then, $\text{cl}_\tau x_1(y) = \bigwedge_{h \in \mathcal{N}_y^\alpha} \bigvee_{0 < h(z)} x_1(z)$, which means for all $h \in \mathcal{N}_y^\alpha$, if $x_1(z) > 0$ whenever $h(z) > 0$, then $\text{cl}_\tau x_1(y) > 0$. From (1), we get that z could not be x with $0 < h(x)$, that is, $\text{cl}_\tau x_1(y) = 0$ for all $y \neq x$. At x, it is clear that $\text{cl}_\tau x_1(x) = 1$. Hence, $\text{cl}_\tau x_1 = x_1$ for all $x \in X$, and (2) is fulfilled.

$(2) \Rightarrow (3)$: For all $x \neq y$ in X, we have $cl_\tau x_1 = x_1$ and $cl_\tau y_1 = y_1$. (2) means that $cl_\tau x_1(y) = x_1(y) = 0 = \bigwedge_{h \in \mathcal{N}_y^\alpha} \bigvee_{0 < h(z)} x_1(z)$, which means for all $h \in \mathcal{N}_y^\alpha$, z could not be x with $0 < h(x)$, that is there is $\alpha \in I_0$ and there is $h \in \mathcal{N}_y^\alpha$ such that $h(x) = 0 < \alpha$ and then, $h(x) < \alpha \leq int_\tau h(y)$. The other case is similar and therefore, (3) is fulfilled.

$(3) \Rightarrow (4)$: As in Theorem 4.1.

$(4) \Rightarrow (1)$: As in Theorem 4.1. $\square$

The next proposition shows that the separation axioms $(\alpha, \beta)T_0$ and $(\alpha, \beta)T_1$ are good extensions in sense of Lowen (1978).

Proposition 4.2 A topological space (X, T) is a T_0-space (T_1-space) if and only if the induced fuzzy topological space $(X, \omega(T))$ is an $(\alpha, \beta)T_0$-space $((\alpha, \beta)T_1$-space).

Proof Let (X, T) be T_0 (T_1) and let $x \neq y$. Then, there is a neighbourhood $\mathcal{O}_y \in T$ such that $x \notin \mathcal{O}_y$. Taking $f \in I^X$ such that $\mathcal{O}_y = s_\alpha f \in T$ for some $\alpha \in I_1$, we get $f(x) \leq \alpha < int_{\omega(T)} f(y)$, That is, $f(x) < \alpha \leq int_{\omega(T)} f(y)$ for some $\alpha \in I_0$. Similarly, if there is a neighbourhood $\mathcal{O}_x \in T$ such that $y \notin \mathcal{O}_x$, we can find $g \in I^X$ such that $\mathcal{O}_x = s_\beta g \in T$ and $g(y) \leq \beta < int_{\omega(T)} g(x)$ for some $\beta \in I_1$, That is, $g(y) < \beta \leq int_{\omega(T)} g(x)$ for some $\beta \in I_0$. Hence, $(X, \omega(T))$ is an $(\alpha, \beta)T_0$-space $((\alpha, \beta)T_1)$.

Conversely, let $(X, \omega(T))$ be an $(\alpha, \beta)T_0$-space $((\alpha, \beta)T_1)$ and $x \neq y$. Then, there exists $f \in I^X$ such that $f(y) < \alpha \leq int_{\omega(T)} f(x)$ for some $\alpha \in I_0$, which means $f(y) \leq \alpha < int_{\omega(T)} f(x)$ for some $\alpha \in I_1$, that is there is $int_{\omega(T)} f \in I^X$ such that $s_\alpha int_{\omega(T)} f = \mathcal{O}_x \in T$ and $y \notin \mathcal{O}_x$. Similarly, the other case is proved. Hence, (X, T) is a T_0-space (T_1). $\square$

Proposition 4.3 Let (X, τ) be an $(\alpha, \beta)T_0$-space $((\alpha, \beta)T_1)$ and let σ be a fuzzy topology on X finer than τ. Then, (X, σ) is also $(\alpha, \beta)T_0$-space $((\alpha, \beta)T_1$-space).

Proof (X, τ) is an $(\alpha, \beta)T_0$-space $((\alpha, \beta)T_1)$ implying that there is $f \in I^X$ such that $\alpha \leq int_\tau f(x)$ and $f(x) < \alpha$ or (and) there is $g \in I^X$ such that $\alpha \leq int_\tau g(x)$ and $g(x) < \alpha$, which implies that $\alpha \leq \tau(f)$ or (and)$\alpha \leq \tau(g)$. Since σ is finer than τ, then $\alpha \leq \sigma(f)$ or (and)$\alpha \leq \sigma(g)$, and thus, $\alpha \leq int_\sigma f(x)$ and $f(x) < \alpha$ or (and) $\alpha \leq int_\sigma g(x)$ and $g(x) < \alpha$. Hence (X, σ) is an $(\alpha, \beta)T_0$-space $((\alpha, \beta)T_1)$. $\square$

5. $(\alpha, \beta)T_2$-spaces
Here, we introduce and study the Hausdorff separation axiom in fuzzy topological spaces.

Definition 5.1 An fuzzy topological space (X, τ) is called an $(\alpha, \beta)T_2$-space if for all $x \neq y$ in X there exist $f \in \mathcal{N}_x^\alpha$ and $g \in \mathcal{N}_y^\beta$ such that $(\alpha \wedge \beta) > \sup(f \wedge g)$; $\alpha, \beta \in I_0$.

Proposition 5.1 Every $(\alpha, \beta)T_2$-space is an $(\alpha, \beta)T_1$-space.

Proof Let (X, τ) be an $(\alpha, \beta)T_2$-space but not $(\alpha, \beta)T_1$-space. That is, for $x \neq y$, we get for all $f \in \mathcal{N}_x^\alpha$ that $f(y) \geq \alpha$ for all $\alpha \in I_0$. Since for any $g \in \mathcal{N}_y^\beta$ we have $g(y) \geq \beta$, then $(f \wedge g)(y) = f(y) \wedge g(y) \geq (\alpha \wedge \beta)$ and thus, $\sup(f \wedge g) \geq (\alpha \wedge \beta)$ which contradicts the axiom $(\alpha, \beta)T_2$. Hence, (X, τ) is an $(\alpha, \beta)T_1$-space. $\square$

Example 5.1   Let $X = \{x, y\}$, and

There are $f = x_1 \in I^X$ and $g = x_{\frac{1}{3}} \vee y_1 \in I^X$ such that, for $\alpha = \frac{1}{5}$ and $\beta = \frac{4}{5}$ in I_0, we get that $f = x_1 \in \mathcal{N}_x^\alpha$ and $g = x_{\frac{1}{3}} \vee y_1 \in \mathcal{N}_y^\beta$ such that $f(y) = x_1(y) = 0 < \frac{1}{5} = \alpha$ and $g(x) = (x_{\frac{1}{3}} \vee y_1)(x) = \frac{1}{3} < \frac{4}{5} = \beta$. That is, (X, τ) is an $(\alpha, \beta)T_1$-space. But for all fuzzy sets $f \in \mathcal{N}_x^\alpha$ and $g \in \mathcal{N}_y^\beta$, we get that $(\alpha \wedge \beta) \leq \sup(f \wedge g)$ and thus, (X, τ) is not $(\alpha, \beta)T_2$-space.

$$\tau(f) = \begin{cases} 1 & \text{at } \overline{0} \text{ or } \overline{1} \\ \frac{1}{5} & \text{at } x_1 \\ \frac{4}{5} & \text{at } x_{\frac{1}{3}} \vee y_1 \\ 0 & \text{otherwise.} \end{cases}$$

THEOREM 5.1   *Let (X, τ) be an fuzzy topological space. Then, the following statements are equivalent.*

(1) (X, τ) is $(\alpha, \beta)T_2$.

(2) For all fuzzy ultrafilter $\mathcal{M}$ on X and for all $x \neq y$, there is $f \in \mathcal{N}_x^\alpha$ such that $\mathcal{M}(f) < \alpha; \alpha \in I_0$ or there is $g \in \mathcal{N}_y^\beta$ such that $\mathcal{M}(g) < \beta; \beta \in I_0$.

(3) For all fuzzy filter $\mathcal{M}$ on X and for all $x \neq y$, there is $f \in \mathcal{N}_x^\alpha$ such that $\mathcal{M}(f) < \alpha; \alpha \in I_0$ or there is $g \in \mathcal{N}_y^\beta$ such that $\mathcal{M}(g) < \beta; \beta \in I_0$.

Proof   (1) $\Rightarrow$ (2): Suppose that there is an fuzzy ultrafilter $\mathcal{M}$ on X such that $\mathcal{M}(f) \geq \alpha$ and $\mathcal{M}(g) \geq \beta$ for all $f \in \mathcal{N}_x^\alpha$ and $g \in \mathcal{N}_y^\beta$. That is, $\mathcal{M}(f \wedge g) = \mathcal{M}(f) \wedge \mathcal{M}(g) \geq \alpha \wedge \beta$, but in common we know that $\mathcal{M}(h) \leq \sup h$ for all $h \in I^X$, which means that for all $f \in \mathcal{N}_x^\alpha$ and $g \in \mathcal{N}_y^\alpha$, we have $\sup(f \wedge g) \geq (\alpha \wedge \beta)$ and therefore, (1) implies (2) is satisfied.

(2) $\Rightarrow$ (3): Since for any fuzzy filter $\mathcal{M}$ on X we find a finer fuzzy ultrafilter $\mathcal{I}$ on X, that is $\mathcal{M}(f) \leq \mathcal{I}(f)$ for all $f \in I^X$, then (2) implies that there is $f \in \mathcal{N}_x^\alpha$ such that $\mathcal{M}(f) \leq \mathcal{I}(f) < \alpha; \alpha \in I_0$ or there is $g \in \mathcal{N}_y^\beta$ such that $\mathcal{M}(g) \leq \mathcal{I}(g) < \beta; \beta \in I_0$. Thus, (3) holds.

(3) $\Rightarrow$ (1): Suppose for all $f \in \mathcal{N}_x^\alpha$ and $g \in \mathcal{N}_y^\beta$; $\alpha, \beta \in I_0$ that $(\alpha \wedge \beta) \leq \sup(f \wedge g)$ and (3) is fulfilled. Then, for all fuzzy filter $\mathcal{M}$ on X, we have $\mathcal{M}(f) < \alpha$ or $\mathcal{M}(g) < \beta$; $\alpha, \beta \in I_0$. Hence, $\mathcal{M}(f \wedge g) < (\alpha \wedge \beta) \leq \sup(f \wedge g)$, which means a contradiction to the common result that $\mathcal{M}(f \wedge g) \leq \sup(f \wedge g)$ and therefore, $\mathcal{M}(f \wedge g) \leq \sup(f \wedge g) < (\alpha \wedge \beta)$. Thus, (1) is satisfied. $\square$

Example 5.2   Let $X = \{x, y\}$, and

$$\tau(f) = \begin{cases} 1 & \text{at } \overline{0} \text{ or } \overline{1} \\ \frac{1}{4} & \text{at } x_{\frac{1}{3}} \\ \frac{1}{4} & \text{at } y_{\frac{1}{3}} \\ 0 & \text{otherwise.} \end{cases}$$

There are $f = x_{\frac{1}{3}} \in I^X$ and $g = y_{\frac{1}{3}} \in I^X$ such that for $\alpha = \frac{1}{4}$ and $\beta = \frac{1}{4}$ in I_0 we get that $f = x_{\frac{1}{3}}$ in $\mathcal{N}_x^\alpha$ and $g = y_{\frac{1}{3}}$ in $\mathcal{N}_y^\beta$ such that $(\alpha \wedge \beta) = \frac{1}{4} > \sup(x_{\frac{1}{3}} \wedge y_{\frac{1}{3}}) = 0$ and thus, (X, τ) is an $(\alpha, \beta)T_2$-space.

Proposition 5.1   A topological space (X, T) is a T_2-space if and only if the induced fuzzy topological space $(X, \omega(T))$ is an $(\alpha, \beta)T_2$-space.

Proof   Let $x \neq y$ in X. Then, there are $\mathcal{O}_x, \mathcal{O}_y \in T$ such that $\mathcal{O}_x \cap \mathcal{O}_y = \emptyset$. Taking $f, g \in I^X$ such that $s_\alpha f = \mathcal{O}_x$, $s_\beta g = \mathcal{O}_y$ for some $\alpha, \beta \in I_1$, then $\text{int}_{\omega(T)} f(x) > \alpha$ and $\text{int}_{\omega(T)} g(y) > \beta$; $\alpha, \beta \in I_1$, that is $\text{int}_{\omega(T)} f(x) \geq \alpha$ and $\text{int}_{\omega(T)} g(y) \geq \beta$; $\alpha, \beta \in I_0$ and then, $f \in \mathcal{N}_x^\alpha$ and $g \in \mathcal{N}_y^\beta$ such that $s_\alpha f \cap s_\beta g = \mathcal{O}_x \cap \mathcal{O}_y = \emptyset$, which means that there is no element $z \in X$ such that

$(f \wedge g)(z) = f(z) \wedge g(z) \geq \text{int}_{\omega(T)} f(z) \wedge \text{int}_{\omega(T)} g(z) > (\alpha \wedge \beta); \alpha, \beta \in I_1$, which means for all $z \in X$, we have $(f \wedge g)(z) \leq (\alpha \wedge \beta); \alpha, \beta \in I_1$. Hence, $\sup(f \wedge g) \leq (\alpha \wedge \beta); \alpha, \beta \in I_1$ and then, $\sup(f \wedge g) < (\alpha \wedge \beta); \alpha, \beta \in I_0$ and thus, $(X, \omega(T))$ is an $(\alpha, \beta)T_2$-space.

Conversely, $x \neq y$ implies that there are $f \in \mathcal{N}_x^\alpha$ and $g \in \mathcal{N}_y^\beta$ such that $\text{int}_{\omega(T)} f(x) \wedge \text{int}_{\omega(T)} g(y) \geq (\alpha \wedge \beta) > \sup(f \wedge g); \alpha, \beta \in I_0$. That is, for $\gamma = \sup(f \wedge g) \in I_1$, we can say $\text{int}_{\omega(T)} f \in \omega(T), x \in s_\gamma \text{int}_{\omega(T)} f$ and $\text{int}_{\omega(T)} g \in \omega(T), y \in s_\gamma \text{int}_{\omega(T)} g$, which means that $s_\gamma \text{int}_{\omega(T)} f = \mathcal{O}_x \in T$, $s_\gamma \text{int}_{\omega(T)} g = \mathcal{O}_y \in T$ and moreover, $\mathcal{O}_x \cap \mathcal{O}_y = \emptyset$ and thus, (X, T) is a T_2-space. (because if there is $z \in (\mathcal{O}_x \cap \mathcal{O}_y)$, then $(f \wedge g)(z) \geq \text{int}_{\omega(T)} f(z) \wedge \text{int}_{\omega(T)} g(z) > \gamma = \sup(f \wedge g)$ which is a contradiction). □

Proposition 5.2 Let (X, τ) be an $(\alpha, \beta)T_2$-space, and let σ be an fuzzy topology on X finer than τ. Then, (X, σ) is also an $(\alpha, \beta)T_2$-space.

Proof Let $x \neq y \in X$. Then, there are $f \in \mathcal{N}_x^\alpha$ and $g \in \mathcal{N}_y^\beta$ such that $(\alpha \wedge \beta) > \sup(f \wedge g); \alpha, \beta \in I_0$, that is $\alpha \leq \text{int}_\tau f(x), \beta \leq \text{int}_\tau g(y)$ and $(\alpha \wedge \beta) > \sup(f \wedge g)$, which means that $\alpha \leq \text{int}_\sigma f(x), \beta \leq \text{int}_\sigma g(y)$ and $(\alpha \wedge \beta) > \sup(f \wedge g); \alpha, \beta \in I_0$ and thus, $f \in \mathcal{N}_x^\alpha$ and $g \in \mathcal{N}_y^\beta$ in (X, σ) such that $(\alpha \wedge \beta) > \sup(f \wedge g); \alpha, \beta \in I_0$. Hence, (X, σ) is an $(\alpha, \beta)T_2$-space. □

6. $(\alpha, \beta)T_3$-spaces and $(\alpha, \beta)T_4$-spaces
In this section, we use fuzzy neighbourhood filters at ordinary sets to define the notions of $(\alpha, \beta)T_3$-spaces and $(\alpha, \beta)T_4$-spaces.

Definition 6.1 A fuzzy topological space (X, τ) is called (α, β) regular if for all $F = \text{cl}_\tau F$ in $P(X)$ and $x \notin F$, there exist $f \in \mathcal{N}_x^\alpha$ and $g \in \mathcal{N}_F^\beta$ such that $(\alpha \wedge \beta) > \sup(f \wedge g); \alpha, \beta \in I_0$.

Definition 6.2 A fuzzy topological space (X, τ) is called $(\alpha, \beta)T_3$-space if it is regular and $(\alpha, \beta)T_1$.

Definition 6.3 A fuzzy topological space (X, τ) is called normal if for all $F_1 = \text{cl}_\tau F_1, F_2 = \text{cl}_\tau F_2 \in P(X)$ with $F_1 \cap F_2 = \emptyset$, there exist $f \in \mathcal{N}_{F_1}^\alpha$ and $g \in \mathcal{N}_{F_2}^\beta$ such that $(\alpha \wedge \beta) > \sup(f \wedge g); \alpha \wedge \beta \in I_0$.

Definition 6.4 A fuzzy topological space (X, τ) is called $(\alpha, \beta)T_4$ if it is normal and $(\alpha, \beta)T_1$.

Proposition 6.1 Every $(\alpha, \beta)T_3$-space is an $(\alpha, \beta)T_2$-space.

Proof Let $x \neq y$ in X. (X, τ) is an $(\alpha, \beta)T_1$-space meaning that $\text{cl}_\tau \{x\} = \{x\}$ for each $x \in X$. Now, $\text{cl}_\tau \{y\} = \{y\}$, $x \notin \{y\}$, and (X, τ) is regular implying that there are $f \in \mathcal{N}_x^\alpha, g \in \mathcal{N}_y^\beta$ such that $(\alpha \wedge \beta) > \sup(f \wedge g); \alpha, \beta \in I_0$. Hence, (X, τ) is an $(\alpha, \beta)T_2$-space. □

THEOREM 6.1 *For each fuzzy topological space (X, τ), the following are equivalent.*

(1) (X, τ) is regular.
(2) For all $y \in F = \text{cl}_\tau F$ and $x \notin F$, we have $\mathcal{N}_x^\alpha \not\subseteq \text{cl}\mathcal{N}_y^\alpha$ and $\mathcal{N}_y^\beta \not\subseteq \text{cl}\mathcal{N}_x^\beta$ for all $y \in F; \alpha, \beta \in I_0$.
(3) For all $x \in X$ and all $\alpha \in I_0$, we have $\text{cl}\mathcal{N}_x^\alpha = \mathcal{N}_x^\alpha$.
(4) For all $x \in X$, for all fuzzy filter $\mathcal{M}$ on X, for all $f \in \mathcal{N}_x^\alpha$, and for all $\alpha \in I_0$, we have $\mathcal{M}(f) \geq \alpha$ implies $\text{cl}\mathcal{M}(f) \geq \alpha$.

Proof (1) ⇒ (2): Let $f \in \mathcal{N}_x^\alpha; \alpha \in I_0$. Suppose that $f \in \text{cl}\mathcal{N}_y^\alpha$ for some $y \in F$, that is, there is $h \in \mathcal{N}_y^\alpha$ with $\text{cl}_\tau h \leq f$, which means that $f(y) \geq \alpha$. Since for all $g \in \mathcal{N}_y^\alpha$, we have $g(y) \geq \alpha$ for all $y \in F; \alpha \in I_0$ then $\sup(f \wedge g) \geq (f \wedge g)(y) \geq \alpha = (\alpha \wedge \alpha)$ for some $f \in \mathcal{N}_x^\alpha$ for all $x \notin F$, and for all $g \in \mathcal{N}_y^\alpha$ for some $y \in F; \alpha \in I_0$, which contradicts (1) and therefore, $f \notin \text{cl}\mathcal{N}_y^\alpha$ for all $y \in F$. Thus, $\mathcal{N}_x^\alpha \not\subseteq \text{cl}\mathcal{N}_y^\alpha$ for all $y \in F$. The other case is similar and hence, (2) is satisfied.

(2) $\Rightarrow$ (3): *From (2) we deduce that for all $f \in \mathcal{N}_x^\alpha$ and $g \in \mathcal{N}_y^\beta$, we have $f \in cl\mathcal{N}_y^\alpha$ or $g \in cl\mathcal{N}_x^\beta$ for all $\alpha, \beta \in I_0$ implies $x = y$. Hence, for all $f \in \mathcal{N}_x^\alpha$, $x \in X$, and all $\alpha \in I_0$, we get that $f \in cl\mathcal{N}_x^\alpha$, which means that $\mathcal{N}_x^\alpha \subseteq cl\mathcal{N}_x^\alpha$, but from that $cl\mathcal{N}_x^\alpha \subseteq \mathcal{N}_x^\alpha$ for all $\alpha \in I_0$ and for all $x \in X$, we get that $cl\mathcal{N}_x^\alpha = \mathcal{N}_x^\alpha$ for all $\alpha \in I_0$ and for all $x \in X$ and thus, (3) holds.*

(3) $\Rightarrow$ (4): *Let $\mathcal{M}$ be a fuzzy filter on X with $\mathcal{M}(f) \geq \alpha$ for all $f \in \mathcal{N}_x^\alpha$ and $\alpha \in I_0$. From (3), $\mathcal{M}(f) \geq \alpha$ for all $f \in cl\mathcal{N}_x^\alpha$ and $\alpha \in I_0$ and then, $cl\mathcal{M}(f) \geq \alpha$ for all $f \in \mathcal{N}_x^\alpha$ and $\alpha \in I_0$ and thus, (4) is fulfilled.*

(4) $\Rightarrow$ (1): *Consider $\mathcal{M} = \mathcal{N}_x$ in (4), we get that $cl\mathcal{N}_x^\alpha = \mathcal{N}_x^\alpha$ for all $x \in X$ and all $\alpha \in I_0$. Now, for $y \in F = cl_\tau F$ and $x \neq y$, we get for all $f \in \mathcal{N}_x^\alpha$ and $g \in \mathcal{N}_y^\beta$ that $f \in cl\mathcal{N}_x^\alpha$ and $g \in cl\mathcal{N}_y^\beta$, which means there are $h \in \mathcal{N}_x^\alpha$ with $cl_\tau h \leq f$ and $k \in \mathcal{N}_y^\beta$ with $cl_\tau k \leq g$. Choose $f = cl_\tau \chi_{F^c} \in \mathcal{N}_x^1$ and $g = cl_\tau (int_\tau \chi_F) \in \mathcal{N}_F^1$, then we can find $h = \chi_{F^c} \in \mathcal{N}_x^1$ and $k = int_\tau \chi_F \in \mathcal{N}_F^1$ such that $(\alpha \wedge \beta) = 1 > 0 = sup(\chi_{F^c} \wedge int_\tau \chi_F) = sup\,(h \wedge k)$, and thus, for all $F = cl_\tau F \in X$ and $x \notin F$, there exist $h \in \mathcal{N}_x^\alpha$ and $k \in \mathcal{N}_F^\beta$ such that $(\alpha \wedge \beta) > sup(h \wedge k)$; $\alpha, \beta \in I_0$ and therefore, (1) is satisfied.* $\square$

THEOREM 6.2   *Let (X, τ) be a fuzzy topological space. Then, the following are equivalent.*

(1) (X, τ) *is normal.*

(2) *For all $F_1 = cl_\tau F_1, F_2 = cl_\tau F_2 \in P(X)$ with $F_1 \cap F_2 = \emptyset$, we have $\mathcal{N}_x^\alpha \nsubseteq cl\mathcal{N}_y^\alpha$ and $\mathcal{N}_y^\beta \nsubseteq cl\mathcal{N}_x^\beta$ for all $x \in F_1$ and $y \in F_2$; $\alpha, \beta \in I_0$.*

(3) *For all $F = cl_\tau F \in P(X)$, and all $\alpha \in I_0$, we have $cl\mathcal{N}_F^\alpha = \mathcal{N}_F^\alpha$.*

(4) *For all $F = cl_\tau F \in P(X)$, for all fuzzy filters $\mathcal{M}$ on X, for all $f \in \mathcal{N}_F^\alpha$, and for all $\alpha \in I_0$, we have $\mathcal{M}(f) \geq \alpha$ implies $cl\mathcal{M}(f) \geq \alpha$.*

Proof   Similar to the Theorem 6.1. $\square$

Proposition 6.2   Every $(\alpha, \beta)T_4$-space is an $(\alpha, \beta)T_3$-space.

Proof   Let $x \notin F = cl_\tau F$ in X. Since (X, τ) is $(\alpha, \beta)T_4$, then it is $(\alpha, \beta)T_1$, which means that $cl_\tau \{x\} - \{x\}$ for all $x \in X$, which implies that we have $F_1 = \{x\} = cl_\tau \{x\}$ and $F_2 = F$ with $F_1 \cap F_2 = \emptyset$. Hence, there are $f \in \mathcal{N}_x^\alpha$ and $g \in \mathcal{N}_F^\beta$ such that $(\alpha \wedge \beta) > sup(f \wedge g)$; $\alpha, \beta \in I_0$ and thus, (X, τ) is regular and it is $(\alpha, \beta)T_1$. Therefore, (X, τ) is $(\alpha, \beta)T_3$. $\square$

Example 6.1   Let $X = \{x, y\}$, and

$$\tau(f) = \begin{cases} 1 & \text{at } \overline{0} \text{ or } \overline{1} \\ \frac{1}{2} & \text{at } x_1 \\ \frac{1}{3} & \text{at } y_{\frac{1}{2}} \\ 0 & \text{otherwise.} \end{cases}$$

We notice that $\{y\}$ is a closed set and $x \notin \{y\}$. Then, there are $f = x_1 \in I^X$ and $g = y_{\frac{1}{2}} \in I^X$ such that for $\alpha = \frac{1}{2}$ and $\beta = \frac{1}{3}$ in I_0, we get that $f = x_1$ in $\mathcal{N}_x^\alpha$ and $g = y_{\frac{1}{2}}$ in $\mathcal{N}_{\{y\}}^\beta$ such that $(\alpha \wedge \beta) = \frac{1}{3} > sup(x_1 \wedge y_{\frac{1}{2}}) = 0$ and thus, (X, τ) is an (α, β) regular space. Also, it is an $(\alpha, \beta)T_1$-space. Hence, (X, τ) is an $(\alpha, \beta)T_3$-space

Example 6.2   Let $X = \{x, y\}$, and

$$\tau(f) = \begin{cases} 1 & \text{at } \overline{0} \text{ or } \overline{1} \\ \frac{1}{2} & \text{at } x_1 \\ \frac{1}{2} & \text{at } y_1 \\ 0 & \text{otherwise.} \end{cases}$$

We see that $\{x\}$ and $\{y\}$ are disjoint closed subsets of X. Then, there are $f = x_1 \in I^X$ and $g = y_1 \in I^X$

such that for $\alpha = \frac{1}{2}$ and $\beta = \frac{1}{2}$ in I_0, we get that $f = x_1$ in $\mathcal{N}^{\alpha}_{\{x\}}$ and $g = y_1$ in $\mathcal{N}^{\beta}_{\{y\}}$ such that $(\alpha \wedge \beta) = \frac{1}{2} > \sup(x_1 \wedge y_1) = 0$ and thus, (X, τ) is an (α, β) normal space. Also, it is an $(\alpha, \beta)T_1$-space. Hence, (X, τ) is an $(\alpha, \beta)T_4$-space

Proposition 6.3   *A topological space (X, T) is T_3 if and only if the induced fuzzy topological space $(X, \omega(T))$ is $(\alpha, \beta)T_3$.*

Proof (X, T) *is T_1 iff $(X, \omega(T))$ is $(\alpha, \beta)T_1$ is proved in Proposition 4.2.*

Let $F = cl_\tau F$ and $x \notin F$ in X. Then, there are $\mathcal{O}_x, \mathcal{O}_F \in T$ such that $\mathcal{O}_x \cap \mathcal{O}_F = \emptyset$. Taking $f = \chi_{F^c}$ and $g = \chi_{\mathcal{O}_F}$ in $\omega(T)$, we get that $int_{\omega(T)}f(x) \wedge int_{\omega(T)}g(F) = 1 > 0 = \sup(f \wedge g)$. Hence, there are $\mathcal{N}^1_x$ and $\mathcal{N}^1_F$ of x and F respectively, such that $1 > \sup(f \wedge g)$ and thus, $(X, \omega(T))$ is an $(\alpha, \beta)T_3$-space.

Conversely, $F = cl_\tau F$ and $x \notin F$ imply there are $f \in \mathcal{N}^{\alpha}_x$ and $g \in \mathcal{N}^{\beta}_y$ for all $y \in F$ such that $int_{\omega(T)}f(x) \wedge int_{\omega(T)}g(F) \geq (\alpha \wedge \beta) > \sup(f \wedge g)$; $\alpha \wedge \beta \in I_0$. That is, $int_{\omega(T)}f \in \omega(T)$, $x \in s_\alpha int_{\omega(T)}f$ and $int_{\omega(T)}g \in \omega(T)$, $F \in s_\beta int_{\omega(T)}g$, which means that $s_\alpha int_{\omega(T)}f = \mathcal{O}_x \in T$, $s_\beta int_{\omega(T)}g = \mathcal{O}_F \in T$ and moreover, $\mathcal{O}_x \cap \mathcal{O}_F = \emptyset$, and thus, (X, T) is a T_3-space.   □

Proposition 6.4   *A topological space (X, T) is T_4 iff the induced fuzzy topological space $(X, \omega(T))$ is an $(\alpha, \beta)T_4$.*

Proof   Similar to Proposition 6.3.   □

Proposition 6.5   *Let (X, τ) be an $(\alpha, \beta)T_3$-space, and let σ be an fuzzy topology on X finer than τ. Then, (X, σ) is also an $(\alpha, \beta)T_3$-space.*

Proof   *Let $x \in X$ and F be a closed subset of X with $x \notin F$. Then, there are $f \in \mathcal{N}^{\alpha}_x$ and $g \in \mathcal{N}^{\beta}_F$ such that $(\alpha \wedge \beta) > \sup(f \wedge g)$; $\alpha, \beta \in I_0$, that is $\alpha \leq int_\tau f(x)$, $\beta \leq int_\tau g(y)$ for all $y \in F$ and $(\alpha \wedge \beta) > \sup(f \wedge g)$, which means that $\alpha \leq int_\sigma f(x)$, $\beta \leq int_\sigma g(y)$ for all $y \in F$ and $(\alpha \wedge \beta) > \sup(f \wedge g)$; $\alpha, \beta \in I_0$ and thus, $f \in \mathcal{N}^{\alpha}_x$ and $g \in \mathcal{N}^{\beta}_F$ in (X, σ) such that $(\alpha \wedge \beta) > \sup(f \wedge g)$; $\alpha, \beta \in I_0$. Hence, (X, σ) is an (α, β) regular space. Proposition 4.3 states that (X, σ) is an $(\alpha, \beta)T_1$-space, and thus, it is an $(\alpha, \beta)T_3$-space.*   □

Proposition 6.6   *Let (X, τ) be an $(\alpha, \beta)T_4$-space and let σ be a fuzzy topology on X finer than τ. Then, (X, σ) is also an $(\alpha, \beta)T_4$-space.*

Proof   Similar to the proof in Proposition 6.5.   □

Funding
The author received no direct funding for this research.

Author details
Ismail Ibedou[1,2]
E-mail: ismail.ibedou@gmail.com
[1] Faculty of Science, Department of Mathematics, Benha University, 13518 Benha, Egypt.
[2] Faculty of Science, Department of Mathematics, Jazan University, KSA.

References
Bayoumi, F., & Ibedou, I. (2001). On GT_i-spaces. *Journal of Institute of Mathematics & Computer Sciences, 14*, 187–199.

Bayoumi, F., & Ibedou, I. (2002a). T_i-spaces I. *The Journal of the Egyptian Mathematical Society, 10*, 179–199.

Bayoumi, F., & Ibedou, I. (2002b). T_i-spaces II. *The Journal of the Egyptian Mathematical Society, 10*, 201–215.

Bayoumi, F., & Ibedou, I. (2004). The relation between the GT_i-spaces and fuzzy proximity spaces, G- compact spaces, fuzzy uniform spaces. *The Journal of Chaos, Solitons and Fractals, 20*, 955–966.

Chang, C. I. (1968). Fuzzy topological spaces. *Journal of Mathematical Analysis and Applications, 24*, 182–190.

DCPE Co-Training for Classification (2012). *Neuro computing, 86*, 75–85.

Eklund, P., & Gähler, W. (1992). Fuzzy filter functors and convergence. *Applications of Category Theory to fuzzy Subsets* (pp. 109–136). Dordrecht: Kluwer.

Fuzzy neuro systems for machine learning for large data sets. (2009, March, 6–7). *Proceedings of the IEEE International Advance Computing Conference, IEEE Explore* (pp. 541–545). Patiala.

Gähler, W. (1995). The general fuzzy filter approach to fuzzy topology I. *Fuzzy Sets and Systems, 76*, 205–224.

Gähler, W. (1995). The general fuzzy filter approach to fuzzy topology II. *Fuzzy Sets and Systems, 76*, 225–246.

Kubiak, T. (1985). *On fuzzy topologies* (PhD Thesis). A. Mickiewicz, Poznan.

Lowen, R. (1978). A comparison of different compactness notions in fuzzy topological spaces. *Journal of Mathematical Analysis and Applications, 64*, 446–454.

Lowen, R. (1977). Initial and final fuzzy topologies and fuzzy Tychonoff theorem. *Journal of Mathematical Analysis and Applications, 58*, 11–21.

Lowen, R. (1979). Convergence in fuzzy topological spaces. *General Topology and Applications, 10*, 147–160.

Söstak, A. P. (1985). On a fuzzy topological structure. *Rendiconti del Circolo Matematico di Palermo - Springer II, 11I*, 89–103.

11

Attractor and self-similar group of generalized fuzzy contraction mapping in fuzzy metric space

R. Uthayakumar[1] and A. Gowrisankar[1]*

*Corresponding author: A. Gowrisankar, Department of Mathematics, The Gandhigram Rural Institute – Deemed University, Gandhigram – 624 302, Dindigul, Tamil Nadu, India

E-mail: gowrisankargri@gmail.com

Reviewing editor: Lishan Liu, Qufu Normal University, China

Abstract: In this paper, we construct a deterministic fractal in fuzzy metric space using generalized fuzzy contraction mapping and its fixed-point theorem in hyperspace of non-empty compact sets. Moreover, we present the self-similar group of $\mathcal{H}$-contraction in fuzzy metric space and prove some familiar results of self-similar group for fuzzy metric space.

Subjects: Advanced Mathematics; Analysis - Mathematics; Applied Mathematics; Chaos Theory; Dynamical Systems; Mathematics & Statistics; Mathematical Analysis; Science

Keywords: attractor; $\mathcal{H}$-contraction; self-similar group; fuzzy metric space

AMS subject classifications: 28A80; 54H11; 47H10

1. Introduction

At the origin, fractal was defined by rough or fragmented geometric shape that can be split into parts where each smaller part is reduced size of the whole. That is, fractal can be defined through self-similar property. According to self-similar property, fractal can be characterized into two types, they are, an object having approximate or statistical self-similarity called random fractal and another one is an object having regular or exact self-similarity called deterministic or regular fractal. Mathematically, sets with non-integral Hausdorff dimension which exceed their topological dimension are called fractals by Mandelbort (1983).

ABOUT THE AUTHOR

A. Gowrisankar was born in Palani, Dindigul, India in 1989. He received the MSc Degree in Mathematics from The Gandhigram Rural Institute - Deemed University, Gandhigram, India in 2012. Currently, he is a Research Scholar in Department of Mathematics, The Gandhigram Rural Institute - Deemed University, Gandhigram, India. His research interests include Iterated Function System, Fractal interpolation theory and fuzzy metric space.

PUBLIC INTEREST STATEMENT

The Euclidean geometry handles regular objects with integer dimension, while the fractal geometry directs irregular objects with noninteger dimension. According to self-similar property, fractal can be characterized into two types, they are, an object having approximate self-similarity is called random fractal and another one is an object having exact self-similarity is called deterministic fractal. Natural objects that approximate fractals to a degree include clouds, mountain ranges, lightning bolts, coastlines, and snowflakes. However, not all self-similar objects are fractals for example, the real line is formally self-similar but fails to have other fractal characteristics. Fractals and Fuzzy spaces play a significant role in the Nonlinear Analysis. Hence, the above studies motivate our direction to investigate the fractal concepts in particular self-similar property in fuzzy setting.

Hutchinson (1981) introduced the formal definition of iterated function systems (IFS) and Barnsley (1993) developed the theory of IFS called the Hutchinson–Barnsley theory (HB Theory) in order to define and construct the fractal as a compact invariant subset of a complete metric space generated by the IFS of Banach contractions. That is, Hutchinson introduced an operator on hyperspace of non-empty compact sets called as Hutchinson–Barnsley operator (HB operator) to define a fractal set as a unique fixed point using the Banach contraction theorem in the complete metric space, in order to generate fractal as a unique fixed point using Banach fixed-point theorem having the aforesaid exact self-similar property. Moreover, these fractal sets have Hausdorff dimension greater than its topological dimension, in such a way that self-similarity is the most fundamental property of the fractals. In order to analyze self-similar sets in depth, we must realize their group structure. In this study, we present the self-similar group in fuzzy setting. Self-similar group is defined through Banach contraction and topological group in the classical metric space, while the fuzzy self-similar group is defined by fuzzy H-contraction and fuzzy topological group in the fuzzy metric space.

Fuzzy set theory was introduced by Zadeh (1965). Kramosil and Michalek (1975) introduced the notion of fuzzy metric space. Many authors have introduced and discussed several notions of fuzzy metric space in different ways and also proved fixed-point theorems with interesting consequent results in the fuzzy metric spaces (Farnoosh, Aghajani, & Azhdari, 2009; George & Veeramani, 1997; Grabiec, 1988; Gregori & Sapena, 2002; Mihet, 2007; Rodriguez-Lopez & Romaguera, 2004; Uthayakumar & Gowrisankar, 2014; Wardowski, 2013). George and Veeramani (1994) imposed some stronger conditions on the fuzzy metric space in order to obtain a Hausdorff topology. Rodriguez-Lopez and Romaguera defined the Hausdorff metric on fuzzy hyperspace and constructed the Hausdorff fuzzy metric space. Besides that, the necessary results of the Hausdorff fuzzy metric on fuzzy hyperspaces are proved in Rodriguez-Lopez and Romaguera (2004). Uthayakumar and Easwaramoorthy (2011), Easwaramoorthy and Uthayakumar (2011), investigated the fuzzy IFS fractals in the fuzzy metric space. On the basis of self-similar group of Banach contraction in classical metric space given by Saltan and Demir (2013), in this paper, we introduce the definition and property of self-similar group and strong self-similar group of $\mathcal{H}$-contraction. If G is a self-similar group (strong self-similar group) of $\mathcal{H}$-contraction, then G is also described as the attractor of a $\mathcal{H}$-IFS and one of the $\mathcal{H}$-contractions of $\mathcal{H}$-IFS is a group homomorphism (isomorphism). The image of G under this $\mathcal{H}$-contraction map is its proper subgroup H being homomorphic (isomorphic) to G. Fractal set can be defined as a self-similar and strong self-similar group in the sense of $\mathcal{H}$-IFS of compact topological space.

The paper is organized into two directions, first one is to construct the fractals in fuzzy metric space using generalized fuzzy contraction mapping. Second direction is that we investigate a fuzzy metric group on self-similar property of fractal set in order to define the topological group with generalized fuzzy contraction. In this paper, we will start with short introduction of deterministic fractals in fuzzy metric space in Section 2 and some of its properties which will be used frequently in the sequel. In Section 3, we present generalization of the fuzzy contraction mappings together with their fixed-point properties. Further, in Section 4, we define the self-similar groups in fuzzy metric space and investigate the properties of these groups. At the end of the paper, two substantial examples are given, which shows the existence of fuzzy self-similar groups.

2. Fuzzy iterated function system
In this section, we recall some pertinent concepts on fuzzy metric spaces in the sense of George and Veeramani. Hausdorff fuzzy metric for a given fuzzy metric space on the set of its non-empty compact subsets as well as Fuzzy IFS Fractals in the fuzzy metric space.

Definition 2.1 (George & Veeramani, 1997, 1994) A binary operation $* : [0, 1] \times [0, 1] \longrightarrow [0, 1]$ is a continuous t-norm, if $([0, 1], *)$ is a topological monoid with unit 1 such that $a * b \leq c * d$ whenever $a \leq c, b \leq d$, and $a, b, c, d \in [0, 1]$.

George and Veeramani modified the Kramosil and Michalek (1975) fuzzy metric space as follows.

Definition 2.2 (George & Veeramani, 1997, 1994) The 3-tuple $(X, M, *)$ is said to be a fuzzy metric space if X is an arbitrary set, $*$ is a continuous t-norm and M is a fuzzy set on $X \times X \times (0, \infty)$ satisfying the following conditions:

(1) $M(x, y, t) > 0$,

(2) $M(x, y, t) = 1$ if and only if $x = y$,

(3) $M(x, y, t) = M(y, x, t)$,

(4) $M(x, y, t) * M(y, z, s) \leq M(x, z, t + s)$,

(5) $M(x, y, \cdot) : (0, \infty) \longrightarrow [0, 1]$ is continuous,

$x, y, z \in X$ and $t, s > 0$.

Definition 2.3 (Gregori & Sapena, 2002; Rodriguez-Lopez & Romaguera, 2004) Let $(X, M, *)$ be a fuzzy metric space. The mapping $f: X \longrightarrow X$ is fuzzy contractive if there exists $k \in (0, 1)$ such that

$$\frac{1}{M(f(x), f(y), t)} - 1 \leq k \left(\frac{1}{M(x, y, t)} - 1 \right)$$

for all $x, y \in X$ and $t > 0$. Here, k is called the fuzzy contractivity ratio of f.

THEOREM 2.1 *(Gregori & Sapena 2002. Fuzzy Banach Contraction Theorem) Let $(X, M, *)$ be a complete fuzzy metric space in which fuzzy contractive sequence are Cauchy. Let $f: X \longrightarrow X$ be a fuzzy contractive mapping with contractivity ratio k. Then f has a unique fixed point.*

Definition 2.4 (Rodriguez-Lopez & Romaguera, 2004) Let $(X, M, *)$ be a fuzzy metric space. Let $\mathcal{H}(X)$ be set of all non-empty compact subsets of X. Define, $M(x, B, t) = \sup_{y \in B} M(x, y, t)$ and $M(A, B, t) = \inf_{x \in A} M(x, B, t)$ for all $x \in X$ and $A, B \in \mathcal{H}(X)$. Then Hausdorff fuzzy metric (H_M) is a function $H_M: \mathcal{H}(X) \times \mathcal{H}(X) \times (0, \infty) \longrightarrow [0, 1]$ defined by

$$H_M(A, B, t) = \min \left\{ M(A, B, t), M(B, A, t) \right\}$$

then H_M is a fuzzy metric on $\mathcal{H}(X)$, and hence $(\mathcal{H}(X), H_M, *)$ is called a Hausdorff fuzzy metric space.

Definition 2.5 (Uthayakumar & Easwaramoorthy, 2011; Easwaramoorthy & Uthayakumar, 2011) Let $(X, M, *)$ be a fuzzy metric space and $f_n: X \longrightarrow X$, $n = 1, 2, 3, \ldots, N$ ($N \in \mathbb{N}$) be N - fuzzy contractive mappings with the corresponding contractivity ratios k_n, $n = 1, 2, 3, \ldots, N$. Then the system $\{X; f_n, \ n = 1, 2, 3, \ldots, N\}$ is called a Fuzzy Iterated Function System (FIFS) of fuzzy contractions in the fuzzy metric space $(X, M, *)$.

Definition 2.6 (Uthayakumar & Easwaramoorthy, 2011; Easwaramoorthy & Uthayakumar, 2011) Let $(X, M, *)$ be a fuzzy metric space. Let $\{X; f_n, \ n = 1, 2, 3, \ldots, N; N \in \mathbb{N}\}$ be a FIFS of fuzzy contractions. Then the Fuzzy Hutchinson–Barnsley operator (FHB operator) of the FIFS of fuzzy contractions is a function $F: \mathcal{H}(X) \longrightarrow \mathcal{H}(X)$ defined by

$$F(B) = \bigcup_{n=1}^{N} f_n(B), \quad \text{for all } B \in \mathcal{H}(X).$$

Definition 2.7 (Uthayakumar & Easwaramoorthy, 2011; Easwaramoorthy & Uthayakumar, 2011) Let $(X, M, *)$ be a complete fuzzy metric space. Let $\{X; f_n, \ n = 1, 2, 3, \ldots, N; N \in \mathbb{N}\}$ be a FIFS of fuzzy contractions and F be the FHB operator of the FIFS of fuzzy contractions. We say that the set $A_\infty \in \mathcal{H}(X)$ is Fuzzy Attractor (Fuzzy Fractal) of the given FIFS of fuzzy contractions, if A_∞ is a unique fixed point of the FHB operator F of fuzzy contractions. Usually, such $A_\infty \in \mathscr{K}(X)$ is also called as Fractal generated by the FIFS of fuzzy contractions.

3. Attractor of generalized fuzzy contraction

In this section, we generate a fractal in fuzzy metric space, which is a generalization of a fractal initiated in the article (Easwaramoorthy & Uthayakumar, 2011). Moreover, we develop the $\mathcal{H}$-IFS theory in order to define and construct the fractal as a compact invariant subset of M-complete fuzzy metric space generated by the fixed-point theorem.

$\mathcal{H}$ denotes a collection of mappings $\eta : (0, 1] \longrightarrow [0, \infty)$ such that η maps $(0, 1]$ onto $[0, \infty)$ and $s > t$ implies $\eta(s) < \eta(t)$ for all $s, t \in (0, 1]$.

Definition 3.1   (Wardowski, 2013) Let $(X, M, *)$ be a fuzzy metric space. A mapping $f : X \longrightarrow X$ is said to be $\mathcal{H}$-contractive with respect to $\eta \in \mathcal{H}$ if there exists $k \in (0, 1)$ such that

$$\eta(M(f(x), f(y), t)) \leq k\eta(M(x, y, t)) \tag{1}$$

for all $x, y \in X$ and $t > 0$.

Remark 3.1   (Wardowski, 2013) Consider a mapping $\eta \in \mathcal{H}$ of the form $\eta(t) = \frac{1}{t} - 1, t \in (0, 1]$. Then the condition (1) reduces to

$$\frac{1}{M(f(x), f(y), t)} - 1 \leq k\left(\frac{1}{M(x, y, t)} - 1\right)$$

for all $x, y \in X$ and $t > 0$.

PROPOSITION 3.1   *(Wardowski, 2013) Let $(X, M, *)$ be a fuzzy metric space and $\eta \in \mathcal{H}$. A sequence $< x_n >_{n \in \mathbb{N}} \subset X$ is M-Cauchy if and only if for given $\epsilon > 0, t > 0$ there exits $n_0 \in \mathbb{N}$ scuh that*

$$\eta(M(x_m, x_n, t)) < \epsilon,$$

for all $m, n \geq n_0$.

PROPOSITION 3.2   *(Wardowski, 2013) Let $(X, M, *)$ be a fuzzy metric space and $\eta \in \mathcal{H}$. A sequence $< x_n >_{n \in \mathbb{N}} \subset X$ is convergent to $x \in X$ if and only if $\lim_{n \rightarrow \infty} \eta(M(x_n, x, t)) = 0$ for all $t > 0$.*

THEOREM 3.1   *Let $(X, M, *)$ be a fuzzy metric space. Let $(\mathscr{K}(X), H_M, *)$ be a corresponding Hausdorff fuzzy metric space. If $f : X \longrightarrow X$ is a fuzzy $\mathcal{H}$-contraction with respect to $\eta \in \mathcal{H}$ on $(X, M, *)$, then f is a fuzzy $\mathcal{H}$-contraction with respect to $\eta \in \mathcal{H}$ on $(\mathscr{K}(X), H_M, *)$.*

Proof   Fix $t > 0$. Let $A, B \in \mathscr{K}(X)$. f is fuzzy $\mathcal{H}$-contraction with respect to $\eta \in \mathcal{H}$ on $(X, M, *)$. Hence, there exlsts $k \in (0, 1)$ such that

$$\eta(M(f(x), f(y), t)) \leq k\eta(M(x, y, t)), \forall x, y \in X$$
$$\eta(\sup_{y \in B} M(f(x), f(y), t)) \leq k\eta(\sup_{y \in B} M(x, y, t)), \forall x \in A, y \in B$$
$$\eta(M(f(x), f(B), t)) \leq k\eta(M(x, B, t)), \forall x \in A, B \in \mathscr{K}(X)$$
$$\eta(\inf_{x \in A} M(f(x), f(B), t)) \leq k\eta(\inf_{x \in A} M(x, B, t)), \forall x \in A, B \in \mathscr{K}(X)$$
$$\eta(M(f(A), f(B), t)) \leq k\eta(M(A, B, t)), \forall A, B \in \mathscr{K}(X)$$
Similarly, $\eta(M(f(B), f(A), t)) \leq k\eta(M(B, A, t)), \forall A, B \in \mathscr{K}(X)$
Hence, $\eta(H_M(f(B), f(A), t)) \leq k\eta(H_M(B, A, t)), \forall A, B \in \mathscr{K}(X)$

THEOREM 3.2   *Let $(X, M, *)$ be a M-complete fuzzy metric space. Let $(\mathscr{K}(X), H_M, *)$ be a corresponding Hausdorff fuzzy metric space and let $f : \mathscr{K}(X) \longrightarrow \mathscr{K}(X)$ be a fuzzy $\mathcal{H}$-contractive with respect to $\eta \in \mathcal{H}$ such that*

(i) $\prod_{i=1}^{k} H_M(A, f(A), t_i) \neq 0$, for all $A \in \mathcal{H}(X), k \in \mathbb{N}$ and any sequence $< t_i >_{i \in \mathbb{N}} \subset (0, \infty), t_i \to 0$,

(ii) $r * s > 0$ implies $\eta(r * s) \leq \eta(r) + \eta(s)$, for all $r, s \in \{H_M(A, f(A), t): A \in \mathcal{H}(X), t > 0\}$,

(iii) $\{H_M(A, f(A), t_i): i \in \mathbb{N}\}$ is bounded for all $A \in \mathcal{H}(X)$ and any sequence $< t_i >_{i \in \mathbb{N}} \subset (0, \infty), t_i \to 0$.

Then f has a unique fixed point A^* and for each A_0, the sequence $< f^n(A_0) >_{n \in \mathbb{N}}$ converges to A^*.

Proof   Fix $A_0 \in \mathcal{H}(X)$. Define $A_1 = f(A_0)$ and $A_n = f(A_{n-1})$ for $n \geq 2$, we have a sequence $< A_n >_{n \in N}$. For $t > 0$,

$$\eta(H_M(A_1, A_2, t)) = \eta(H_M(f(A_0), f(A_1), t)) \leq k\eta(H_M(A_0, A_1, t)),$$

$$\eta(H_M(A_2, A_3, t)) = \eta(H_M(f(A_1), f(A_2), t)) \leq k\eta(H_M(A_1, A_2, t)) \leq k^2\eta(H_M(A_0, A_1, t)),$$

$$\eta(H_M(A_n, A_{n+1}, t)) = \eta(H_M(f(A_{n-1}), f(A_n), t)) \leq k\eta(H_M(A_{n-1}, A_n, t)) \leq \ldots \leq k^n\eta(H_M(A_0, A_1, t)).$$

Hence, $\eta(H_M(A_n, A_{n+1}, t)) \leq k^n\eta(H_M(A_0, A_1, t))$ for all $n \geq 1$.

Clearly, $H_M(A_n, A_{n+1}, t) \geq \eta(H_M(A_0, A_1, t)$ for all $n \geq 1$ and $t > 0$. For $m, n \in \mathbb{N}, m < n, t < 0$ and let $< a_i >_{i \in \mathbb{N}}$ be a strictly decreasing sequence of positive number such that $\sum_{i=1}^{\infty} a_i = 1$.

$$H_M(A_m, A_n, t) = H_M\left(A_m, A_n, t + \sum_{i=m}^{n-1} a_i t - \sum_{i=m}^{n-1} a_i t\right)$$

$$\geq H_M\left(A_m, A_m, t - \sum_{i=m}^{n-1} a_i t\right) * H_M\left(A_m, A_n, \sum_{i=m}^{n-1} a_i t\right)$$

$$= 1 * H_M\left(A_m, A_n, \sum_{i=m}^{n-1} a_i t\right)$$

$$\geq \prod_{i=m}^{n-1} H_M(A_i, A_{i+1}, a_i t)$$

$$\geq \prod_{i=m}^{n-1} H_M(A_0, A_1, a_i t) < 0$$

$H_M(A_m, A_n, t) \geq \prod_{i=m}^{n-1} H_M(A_i, A_i, a_i t) \implies \eta(H_M(A_m, A_n, t)) \leq \eta(\prod_{i=m}^{n-1} H_M(A_i, A_i, a_i t)) \leq \sum_{i=m}^{n-1} k^i \eta (H_M(A_i, A, a_i t)). \, \eta(H_M(A_m, A_n, t)) \leq \sum_{i=m}^{n-1} k^i \eta(H_M(A_i, A_i, a_i t))$.

Easy to verify that a sequence $< \eta(H_M(A_0, A_1, a_i t)) <_{i \in \mathbb{N}}$ is non-decreasing and, by (c), bounded, hence we have a convergence of the series $\sum_{i=1}^{\infty} \eta(H_M(A_i, A_i, a_i t))$. Consquently, for any $\epsilon > 0$, there exist $k \in \mathbb{N}$ such that $\sum_{i=m}^{n-1} \eta(H_M(A_i, A_i, a_i t)) < \epsilon$ for all $m, n \geq N, m < n$. Thus, by Proposition 3.1, $< A_n <_{i \in \mathbb{N}}$ is an M-Cauchy sequence. By the M-completeness of X, there exists $A^* \in X$ such that $\lim_{n \to \infty} A_n = A^*$. Due to Proposition 3.2, $\lim_{n \to \infty} \eta(H_M(A_n, A^*, t)) = 0$ for each $t > 0$. Hence for all $t > 0$, we obtain $\eta(H_M(f(A^*), A_{n+1}, t)) \leq k\eta(H_M(A^*, A_n, t)) \longrightarrow 0$ as $n \longrightarrow \infty$. Finally, from the Proposition 3.2, we have $A^* = \lim_{n \to \infty} A_{n+1} = f(A^*)$.

Suppose that there exists $A' \in \mathcal{H}(X), A' \neq A^*$ such that $f(A') = A'$. Then, any $t > 0$,

$$\eta(H_M(A', A^*, t)) = \eta(H_M(f(A'), f(A^*), t)) \leq k\eta(H_M(A', A^*, t)).$$

Since $H_M(A', A^*, t) \neq 1$, it is contradiction to the definition of η. Hence, $A' = A^*$, A^* is a unique fixed point of f.

Definition 3.2   Let $(X, M, *)$ be a fuzzy metric space and $f_n : X \longrightarrow X, n = 1, 2, 3, \ldots, N (N \in \mathbb{N})$ be N-$\mathcal{H}$ contractive mappings. Then the system $\{X; f_n, \ n = 1, 2, 3, \ldots, N\}$ is called a $\mathcal{H}$-Iterated Function System ($\mathcal{H}$-IFS) of $\mathcal{H}$-contractions in $(X, M, *)$. The Hutchinson–Barnsley operator (HB operator) of the $\mathcal{H}$-IFS is a function $F : \mathcal{K}(X) \longrightarrow \mathcal{K}(X)$ defined by

$$F(B) = \bigcup_{n=1}^{N} f_n(B), \quad \text{for all} \quad B \in \mathcal{K}(X)$$

Definition 3.3   Let $(X, M, *)$ be a complete fuzzy metric space. Let $\{X; f_n, \ n = 1, 2, 3, \ldots, N; N \in \mathbb{N}\}$ be a $\mathcal{H}$-IFS and F be the HB operator of the $\mathcal{H}$-IFS. If F has a unique fixed point A^* in $(X, M, *)$, then the set $A^* \in \mathcal{K}(X)$ is called the Attractor (or Fractal) generated by the $\mathcal{H}$-IFS of $\mathcal{H}$-contractions.

THEOREM 3.3   Let $\{X; f_0, f_1, \ldots, f_n\}$ be a $\mathcal{H}$-IFS with attractor A. If the $\mathcal{H}$-contraction mappings $f_0, f_1, \ldots, f_n$ are one-to-one on A and

$$f_i(A) \cap f_j(A) = \phi \text{ for all } i, j \in \{0, 1, 2, \ldots, n\} \text{ with } \pi \neq j$$

then A is totally disconnected set.

Proof   Suppose that there exist a connected subset B of A contains more than two points. A is an attractor of given $\mathcal{H}$-IFS, therefore $f_0(A) \cup f_2(A) \cup \cdots \cup f_n(A) = A$. $\mathcal{H}$-contraction mappings are t-uniform continuous, hence $f_i(B)$ is connected and $f_i(B) \cap f_j(B) = B$ for all $i, j \in \{0, 1, 2, \ldots, n\}$. Clearly, it gives the contradiction to $f_0, f_2, \ldots, f_n$ are one-to-one on A. Therefore, only connected subset of A is single point set, there are no other connected subsets in A. Hence, A is totally disconnected.

4. Fuzzy self-similar group

Romaguera and Sanchis (2001) extended the classical theorems on metric groups to the fuzzy setting. According to the definition of self-similar group of Banach contraction in classical metric space given by Saltan and Demir (2013), in this section, we introduce the definition and property of fuzzy self-similar group and strong fuzzy self-similar group of fuzzy contraction in fuzzy metric space. Then, we investigate some properties of strong fuzzy self-similar and fuzzy profinite groups.

Topological groups can be defined concisely as group objects in the category of topological spaces, in the same way that ordinary groups are group objects in the category of sets. Now we recall the definition of self-similar group in compact topological space and profinite group.

Definition 4.1   (Demir & Saltan, 2012; Saltan & Demir, 2013) Let (G, d) be a compact topological group with a translation-invariant metric d. G is called a self-similar group, if there exists a proper subgroup H of finite index and a surjective homomorphism $\phi : G \longrightarrow H$, which is a contraction with respect to d.

Definition 4.2   (Saltan & Demir, 2013) Let (G, d) be a compact topological group with a translation-invariant metric d. G is called a strong self-similar group, if there exists a proper subgroup H of finite index and a group isomorphism $\phi : G \longrightarrow H$, which is a contraction with respect to d.

Definition 4.3   (Dixon, Du Sautoy, Mann, & Segal, 1999; Saltan & Demir, 2013) A topological group G is profinite, if it is topologically isomorphic to an inverse limit of finite discrete topological groups. Equivalently, a profinite group is a compact, Hausdorff, and totally disconnected topological group.

A fuzzy metric group is a 4-tuple $(G, ., M, *)$ such that $(G, M, *)$ is a fuzzy metric space and $(G, ., \tau_M)$ is a topological group, where τ_M is a topology induced by the fuzzy metric $(M, *)$.

Definition 4.4   Let $(G, ., M, *)$ be a compact topological fuzzy metric group (simply fuzzy group) with a translation-invariant fuzzy metric $(M, *)$. Then G is called a self-similar group of $\mathcal{H}$-contraction, if there exists a proper subgroup H of finite index and a surjective homomorphism $\phi:G \longrightarrow H$, which is a $\mathcal{H}$-contraction with respect to $\eta \in \mathcal{H}$ on fuzzy metric $(M, *)$.

Definition 4.5 Let $(G, ., M, *)$ be a compact fuzzy topological group with a translation-invariant fuzzy metric $(M, *)$. Then G is called a strong self-similar group of $\mathcal{H}$-contraction, if there exists a proper subgroup H of finite index and a group isomorphism $\phi:G \longrightarrow H$, which is a $\mathcal{H}$-contraction with respect to $\eta \in \mathcal{H}$ on fuzzy metric $(M, *)$.

Definition 4.6 A fuzzy topological group G is fuzzy profinite, if it is topologically isomorphic to an inverse limit of finite discrete topological fuzzy groups.

THEOREM 4.1 *If $(G, ., M, *)$ is a fuzzy profinite topological group, then G is profinite if and only if it is Hausdorff compact and totally disconnected.*

Proof Assume that G is profinite group, then G is Hausdorff, compact, and totally disconnected. Since every topological group is Hausdorff, and finite discrete groups are compact and totally disconnected.

Conversely, Let G be Hausdorff, compact, and totally disconnected. Since all components, i.e. all points of G, are closed and $e = comp\{e\}$ is the intersection of all open–closed neighborhoods of e. It is easy to show that every open–closed neighborhood of e contains an open normal subgroup, which implies G has a topological isomorphic to an inverse limit of finite discrete topological group.

PROPOSITION 4.1 *A strong self-similar group of $\mathcal{H}$-contraction is the attractor of $\mathcal{H}$-IFS.*

Proof Let $(X, ., M, *)$ be a strong self-similar group of $\mathcal{H}$-contraction. Hence, there is a proper subgroup H of X with $[X:H] = n$ such that the mapping $\phi_o:X \longrightarrow H$ is a group isomorphism and is a $\mathcal{H}$-contraction with respect to $\eta \in \mathcal{H}$. Let $x_o = e$ be the identity element of X. For all $i,j \in \{0, 1, 2 \ldots, n-1\}$ and $i \neq j$, there are cosets of H in X such that $(H.x_i) \cap (H.x_j) = \phi$ and $X = H \cup (H.x_1) \cup (H.x_2) \cup \cdots \cup (H.x_n - 1)$. Define $\phi_i:X \longrightarrow X$ by $\phi_i(g) = \phi_o(g).x_i$, $i = 1, 2, 3, \ldots, n-1$. Clearly,

$$\phi_i(X) = H.x_i$$

because of ϕ_o is surjective. Since ϕ_o is a $\mathcal{H}$-contraction mapping with respect to η and $(M, *)$ is a translation invariant fuzzy metric, we obtain that

$$M((\phi_i(g), \phi_i(h)), t) = M(\phi_o(g).x_i, \phi_o(h).x_i, t)$$
$$= M(\phi_o(g), \phi_o(h), t)$$
$$\eta(M((\phi_i(g), \phi_i(h)), t)) = \eta(M(\phi_o(g), \phi_o(h), t))$$
$$\leq k\eta(M(g, h, t)),$$

for all $g, h \in X$. Therefore, ϕ_i is a $\mathcal{H}$-contraction mapping with respect to η for $i = 1, 2, 3, \ldots, n-1$ and

$$X = H \cup (H.x_1) \cup (H.x_2) \cup \cdots \cup (H.x_{n-1})$$
$$= \phi_o(X) \cup \phi_1(X) \cup \phi_2(X) \cup \cdots \cup \phi_{n-1}(X)$$
$$X = \bigcup_{n=0}^{n-1} \phi_i(X)$$

Thus, X is the attractor of the $\mathcal{H}$-IFS $\{X; \phi_o, \phi_1, \ldots, \phi_{n-1}\}$.

THEOREM 4.2 *Let $(G, ., M, *)$ and $(G', ., M', *')$ be compact fuzzy topological groups. If G is a strong self-similar group of $\mathcal{H}$-contraction and $f:G \longrightarrow G'$ is both an isometry map and a group isomorphism, so is G'.*

Proof First, we show that M' is a translation-invariant fuzzy metric. There exists $x, y, z \in G$ such that $f(x) = x', f(y) = y'$, and $f(z) = z'$ for all $x', y', z' \in G$, since f is surjective and isometric, M is a translation-invariant fuzzy metric, hence we get

$$M'(x'.'z', y'.'z', t) = M'(f(x).'f(z), f(y).'f(z), t)$$
$$= M'(f(x.z), f(y.z), t)$$
$$= M(x.z, y.z, t)$$
$$= M(x, y, t)$$
$$= M'(f(x), f(y), t)$$
$$= M'(x', y', t).$$

G is a strong self-similar group of $\mathcal{H}$-contraction, there exists a subgroup H with $[G:H] = n$ and a group isomorphism $\phi: G \longrightarrow H$. Let $f(H) = H'$. f is a group isomorphism on G, hence f maps subgroup H of G into subgroup H' of G' with $[G':H'] = n$.

Define $\phi':G' \longrightarrow H'$ by $\phi'(g) = f_{|H} \circ \phi \circ f^{-1}(g)$, where $f_{|H}$ is a function from H to G' defined by $f_{|H}(g) = f(g), \forall x \in H$.

$f, f_{|H}$ and ϕ are group isomorphisms, it is clear that ϕ' is also a group isomorphism. Further, ϕ is a $\mathcal{H}$-contraction mapping with respect to $\eta \in \mathcal{H}$ and $f, f_{|H}$ are isometries, hence

$$M'(\phi'(g'), \phi'(h'), t) = M'(f_{|H} \circ \phi \circ f^{-1}(g'), f_{|H} \circ \phi \circ f^{-1}(h'), t)$$
$$= M'(f_{|H}(\phi \circ f^{-1})(g'), f_{|H}(\phi \circ f^{-1})(h'), t)$$
$$= M(\phi \circ f^{-1}(g'), \phi \circ f^{-1}(h'), t)$$
$$\eta(M'(\phi'(g'), \phi'(h'), t)) = \eta(M(\phi \circ f^{-1}(g'), \phi \circ f^{-1}(h'), t))$$
$$\leq k\, \eta((M(f^{-1}(g'), f^{-1}(h'), t))$$
$$= k\, \eta(M'(g', h', t))$$

for all $g', h' \in G'$. It gives that ϕ' is $\mathcal{H}$-contraction mapping on G'. Therefore, there exists a $\phi':G' \longrightarrow H'$ such that ϕ' is a group isomorphism and $\mathcal{H}$-contraction on G', hence G' is a strong self-similar group.

THEOREM 4.3 *If $G_1, G_2, \ldots, G_n$ are strong fuzzy self-similar groups of $\mathcal{H}$-contraction, so is $G_1 \times G_2 \times \cdots \times G_n$.*

Proof Consider the product fuzzy metric $M_p(X, Y, t) = M_1(x_1, y_1, t) * M_2(x_2, y_2, t) * \cdots * M_n(x_n, y_n, t)$ for all $t > 0$, $X, Y \in G_1 \times G_2 \times \cdots \times G_n$, $(G_1, M_1, *, .), (G_2, M_2, *, .), \ldots, (G_n, M_n, *, .)$ are compact fuzzy topological groups, hence $G_1 \times G_2 \times \cdots \times G_n$ is a compact fuzzy topological group. Moreover, there are subgroups $H_1, H_2, \ldots, H_n$ of $G_1, G_2, \ldots, G_n$ respectively, such that $[G_i:H_i] = m_i$ and the mappings

$$\varphi_i: G_i \longrightarrow H_i$$

are $\mathcal{H}$-contraction with respect to η and group isomorphisms for $i = 1, 2, \ldots, n$, since these groups are strong self-similar groups of $\mathcal{H}$-contraction. Define the mapping $\phi:G_1 \times G_2 \times \cdots \times G_n \longrightarrow H_1 \times H_2 \times \cdots \times H_n$ by $\phi(g_1, g_2, \ldots, g_n) = (\varphi_1(g_1), \varphi_2(g_2), \ldots, \varphi_n(g_n))$. Clearly, $H_1 \times H_2 \times \cdots \times H_n$ is a subgroup of $G_1 \times G_2 \times \cdots \times G_n$ and $[G_1 \times G_2 \times \cdots \times G_n:H_1 \times H_2 \times \cdots \times H_n] = m_1 m_2 \ldots m_n$. $\varphi_1, \varphi_2, \ldots, \varphi_n$ are group homomorphisms, hence

$$\phi(g.h) = \phi((g_1, g_2, \ldots, g_n).(h_1, h_2, \ldots, h_n))$$
$$= \phi((g_{1\cdot1}h_1, g_{2\cdot2}h_2, \ldots, g_{n\cdot n}h_n))$$
$$= (\varphi_1(g_{1\cdot1}h_1), \varphi_2(g_{2\cdot2}h_2), \ldots, \varphi_n(g_{n\cdot n}h_n))$$
$$= (\varphi_1(g_1)._1\varphi_1(h_1), \varphi_2(g_2)._2\varphi_2(h_2), \ldots, \varphi_n(g_n)._n\varphi_n(h_n))$$
$$= (\varphi_1(g_1), \ldots, \varphi_n(g_n)).(\varphi_1(h_1), \ldots, \varphi_n(h_n))$$
$$= \phi(g_1, g_2, \ldots, g_n).\phi(h_1, h_2, \ldots, h_n)$$
$$= \phi(g).\phi(h).$$

It gives that ϕ is group homomorphism. $\varphi_1, \varphi_2, \ldots, \varphi_n$ are bijective implies ϕ is bijective. Take $\mathcal{H}$-contraction ratio b_i of $\mathcal{H}$-contraction mappings φ_i for $i = 1, 2, \ldots, n$ and choose $b = max\{b_1, b_2, \ldots, b_n\}$. Then,

$$\eta(M_p(\phi(g), \phi(h), t)) = \eta(M_p(\phi(g_1, g_2, \ldots, g_n), \phi(h_1, h_2, \ldots, h_n), t))$$
$$= \eta(M_p((\varphi_1(g_1), \ldots, \varphi_n(g_n)), (\varphi_1(h_1), \ldots, \varphi_n(h_n)), t))$$
$$= \eta(M_1(\varphi_1(g_1), \varphi_1(h_1), t)).\eta(M_2(\varphi_2(g_2), \varphi_2(h_2), t)). \cdots .\eta(M_n(\varphi_n(g_n), \varphi_n(h_n), t))$$
$$\leq b_1\eta(M_1(g_1, h_1, t)).b_2\eta(M_2(g_2, h_2, t)). \cdots .b_n\eta(M_n(g_n, h_n, t))$$
$$\leq b\eta(M_1(g_1, h_1, t)).b\eta(M_2(g_2, h_2, t)). \cdots .b\eta(M_n(g_n, h_n, t))$$
$$= b\eta(M_p((g_1, \ldots, g_n), (h_1, \ldots, h_n), t))$$
$$= b\eta(M_p(g, h, t))$$

Hence, ϕ is $\mathcal{H}$-contraction mapping with respect to $\eta \in \mathcal{H}$. It gives $G_1 \times G_2 \times \cdots \times G_n$ is a strong self-similar group of $\mathcal{H}$-contraction.

The above Theorem 4.3 shows that, finite product of strong self-similar groups of $\mathcal{H}$-contraction is also a strong self-similar group of $\mathcal{H}$-contraction.

PROPOSITION 4.2   *A self-similar group of $\mathcal{H}$-contraction is a disconnected set.*

Proof Let G be a self-similar group of $\mathcal{H}$-contraction. Then G is a topological fuzzy group. Proposition 4.1 shows that G is the attractor of the $\mathcal{H}$-IFS $\{\phi_0, \ldots, \phi_{n-1}\}$. Hence,

$$G = \phi_0(G) \cup \phi_1(G) \cup \cdots \cup \phi_{n-1}(G)$$
$$\Phi = \phi_i(G) \cap \phi_j(G)$$

for all $i, j \in \{0, 1, 2, \ldots, n-1\}$ and $i \neq j$. For every $i = 1, 2, \ldots, n-1$, the mappings $\phi_i: G \longrightarrow \phi_i(G)$ are $\mathcal{H}$-contraction with respect to $\eta \in \mathcal{H}$. $\mathcal{H}$-contraction mapping is t-uniform continuous. Further, t-uniformly continuous function maps compact set into compact set. Hence, $\phi_i(G)$ is compact subspace in $(G, M, *)$. $\phi_i(G)$ is a closed set in $(G, M, *)$ for all $i \in \{0, 1, 2, \ldots, n-1\}$, since $(G, M, *)$ is Hausdroff space. Due to the fact that

$$G = \phi_0(G) \cup [\phi_1(G) \cup \cdots \cup \phi_{n-1}(G)]$$
$$\Phi = \phi_0(G) \cap [\phi_1(G) \cup \cdots \cup \phi_{n-1}(G)])$$

hence G can be written as disjoint union of non-empty closed sets $\phi_0(G), [\phi_1(G) \cup \cdots \cup \phi_{n-1}(G)]$. That is, G is a disconnected set.

PROPOSITION 4.3   *A strong self-similar group of $\mathcal{H}$-contraction is a totally disconnected set.*

Proof Let G be a strong self-similar group of $\mathcal{H}$-contraction. Proposition 4.1 shows that G is the attractor of a $\mathcal{H}$-IFS $\{\phi_0, \ldots, \phi_{n-1}\}$. Since $\phi_0: G \longrightarrow H$ is one-to-one, we get

$$\phi_i(g) = \phi_i(h)$$
$$\phi_0(g) * x_i = \phi_0(h) * x_i$$
$$\phi_0(g) = \phi_0(h)$$
$$g = h$$

for all $g, h \in G$. This shows that ϕ_i is one-to-one for $i = 1, 2, \ldots, n-1$. In addition that, $\phi_i(G) \cap \phi_j(G) = \Phi$, for all $i, j \in \{0, 1, 2, \ldots, n-1\}$ and $i \neq j$. As per the Theorem 3.3, G is a totally disconnected set.

The following theorem illuminates connections between fuzzy profinite group and strong self-similar group of $\mathcal{H}$-contraction.

THEOREM 4.4   *A strong self-similar group of $\mathcal{H}$-contraction is a fuzzy profinite group.*

Proof   Let A be a strong self-similar group of $\mathcal{H}$-contraction. By the Definition 4.5, A is a compact topological fuzzy group and also A is Hausdorff since every fuzzy metric space is Hausdorff. Proposition 4.3 shows that A is a totally disconnected set. Finally, we get A is compact, Hausdorff, and totally disconnected. Thus, we have the properties which characterize profinite groups. This shows that a strong self-similar group of $\mathcal{H}$-contraction is a profinite group.

Remark 4.1   As per the Remark 3.1, $\mathcal{H}$-contraction is the generalization of fuzzy contraction. Hence, the above theorems and propositions are all true in the sense of the self-similar group of fuzzy contraction, in this case, $\eta \in \mathcal{H}$ of the form $\eta(t) = \frac{1}{t} - 1, t \in (0, 1]$. We illustrate our result by the following examples.

Example 4.1   Consider the continuous t-norm $a * b = ab$. Given a finite group G with fuzzy metric induced by discrete metric. Since G is a discrete topological space, it is a compact topological group with respect to the fuzzy metric. Let $|G| = m$ and $H = \{e\}$, where e is the identity element of G. So, it is clear that $[G:H] = m$. Moreover, the mapping

$$\phi: G \to H$$
$$g \mapsto e$$

is a surjective group homomorphism. If $\eta(t) = \frac{1}{t} - 1, t \in (0, 1]$, then ϕ is fuzzy contraction. As a result, G is a self-similar group but not a strong self-similar group. Since, no finite group is isomorphic to its proper subgroup.

4.1. Construction of fuzzy self-similar group

Example 4.2   Self-similar group of fuzzy contraction on Cantor set
Consider the direct product group

$$G = \mathbb{Z}/2\mathbb{Z} \times \mathbb{Z}/2\mathbb{Z} \times \mathbb{Z}/2\mathbb{Z} \times \cdots$$

Define the fuzzy metric on G by

$$M_d(x, y, t) = \frac{t}{t + d(x, y)}$$

where $d(x, y) = 2 \left| \sum_{i=1}^{\infty} \frac{x_i - y_i}{3^i} \right|$ for all $x, y \in G$ and $t > 0$. G is compact topological group with respect to the fuzzy metric defined above. Consider $\eta(t) = \frac{1}{t} - 1$, define the $\mathcal{H}$-contraction mappings with respect to η as follows:

$$f_0 : \mathbb{Z}/2\mathbb{Z} \times \mathbb{Z}/2\mathbb{Z} \times \mathbb{Z}/2\mathbb{Z} \times \cdots \to \mathbb{Z}/2\mathbb{Z} \times \mathbb{Z}/2\mathbb{Z} \times \mathbb{Z}/2\mathbb{Z} \times \cdots$$
$$(x_1, x_2, x_3, \ldots) \mapsto (0, 0, x_2, x_3, \ldots)$$
$$f_1 : \mathbb{Z}/2\mathbb{Z} \times \mathbb{Z}/2\mathbb{Z} \times \mathbb{Z}/2\mathbb{Z} \times \cdots \to \mathbb{Z}/2\mathbb{Z} \times \mathbb{Z}/2\mathbb{Z} \times \mathbb{Z}/2\mathbb{Z} \times \cdots$$
$$(x_1, x_2, x_3, \ldots) \mapsto (0, 1, x_2, x_3, \ldots)$$
$$f_2 : \mathbb{Z}/2\mathbb{Z} \times \mathbb{Z}/2\mathbb{Z} \times \mathbb{Z}/2\mathbb{Z} \times \cdots \to \mathbb{Z}/2\mathbb{Z} \times \mathbb{Z}/2\mathbb{Z} \times \mathbb{Z}/2\mathbb{Z} \times \cdots$$
$$(x_1, x_2, x_3, \ldots) \mapsto (1, 0, x_2, x_3, \ldots)$$
$$f_3 : \mathbb{Z}/2\mathbb{Z} \times \mathbb{Z}/2\mathbb{Z} \times \mathbb{Z}/2\mathbb{Z} \times \cdots \to \mathbb{Z}/2\mathbb{Z} \times \mathbb{Z}/2\mathbb{Z} \times \mathbb{Z}/2\mathbb{Z} \times \cdots$$
$$(x_1, x_2, x_3, \ldots) \mapsto (1, 1, x_2, x_3, \ldots)$$

Then, G is the attractor of the FIFS $\{G; f_0, f_1, f_2, f_3\}$.

$$f_0(G) = \{0\} \times \{0\} \times \mathbb{Z}/2\mathbb{Z} \times \cdots \times \mathbb{Z}/2\mathbb{Z} \times \cdots$$

f_0 is a surjective group homomorphism but it is not one-to-one. Therefore, G is a fuzzy self-similar group of $\mathcal{H}$-contraction. The $\mathcal{H}$-contraction mappings f_0, f_1, f_2, f_3 are defined on the Cantor set C by

$$f_0 : C \to C$$

$$x \mapsto f_0(x) = \begin{cases} \frac{x}{3} & \text{for } 0 \le x \le \frac{1}{3} \\ \frac{1}{3}\left(x - \frac{2}{3}\right) & \text{for } \frac{2}{3} \le x \le 1 \end{cases}$$

$$f_1 : C \to C$$

$$x \mapsto f_1(x) = \begin{cases} \frac{x}{3} + \frac{2}{9} & \text{for } 0 \le x \le \frac{1}{3} \\ \frac{1}{3}\left(x - \frac{2}{3}\right) + \frac{2}{9} & \text{for } \frac{2}{3} \le x \le 1 \end{cases}$$

$$f_2 : C \to C$$

$$x \mapsto f_2(x) = \begin{cases} \frac{x}{3} + \frac{2}{3} & \text{for } 0 \le x \le \frac{1}{3} \\ \frac{1}{3}\left(x - \frac{2}{3}\right) + \frac{2}{3} & \text{for } \frac{2}{3} \le x \le 1 \end{cases}$$

$$f_3 : C \to C$$

$$x \mapsto f_3(x) = \begin{cases} \frac{x}{3} + \frac{8}{9} & \text{for } 0 \le x \le \frac{1}{3} \\ \frac{1}{3}\left(x - \frac{2}{3}\right) + \frac{8}{9} & \text{for } \frac{2}{3} \le x \le 1 \end{cases}$$

That is, the mappings $f_0, f_1, f_2,$ and f_3 are fuzzy $\mathcal{H}$-contractions with contraction constant $\frac{1}{3}$ and translations $0, \frac{2}{9}, \frac{2}{3}, \frac{8}{9}$, respectively.

Example 4.3   Strong self-similar group of fuzzy contraction on Cantor set

Consider the compact topological group $G = \mathbb{Z}/2\mathbb{Z} \times \mathbb{Z}/2\mathbb{Z} \times \cdots$ with the fuzzy metric defined in the previous example.

Define $f_0 : G \to G$ by $(x_1, x_2, x_3, \ldots) \mapsto (0, x_1, x_2, \ldots)$

$f_1 : G \to G$ by $(x_1, x_2, x_3, \ldots) \mapsto (1, x_1, x_2, \ldots)$. f_0, f_1 are fuzzy $\mathcal{H}$-contractions with respect to $\eta(t) = \frac{1}{t} - 1$ and G is the attractor of the IFS $\{G; f_0, f_1\}$. Moreover,

$$f_0(G) = \{0\} \times \mathbb{Z}/2\mathbb{Z} \times \cdots \times \mathbb{Z}/2\mathbb{Z} \times \cdots$$

is a subgroup of G and f_0 is an isomorphism. This shows that G is a strong fuzzy self-similar group.

Funding

The research work has been supported by University Grants Commission, Government of India, New Delhi, India under the schemes of UGC—Major Research Project with [grant number F.No. 42-21/2013] (SR)/dated 12.03.2013 and UGC-SAP.

Author details

R. Uthayakumar[1]
E-mail: uthayagri@gmail.com
A. Gowrisankar[1]
E-mail: gowrisankargri@gmail.com
[1] Department of Mathematics, The Gandhigram Rural Institute – Deemed University, Gandhigram – 624 302, Dindigul, Tamil Nadu, India.

References

Barnsley, M. F. (1993). *Fractals everywhere* (2nd ed.). Boston, MA: Academic Press.

Demir, B., & Saltan, M. (2012). A self-similar group in the sense of iterated function system. *Far East Journal of Mathematical Science, 60*, 83–99.

Dixon, J. D., Du Sautoy, M. P. F., Mann, A., & Segal, D. (1999). *Analytic Pro-p groups*. Cambridge: Cambridge University Press.

Easwaramoorthy, D., & Uthayakumar, R. (2011). Analysis on fractals in Fuzzy metric spaces, fractals. *World Scientific, 19*, 379–386.

Farnoosh, R., Aghajani, A., & Azhdari, P. (2009). Contraction theorems in fuzzy metric space. *Chaos, Solitons and Fractals 41*

George, A., & Veeramani, P. (1994). On some result in fuzzy metric spaces. *Fuzzy Sets and Systems, 64*, 395–399.

George, A., & Veeramani, P. (1997). On some results of analysis for fuzzy metric spaces. *Fuzzy Sets and Systems, 90*, 365–368.

Grabiec, M. (1988). Fixed points in fuzzy metric spaces. *Fuzzy Sets and Systems, 27*, 385–389.

Gregori, V., & Sapena, A. (2002). On fixed -- point theorems in fuzzy metric spaces. *Fuzzy Sets and Systems, 125*, 245–252.

Hutchinson, J. E. (1981). Fractals and self similarity. *Indiana University Mathematics Journal, 30*, 713–747.

Kramosil, O., & Michalek, J. (1975). Fuzzy metric and statistical metric space. *Kybernetika, 11*, 326–334.

Mandelbrot, B. B. (1983). *The fractal geometry of nature*. New York, NY: W.H. Freeman and Company.

Mihet, D. (2007). On fuzzy contractive mappings in fuzzy metric spaces. *Fuzzy Sets and Systems, 158*, 915–921.

Rodriguez-Lopez, J., & Romaguera, S. (2004). The Hausdorff fuzzy metric on compact sets. *Fuzzy Sets and Systems, 147*, 273–283.

Romaguera, S., & Sanchis, M. (2001). On fuzzy metric groups. *Fuzzy Sets and Systems, 124*, 109–115.

Saltan, M., & Demir, B. (2013). Self-similar groups in the sense of an iterated function system and their properties. *Jouranl of Mathematical Analysis and Applications, 408*, 694–704.

Uthayakumar, R., & Easwaramoorthy, D. (2011). Hutchinson--Barnsley Operator in Fuzzy metric spaces. *World Academy of Science, Engineering and Technology, 56*, 1372–1376.

Uthayakumar, R., & Gowrisankar, A. (2014). Fractals in product fuzzy metric space, fractals, wavelets, and their applications. *Springer proceedings in Mathematics & Statistics, 92*, 157–164.

Wardowski, D. (2013). Fuzzy contractive mapping and fixed points in fuzzy metric spaces. *Fuzzy Sets and Systems, 222*, 108–114.

Zadeh, L. A. (1965). Fuzzy sets. *Information and control, 8*, 338–353.

12

Direct product of general intuitionistic fuzzy sets of subtraction algebras

Muhammad Gulistan[1]*, Shah Nawaz[1] and Syed Zaheer Abbas[1]

*Corresponding author: Muhammad Gulistan, Department of Mathematics, Hazara University, Mansehra, Pakistan
E-mail: gulistanmath@hu.edu.pk

Reviewing editor: Hari M. Srivastava, University of Victoria, Canada

Abstract: We define direct product of $(\in, \in \vee q_k)$-intuitionistic fuzzy sets and direct product of $(\in, \in \vee q_k)$-intuitionistic fuzzy soft sets of subtraction algebras and investigate some related properties.

Subjects: Advanced Mathematics; Algebra; Mathematics & Statistics; Science

Keywords: subtraction algebras; direct product; an $(\in, \in \vee q_k)$-intuitionistic fuzzy soft subalgebras; an $(\in, \in \vee q_k)$-intuitionistic fuzzy soft ideals

AMS Subject Classifications: 06F35; 03G25; 08A72

1. Introduction

The system $(X; \circ, \backslash)$ by Schein (1992), is a set of functions closed under the composition "∘" under the composition of function(and hence $(X, \circ)$ is a function semigroup) and the set theoretical subtraction "$\backslash$" (and hence $(X, \backslash)$ is a subtraction algebra in the sense of Abbot (1969)). He proved that every subtraction semigroup is isomorphic to a difference semigroup of invertible functions. Zelinka (1995) discussed a problem proposed by B. M. Schein concerning the structure of multiplication in a subtraction semigroup. He solved the problem for subtraction algebras of a special type called the atomic subtraction algebras. Jun, Kim, and Roh (2005) introduced the notion of ideals in subtraction algebras and discussed characterization of ideals. To study more about subtraction algebras see Ceven (2009), Jun and Kim (2007). The fuzzifications of ideals in subtraction algebras were discussed in Lee and Park (2007).

ABOUT THE AUTHORS

Dr Muhammad Gulistan, Works as an assistant professor in the Department of Mathematics, Hazara University Mansehra, Pakistan. He published more than 25 research papers in the field of fuzzy sets, cubic sets and hyper structures.

Shah Nawaz is the M.Phil's student of Dr Muhammad Gulistan. He is working in the same filed.

Syed Zaheer Abbas works as an assistant professor in the Department of Mathematics, Hazara University Mansehra, Pakistan. His field of interest is fuzzy sets and abstract algebras.

PUBLIC INTEREST STATEMENT

Real world is featured with complex phenomenon. As vulnerability is unavoidably included in issues emerge in different fields of life and traditional techniques neglected to handle these sorts of issues. Managing with loose, unverifiable, or defective data was a major assignment for a long time. Numerous models were introduced with a specific end goal to appropriately join instability into framework portrayal; L.A. Zadeh in 1965 presented the thought of a fuzzy set. Zadeh supplanted traditional trademark capacity of established fresh sets which tackles its qualities in {0, 1} by enrollment capacity which tackles its values in shut interim [0, 1]. Be that as it may, it is by all accounts the restricted case so this was summed up by K.T. Atanassov in 1986. Soft sets are additionally considered an exceptionally convenient device with a specific end goal to handle loose data. Here, we used a combination of soft and intutionistic fuzzy sets.

Bhakat and Das (1996), introduced a new type of fuzzy subgroups, that is, the $(\in, \in \vee q)$ fuzzy subgroups. In fact, the $(\in, \in \vee q_k)$ fuzzy subgroup is an important generalization of Rosenfeld's fuzzy subgroup. Shabir et al. characterized semigroups by $(\in, \in \vee q_k)$-fuzzy ideals in Shabir (2010). Gulistan, Shahzad, and Yaqoob (2014) studied $(\in, \in \vee q_k)$-fuzzy KU-ideals of KU-algebras.

Molodtsov (1999) introduced the concept of soft set as a new mathematical tool for dealing with uncertainties that are free from the difficulties that have troubled the usual theoretical approaches. Molodtsov pointed out several directions for the applications of soft sets. At present, works on the soft set theory are progressing rapidly. Maji, Roy, and Biswas (2002) described the application of soft set theory to a decision-making problem. Maji, Biswas, and Roy (2003) also studied several operations on the theory of soft sets. The most appropriate theory for dealing with uncertainties is the theory of fuzzy sets developed by Zadeh (1965).

The notion of fuzzy soft sets, as a generalization of the standard soft sets, was introduced in Maji, Biswas, and Roy (2001a), and an application of fuzzy soft sets in a decision-making problem was presented. In Ahmad and Athar (2009), have introduced arbitrary fuzzy soft union and fuzzy soft intersection. Aygunoglu and Aygun introduced the notion of fuzzy soft group and studied its properties. In Jun, Lee, and Park (2010) have introduced the notion of fuzzy soft BCK/BCI-algebras and (closed) fuzzy soft ideals, and then derived their basic properties. Recently, Yang (2011) have studied fuzzy soft semigroups and fuzzy soft (left, right) ideals, and have discussed fuzzy soft image and fuzzy soft inverse image of fuzzy soft semigroups (ideals)in detail. Recently, Khan et al. gave the idea of $(\in, \in \vee q_k)$-intuitionistic (fuzzy ideals, fuzzy soft ideals) of subtraction algebras (Khan, Davvaz, Yaqoob, Gulistan, & Khalaf, 2015). Gulistan, Khan, Yaqoob, Shahzad, and Ashraf (2016) defined direct product of generalized cubic sets in Hv-LA-semigroups. Yaqoob (yaqoob), studied interval-valued intuitionistic fuzzy ideals of regular LA-semigroups. In Akram and Yaqoob (2013) and Yaqoob, Akram, and Aslam (2013), the authors applied the concept of intuitionistic fuzzy soft sets to ordered ternary semigroups and groups. Also see Aslam, Abdullah, Davvaz, and Yaqoob (2012), Khan, Jun, Gulistan, and Yaqoob (2015), Khan, Yousafzai, Khan, and Yaqoob (2013), Yaqoob, Aslam, Davvaz, and Ghareeb (2014), Yaqoob, Chinram, Ghareeb, and Aslam (2013), Yaqoob, Mostafa, and Ansari (2013), Yousafzai, Yaqoob, and Ghareeb (yousafzai), Yousafzai, Yaqoob, and Hila (2012).

The aim of this article is to study the concept of Direct product of $(\in, \in \vee q_k)$-intuitionistic fuzzy sets and Direct product of $(\in, \in \vee q_k)$-intuitionistic fuzzy soft sets of subtraction algebras and investigate some related properties.

2. Preliminaries
In this section we recall some of the basic concepts of subtraction algebra which will be very helpful in further study of the paper. Throughout the paper X denotes the subtraction algebra unless otherwise specified.

Definition 2.1   (Aygunoglu & Aygun, 2009) A nonempty set X together with a binary operation "-" is said to be a subtraction algebra if it satisfies the following:

(S_1) $x - (y - x) = x$,
(S_2) $x - (x - y) = y - (y - x)$,
(S_3) $(x - y) - z = (x - z) - y$, for all $x, y, z \in X$.

The last identity permits us to omit parentheses in expression of the form $(x - y) - z$. The subtraction determines an order relation on X : $a \leq b \Leftrightarrow a - b = 0$, where $0 = a - a$ is an element that does not depend upon the choice of $a \in X$. The ordered set $(X; \leq)$ is a semi-Boolean algebra in the sense of Abbot (1969), that is, it is a meet semi lattice with zero, in which every interval $[0, a]$ is a boolean algebra with respect to the induced order. Here $a \wedge b = a - (a - b)$; the complement of an element $b \in [0, a]$ is $a - b$ and is denoted by b'; and if $b, c \in [0, a]$; then

$b \vee c = \left(b' \wedge c'\right)' = \left((a - b) \wedge (a - c)\right) = a - (a - b) - \left((a - b) - (a - c)\right)$. In a subtraction algebra, the following are true (see Aygunoglu & Aygun, 2009):

$(a1)$ $(x - y) - y = x - y$,
$(a2)$ $x - 0 = x$ and $0 - x = 0$,
$(a3)$ $(x - y) - x = 0$,
$(a4)$ $x - (x - y) \leq y$,
$(a5)$ $(x - y) - (y - x) = x - y$,
$(a6)$ $x - (x - (x - y)) = x - y$,
$(a7)$ $(x - y) - (z - y) \leq x - z$,
$(a8)$ $x \leq y$ if and only if $x = y - w$ for some $w \in X$,
$(a9)$ $x \leq y$ implies $x - z \leq y - z$ and $z - y \leq z - x$, for all $z \in X$,
$(a10)$ $x, y \leq z$ implies $x - y = x \wedge (z - y)$,
$(a11)$ $(x \wedge y) - (x \wedge z) \leq x \wedge (y - z)$,
$(a12)$ $(x - y) - z = (x - z) - (y - z)$.

Definition 2.2   (Aygunoglu & Aygun, 2009) A nonempty subset A of a subtraction algebra X is called an ideal of X, denoted by $A \lhd X$: if it satisfies:

$(b1)$ $a - x \in A$ for all $a \in A$ and $x \in X$,
$(b2)$ for all $a, b \in A$, whenever $a \vee b$ exists in X then $a \vee b \in A$.

Proposition 2.3   (Aygunoglu & Aygun, 2009) A nonempty subset A of a subtraction algebra X is called an ideal of X, if and only if it satisfies:

$(b3)$ $0 \in A$,
$(b4)$ for all $x \in X$ and for all $y \in A, x - y \in A \Rightarrow x \in A$.

Proposition 2.4   (Aygunoglu & Aygun, 2009) Let X be a subtraction algebra and $x, y \in X$. If $w \in X$ is an upper bound for x and y, then the element $x \vee y = w - ((w - y) - x)$ is the least upper bound for x and y.

Definition 2.5   (Aygunoglu & Aygun, 2009) Let Y be a nonempty subset of X; then, Y is called a subalgebra of X if $x - y \in Y$, whenever $x, y \in Y$.

Definition 2.6   (Lee & Park, 2007) Let f be a fuzzy set of X. Then f is called a fuzzy subalgebra of X if it satisfies $(FS) f(x - y) \geq \min\{f(x), f(y)\}$, whenever $x, y \in X$.

Definition 2.7   (Lee & Park, 2007) A fuzzy set f is said to be a fuzzy ideal of X if it satisfies:

$(FI1)$ $f(x - y) \geq f(x)$,
$(FI2)$ If there exists $x \vee y$, then $f(x \vee y) \geq \min\{f(x), f(y)\}$, for all $x, y \in X$.

   Here we mentioned some of the related definitions and results which are directly used in our work. For details we refer the reader Khan et al. (2015).

Definition 2.8   Atanassov (1986) An intuitionistic fuzzy set A in X is an object of the form $A = \{(x, \mu_A(x), \gamma_A(x)): x \in X\}$, where the function $\mu_A: X \rightarrow [0, 1]$ and $\gamma_A: X \rightarrow [0, 1]$ denote the degree of membership and degree of non-membership of each element $x \in X$, and $0 \leq \mu_A(x) + \gamma_A(x) \leq 1$ for all $x \in X$. For simplicity, we will use the symbol $A = (\mu_A, \gamma_A)$ for the intuitionistic fuzzy set $A = \{(x, \mu_A(x), \gamma_A(x)): x \in X\}$. We define $0(x) = 0$ and $1(x) = 1$ for all $x \in X$.

Definition 2.9   (Khan et al., 2015) Let X be a subtraction algebra. An intuitionistic fuzzy set

$A = \{(x, \mu_A(x), \gamma_A(x)): x \in X\}$, of the form

$$x_{(\alpha,\beta)}y = \begin{cases} (\alpha, \beta) & \text{if } y = x \\ (0, 1) & \text{if } y \neq x \end{cases},$$

is said to be an intuitionistic fuzzy point with support x and value (α, β) and is denoted by $x_{(\alpha,\beta)}$. A intuitionistic fuzzy point $x_{(\alpha,\beta)}$ is said to intuitionistic belongs to (resp., intuitionistic quasi-coincident) with intuitionistic fuzzy set $A = \{(x, \mu_A(x), \gamma_A(x)): x \in X\}$ written $x_{(\alpha,\beta)} \in A$ (resp:$x_{(\alpha,\beta)}qA$) if $\mu_A(x) \geq \alpha$ and $\gamma_A(x) \leq \beta$ (resp., $\mu_A(x) + \alpha < 1$ and $\gamma_A(x) + \beta < 1$). By the symbol $x_{(\alpha,\beta)}q_kA$ we mean $\mu_A(x) + \alpha + k > 1$ and $\gamma_A(x) + \beta + k < 1$, where $k \in (0, 1)$.

We use the symbol $x_t \in \mu_A$ implies $\mu_A(x) \geq t$ and $\frac{t}{x}[\in]\gamma_A$ implies $\gamma_A(x) \leq t$, in the whole paper.

Definition 2.10 (Khan et al., 2015) An intuitionistic fuzzy set $A = (\mu_A, \gamma_A)$ of X is said to be an $(\in, \in \vee q_k)$ -intuitionistic fuzzy subalgebra of X if

$$x_{(t_1, t_3)} \in A, y_{(t_2, t_4)} \in A \Rightarrow (x - y)_{(t_1 \wedge t_2, t_3 \vee t_4)} \in \vee q_kA, \text{ for all } x, y \in X, t_1, t_2, t_3, t_4, k \in (0, 1).$$

Definition 2.11 (Khan et al. 2015) An intuitionistic fuzzy set $A = (\mu_A, \gamma_A)$ of X is said to be an $(\in, \in \vee q_k)$ -intuitionistic fuzzy ideal of X if it satisfies the following conditions,

(i) $x_{(t,t)} \in A, y \in X \Rightarrow (x - y)_t \in \vee q_kA$,

(ii) If there exist $x \vee y$, then $x_{(t_1, t_3)} \in A, y_{(t_2, t_4)} \in A \Rightarrow (x \vee y)_{(t_1 \wedge t_2, t_3 \vee t_4)} \in \vee q_kA$, for all $x, y \in X$, $t, t_1, t_2, t_3, t_4, k \in (0, 1)$.

Molodtsov defined the notion of a soft set as follows.

Definition 2.12 (Molodtsov, 1999) A pair (F, A) is called a soft set over U, where F is a mapping given by $F: A \longrightarrow P(U)$. In other words a soft set over U is a parametrized family of subsets of U.

The class of all intuitionistic fuzzy sets on X will be denoted by IF(X).

Definition 2.13 (Maji, Biswas, & Roy, 2001b, 2004) Let U be an initial universe and E be the set of parameters. Let $A \subseteq E$. A pair $(\tilde{F}, A)$ is called an intuitionistic fuzzy soft set over U, where $\tilde{F}$ is a mapping given by $\tilde{F}: A \longrightarrow IF(U)$.

In general, for every $\epsilon \in A$. $\tilde{F}[\epsilon] = \langle \mu_{\tilde{F}[\epsilon]}, \gamma_{\tilde{F}[\epsilon]} \rangle$ is an intuitionistic fuzzy set in U and it is called intuitionistic fuzzy value set of parameter ϵ.

Definition 2.14 (Khan et al., 2015) An intuitionistic fuzzy soft set $\langle \tilde{F}, A \rangle$ of X is called an $(\in, \in \vee q_k)$ -intuitionistic fuzzy soft subalgebra of X, if for all $\epsilon \in A, \tilde{F}[\epsilon] = \langle \mu_{\tilde{F}[\epsilon]}, \gamma_{\tilde{F}[\epsilon]} \rangle$ is an $(\in, \in \vee q_k)$-intuitionistic fuzzy subalgebra of X, if

(i) $\mu_{\tilde{F}[\epsilon]}(x - y) \geq \min\{\mu_{\tilde{F}[\epsilon]}(x), \mu_{\tilde{F}[\epsilon]}(y), \frac{1-k}{2}\}$,

(ii) $\gamma_{\tilde{F}[\epsilon]}(x - y) \leq \max\{\gamma_{\tilde{F}[\epsilon]}(x), \gamma_{\tilde{F}[\epsilon]}(y), \frac{1-k}{2}\}$, for all $x, y \in X$.

Definition 2.15 (Khan et al., 2015) An intuitionistic fuzzy soft set $\langle \tilde{F}, A \rangle$ of X is called an $(\in, \in \vee q_k)$ -intuitionistic fuzzy soft ideal of X, if for all $\epsilon \in A, \tilde{F}[\epsilon] = \langle \mu_{\tilde{F}[\epsilon]}, \gamma_{\tilde{F}[\epsilon]} \rangle$ is an $(\in, \in \vee q_k)$-intuitionistic fuzzy soft ideal of X, if

(i) $\mu_{\tilde{F}_{[e]}}(x-y) \geq \min\{\mu_{\tilde{F}_{[e]}}(x), \frac{1-k}{2}\}$,

(ii) $\gamma_{\tilde{F}_{[e]}}(x-y) \leq \max\{\gamma_{\tilde{F}_{[e]}}(x), \frac{1-k}{2}\}$,

(iii) $\mu_{\tilde{F}_{[e]}}(x \vee y) \geq \min\{\mu_{\tilde{F}_{[e]}}(x), \mu_{\tilde{F}_{[e]}}(y), \frac{1-k}{2}\}$,

(iv) $\gamma_{\tilde{F}_{[e]}}(x \vee y) \leq \max\{\gamma_{\tilde{F}_{[e]}}(x), \gamma_{\tilde{F}_{[e]}}(y), \frac{1-k}{2}\}$, for all $x, y \in X$.

3. Direct product of an $(\in, \in \vee q_k)$-intuitionistic fuzzy subalgebra/ideals

In this section, we define Direct product of $(\in, \in \vee q_k)$-intuitionistic fuzzy sets and investigate some related properties.

Definition 3.1   Let $A = (\mu_A(x), \gamma_A(x))$ and $B = (\mu_B(x), \gamma_B(x))$ be two $(\in, \in \vee q_k)$-intuitionistic fuzzy sets of X_1 and X_2, respectively. Then the Direct product of $A = (\mu_A(x), \gamma_A(x))$ and $B = (\mu_B(x), \gamma_B(x))$ is defined as $A \times B = \langle \mu_{A \times B}, \gamma_{A \times B} \rangle$, where $\mu_{A \times B}(x, y) = \mu_A(x) \wedge \mu_B(y)$ and $\gamma_{A \times B}(x, y) = \gamma_A(x) \vee \gamma_B(y)$ for all $(x, y) \in X_1 \times X_2$.

Definition 3.2   An intuitionistic fuzzy set $A \times B$ of $X_1 \times X_2$ is called an $(\in, \in \vee q_k)$-intuitionistic fuzzy subalgebra of $X_1 \times X_2$ if it satisfies,

(i) $\mu_{A \times B}((x_1, y_1) - (x_2, y_2)) \geq \min\{\mu_{A \times B}(x_1, y_1), \mu_{A \times B}(x_2, y_2), \frac{1-k}{2}\}$,

(ii) $\gamma_{A \times B}((x_1, y_1) - (x_2, y_2)) \leq \max\{\gamma_{A \times B}(x_1, y_1), \gamma_{A \times B}(x_2, y_2), \frac{1-k}{2}\}$.

Example 3.3   Let $X_1 = \{0, a, b\}$ and $X_2 = \{0, a, b, c\}$ be two subtraction algebras with the following Cayley tables

−	0	a	b
0	0	0	0
a	a	0	a
b	b	b	0

and

−	0	a	b	c
0	0	0	0	0
a	a	0	a	0
b	b	b	0	0
c	c	b	a	0

.

Let us define the intuitionistic fuzzy sets $A = (\mu_A(x), \gamma_A(x))$ of X_1 and $B = (\mu_B(x), \gamma_B(x))$ of X_2 as follows

X	0	a	b
$\mu_A(x)$	0.5	0.6	0.7
$\gamma_A(x)$	0.1	0.2	0.21

and

X	0	a	b	c
$\mu_B(x)$	0.6	0.5	0.4	0.3
$\gamma_B(x)$	0.2	0.21	0.23	0.24

.

Then $X_1 \times X_2 = \{(0,0), (0,a), (0,b), (0,c), (a,0), (a,a), (a,b), (a,c), (b,0), (b,a), (b,b), (b,c)\}$ is a subtraction algebra. Now define the direct product $A \times B$ on $X_1 \times X_2$ as $A \times B = \langle \mu_{A \times B}, \gamma_{A \times B} \rangle$ where $\mu_{A \times B}: X_1 \times X_2 \to [0,1]$ and $\gamma_{A \times B}: X_1 \times X_2 \to [0,1]$,

$X_1 \times X_2$	$\mu_{A \times B}$	$\gamma_{A \times B}$
(0, 0)	0.5	0.2
(0, a)	0.5	0.21
(0, b)	0.4	0.23
(0, c)	0.3	0.24
(a, 0)	0.6	0.2
(a, a)	0.5	0.21
(a, b)	0.4	0.23

$X_1 \times X_2$	$\mu_{A\times B}$	$\gamma_{A\times B}$
(a, c)	0.3	0.24
$(b, 0)$	0.6	0.21
(b, a)	0.5	0.21
(b, b)	0.4	0.23
(b, c)	0.3	0.24

then $A \times B$ is an $(\in, \in \vee q_{0.4})$–intuitionistic fuzzy subalgebra of $X_1 \times X_2$.

Definition 3.4 An intuitionistic fuzzy set $A \times B$ of $X_1 \times X_2$ is called an $(\in, \in \vee q_k)$-intuitionistic fuzzy ideal of $X_1 \times X_2$ if it satisfies

(i) $\mu_{A\times B}((x_1, y_1) - (x_2, y_2)) \geq \min\{\mu_{A\times B}(x_1, y_1), \frac{1-k}{2}\}$,

(ii) $\gamma_{A\times B}((x_1, y_1) - (x_2, y_2)) \leq \max\{\gamma_{A\times B}(x_1, y_1), \frac{1-k}{2}\}$,

(iii) $\mu_{A\times B}((x_1, y_1) \vee (x_2, y_2)) \geq \min\{\mu_{A\times B}(x_1, y_1), \mu_{A\times B}(x_2, y_2), \frac{1-k}{2}\}$,

(iv) $\gamma_{A\times B}((x_1, y_1) \vee (x_2, y_2)) \leq \max\{\gamma_{A\times B}(x_1, y_1), \gamma_{A\times B}(x_2, y_2), \frac{1-k}{2}\}$.

Theorem 3.5 Let A and B be two $(\in, \in \vee q_k)$-intuitionistic fuzzy subalgebras of X_1 and X_2, respectively. Then the Direct product $A \times B$ is an $(\in, \in \vee q_k)$-intuitionistic fuzzy subalgebra of $X_1 \times X_2$.

Proof Let A and B be two $(\in, \in \vee q_k)$-intuitionistic fuzzy subalgebras of X_1 and X_2, respectively. For any $(x_1, y_1), (x_2, y_2) \in X_1 \times X_2$. We have

$$\mu_{A\times B}((x_1, y_1) - (x_2, y_2)) = \mu_{A\times B}(x_1 - x_2, y_1 - y_2)$$
$$= \mu_A(x_1 - x_2) \wedge \mu_B(y_1 - y_2)$$
$$\geq \min\{\mu_A(x_1), \mu_A(x_2), \frac{1-k}{2}\} \wedge \min\{\mu_B(y_1), \mu_B(y_2), \frac{1-k}{2}\}$$
$$= \min\{\mu_A(x_1), \mu_B(y_1), \frac{1-k}{2}\} \wedge \min\{\mu_A(x_2), \mu_B(y_2), \frac{1-k}{2}\}$$
$$= \{\mu_{A\times B}(x_1, y_1), \mu_{A\times B}(x_2, y_2), \frac{1-k}{2}\}.$$

Also

$$\gamma_{A\times B}((x_1, y_1) - (x_2, y_2)) = \gamma_{A\times B}(x_1 - x_2, y_1 - y_2)$$
$$= \gamma_A(x_1 - x_2) \vee \gamma_B(y_1 - y_2)$$
$$\leq \max\{\gamma_A(x_1), \gamma_A(x_2), \frac{1-k}{2}\} \vee \{\gamma_B(y_1), \gamma_B(y_2), \frac{1-k}{2}\}$$
$$= \max\{\gamma_A(x_1), \gamma_B(y_1), \frac{1-k}{2}\} \vee \max\{\gamma_A(x_2), \gamma_B(y_2), \frac{1-k}{2}\}$$
$$= \{\gamma_{A\times B}(x_1, y_1), \gamma_{A\times B}(x_2, y_2), \frac{1-k}{2}\}.$$

Hence this shows that $A \times B$ is an $(\in, \in \vee q_k)$-intuitionistic fuzzy subalgebra of $X_1 \times X_2$. □

THEOREM 3.6 Let A and B be two $(\in, \in \vee q_k)$-intuitionistic fuzzy ideals of X_1 and X_2, respectively. Then the Direct product $A \times B$ is an $(\in, \in \vee q_k)$-intuitionistic fuzzy ideal of $X_1 \times X_2$.

Proof Straightforward. □

Proposition 3.7   Every an $(\in, \in \vee q_k)$-intuitionistic fuzzy ideal $A \times B = (\mu_{A \times B}(x), \gamma_{A \times B}(x))$ of $X_1 \times X_2$ satisfies the following,

(i) $\mu_{A \times B}(0, 0) \geq \min\{\mu_{A \times B}(x_1, y_1), \frac{1-k}{2}\}$,

(ii) $\gamma_{A \times B}(0, 0) \leq \max\{\gamma_{A \times B}(x_1, y_1), \frac{1-k}{2}\}$.

Proof   By letting $(x_1, y_1) = (x_2, y_2)$ in conditions (*i*) and (*ii*) in Definition , we get the required proof. □

LEMMA 3.8   *If an* $(\in, \in \vee q_k)$-*intuitionistic fuzzy set* $A \times B = (\mu_{A \times B}(x), \gamma_{A \times B}(x))$ *of* $X_1 \times X_2$ *satisfies the following,*

(i) $\mu_{A \times B}(0, 0) \geq \min\{\mu_{A \times B}(x_1, y_1), \frac{1-k}{2}\}$,

(ii) $\gamma_{A \times B}(0, 0) \leq \max\{\gamma_{A \times B}(x_1, y_1), \frac{1-k}{2}\}$,

(iii) $\mu_{A \times B}((x_1, y_1) - (x_3, y_3)) \geq \min\{\mu_A(((x_1, y_1) - (x_2, y_2)) - (x_3, y_3)), \mu_A(x_2, y_2), \frac{1-k}{2}\}$,

(iv) $\gamma_{A \times B}((x_1, y_1) - (x_3, y_3)) \leq \max\{\gamma_{A \times B}(((x_1, y_1) - (x_2, y_2)) - (x_3, y_3)), \gamma_{A \times B}(x_2, y_2), \frac{1-k}{2}\}$, *then we have* $(x_1, y_1) \leq (a, b) \Rightarrow \mu_{A \times B}(x_1, y_1) \geq \min\{\mu_{A \times B}(a, b), \frac{1-k}{2}\}$ *and* $\gamma_{A \times B}(x) \leq \max\{\gamma_{A \times B}(a, b), \frac{1-k}{2}\}$ *for all* $(a, b), (x_1, y_1), (x_2, y_2), (x_3, y_3) \in X_1 \times X_2$.

Proof   Let $(a, b), (x_1, y_1) \in \in X_1 \times X_2$ and $(x_1, y_1) \leq (a, b)$.

Consider
$$\mu_{A \times B}(x_1, y_1) = \mu_{A \times B}((x_1, y_1) - (0, 0)),$$
$$\geq \min\{\mu_{A \times B}(((x_1, y_1) - (a, b)) - (0, 0)), \mu_{A \times B}(a, b), \frac{1-k}{2}\} \text{ by } (iii),$$
$$= \min\{\mu_{A \times B}(0, 0), \mu_{A \times B}(a, b), \frac{1-k}{2}\},$$
$$= \min\{\mu_{A \times B}(a, b), \frac{1-k}{2}\} \text{ by } (i).$$

Also consider
$$\gamma_{A \times B}(x_1, y_1) = \gamma_{A \times B}((x_1, y_1) - (0, 0)),$$
$$\leq \max\{\gamma_{A \times B}(((x_1, y_1) - (a, b)) - (0, 0)), \gamma_{A \times B}(a, b), \frac{1-k}{2}\} \text{ by}(iv),$$
$$= \max\{\gamma_{A \times B}(0, 0), \gamma_{A \times B}(a, b), \frac{1-k}{2}\},$$
$$= \max\{\gamma_{A \times B}(a, b), \frac{1-k}{2}\} \text{ by}(ii).$$

Definition 3.9   Let $A = (\mu_A(x), \gamma_A(x))$ and $B = (\mu_B(x), \gamma_B(x))$ be intuitionistic fuzzy sets of X_1 and X_2, respectively. Define the intuitionistic level set for the $A \times B$ as $(A \times B)_{(\alpha, \beta)} = \{(x, y) \in X_1 \times X_2 | \mu_{A \times B}(x, y) \geq \alpha, \gamma_{A \times B}(x, y) \leq \beta\}$ where $\alpha \in (0, \frac{1-k}{2}], \beta \in [\frac{1-k}{2}, 1)\}$.

THEOREM 3.10   *Let A and B be two* $(\in, \in \vee q_k)$-*intuitionistic fuzzy subalgebras of* X_1 *and* X_2, *respectively. Then the Direct product* $A \times B$ *is an* $(\in, \in \vee q_k)$ -*intuitionistic fuzzy subalgebra of* $X_1 \times X_2$ *if and only if* $(A \times B)_{(\alpha, \beta)} \neq \Phi$ *is a subalgebra of* $X_1 \times X_2$.

Proof   Straightforward.   □

THEOREM 3.11   *Let A and B be two* $(\in, \in \vee q_k)$-*intuitionistic fuzzy ideals of* X_1 *and* X_2, *respectively. Then the Direct product* $A \times B$ *is an* $(\in, \in \vee q_k)$-*intuitionistic fuzzy ideal of* $X_1 \times X_2$ *if and only if* $(A \times B)_{(\alpha, \beta)} \neq \Phi$ *is an ideal of* $X_1 \times X_2$.

Proof Straightforward. □

4. Direct product of $(\in, \in \vee q_k)$-intuitionistic fuzzy soft subalgebras

In this section, we define Direct product of $(\in, \in \vee q_k)$-intuitionistic fuzzy soft sets and investigate some related properties.

Definition 4.1 Let $\langle \widetilde{F}, A \rangle$ and $\langle \widetilde{G}, B \rangle$ be two $(\in, \in \vee q_k)$-intuitionistic fuzzy soft sets of X_1 and X_2, respectively. Then the Direct product of $(\in, \in \vee q_k)$-intuitionistic fuzzy soft sets $\langle \widetilde{F}, A \rangle$ and $\langle \widetilde{G}, B \rangle$ is defined as

$$\langle \widetilde{F}, A \rangle \otimes \langle \widetilde{G}, B \rangle = (\widehat{\mathcal{U}}, A \times B),$$

where $\widehat{\mathcal{U}}[\epsilon, \varepsilon] = \widetilde{F}[\epsilon] \times \widetilde{G}[\varepsilon] = \left\langle \mu_{\widetilde{F}[\epsilon] \times \widetilde{G}[\varepsilon]}, \gamma_{\widetilde{F}[\epsilon] \times \widetilde{G}[\varepsilon]} \right\rangle$ for all $[\epsilon, \varepsilon] \in A \times B.$

Here $\mu_{\widetilde{F}[\epsilon] \times \widetilde{G}[\varepsilon]}(x, y) = \mu_{\widetilde{F}[\epsilon]}(x) \wedge \mu_{\widetilde{G}[\varepsilon]}(y)$ and $\gamma_{\widetilde{F}[\epsilon] \times \widetilde{G}[\varepsilon]}(x, y) = \gamma_{\widetilde{F}[\epsilon]}(x) \vee \gamma_{\widetilde{G}[\varepsilon]}(y)$ for all $(x, y) \in X_1 \times X_2$ and $[\epsilon, \varepsilon] \in A \times B.$

Definition 4.2 An $(\in, \in \vee q_k)$-intuitionistic fuzzy soft set $\langle \widetilde{F}, A \rangle \otimes \langle \widetilde{G}, B \rangle$ of $X_1 \times X_2$ is called an an $(\in, \in \vee q_k)$-intuitionistic fuzzy soft subalgebra of $X_1 \times X_2$ if it satisfies

(i) $\mu_{\widetilde{F}[\epsilon] \times \widetilde{G}[\varepsilon]}((x_1, y_1) - (x_2, y_2)) \geq \min\{\mu_{\widetilde{F}[\epsilon] \times \widetilde{G}[\varepsilon]}(x_1, y_1), \mu_{\widetilde{F}[\epsilon] \times \widetilde{G}[\varepsilon]}(x_2, y_2), \frac{1-k}{2}\};$

(ii) $\gamma_{\widetilde{F}[\epsilon] \times \widetilde{G}[\varepsilon]}((x_1, y_1) - (x_2, y_2)) \leq \max\{\gamma_{\widetilde{F}[\epsilon] \times \widetilde{G}[\varepsilon]}(x_1, y_1), \gamma_{\widetilde{F}[\epsilon] \times \widetilde{G}[\varepsilon]}(x_2, y_2), \frac{1-k}{2}\}.$

Definition 4.3 An $(\in, \in \vee q_k)$-intuitionistic fuzzy soft set $\langle \widetilde{F}, A \rangle \otimes \langle \widetilde{G}, B \rangle$ of $X_1 \times X_2$ is called an an $(\in, \in \vee q_k)$-intuitionistic fuzzy soft ideal of $X_1 \times X_2$ if it satisfies

(i) $\mu_{\widetilde{F}[\epsilon] \times \widetilde{G}[\varepsilon]}((x_1, y_1) - (x_2, y_2)) \geq \min\{\mu_{\widetilde{F}[\epsilon] \times \widetilde{G}[\varepsilon]}(x_1, y_1), \frac{1-k}{2}\},$

(ii) $\gamma_{\widetilde{F}[\epsilon] \times \widetilde{G}[\varepsilon]}((x_1, y_1) - (x_2, y_2)) \leq \max\{\gamma_{\widetilde{F}[\epsilon] \times \widetilde{G}[\varepsilon]}(x_1, y_1), \frac{1-k}{2}\},$

(iii) $\mu_{\widetilde{F}[\epsilon] \times \widetilde{G}[\varepsilon]}((x_1, y_1) \vee (x_2, y_2)) \geq \min\{\mu_{\widetilde{F}[\epsilon] \times \widetilde{G}[\varepsilon]}(x_1, y_1), \mu_{\widetilde{F}[\epsilon] \times \widetilde{G}[\varepsilon]}(x_2, y_2), \frac{1-k}{2}\},$

(iv) $\gamma_{\widetilde{F}[\epsilon] \times \widetilde{G}[\varepsilon]}((x_1, y_1) \vee (x_2, y_2)) \leq \max\{\gamma_{\widetilde{F}[\epsilon] \times \widetilde{G}[\varepsilon]}(x_1, y_1), \gamma_{\widetilde{F}[\epsilon] \times \widetilde{G}[\varepsilon]}(x_2, y_2), \frac{1-k}{2}\}.$

THEOREM 4.4 Let $\langle \widetilde{F}, A \rangle$ and $\langle \widetilde{G}, B \rangle$ be two $(\in, \in \vee q_k)$-intuitionistic fuzzy soft subalgebras of X_1 and X_2, respectively. Then the Direct product $\langle \widetilde{F}, A \rangle \otimes \langle \widetilde{G}, B \rangle$ is an $(\in, \in \vee q_k)$-intuitionistic fuzzy soft subalgebra of $X_1 \times X_2$.

Proof Let $\langle \widetilde{F}, A \rangle$ and $\langle \widetilde{G}, B \rangle$ be two $(\in, \in \vee q_k)$-intuitionistic fuzzy soft groups of X_1 and X_2, respectively. For any $(x_1, y_1), (x_2, y_2) \in X_1 \times X_2$ and $[\epsilon, \varepsilon] \in A \times B$. We have

$$\mu_{\widetilde{F}[\epsilon] \times \widetilde{G}[\varepsilon]}((x_1, y_1) - (x_2, y_2)) = \mu_{\widetilde{F}[\epsilon] \times \widetilde{G}[\varepsilon]}(x_1 - x_2, y_1 - y_2)$$

$$= \mu_{\widetilde{F}[\epsilon]}(x_1 - x_2) \wedge \mu_{\widetilde{G}[\varepsilon]}(y_1 - y_2)$$

$$\geq \min\{\mu_{\widetilde{F}[\epsilon]}(x_1), \mu_{\widetilde{F}[\epsilon]}(x_2), \frac{1-k}{2}\} \wedge \min\{\mu_{\widetilde{G}[\varepsilon]}(y_1), \mu_{\widetilde{G}[\varepsilon]}(y_2), \frac{1-k}{2}\}$$

$$= \min\{\mu_{\widetilde{F}[\epsilon]}(x_1), \mu_{\widetilde{G}[\varepsilon]}(y_1), \frac{1-k}{2}\} \wedge \min\{\mu_{\widetilde{F}[\epsilon]}(x_2), \mu_{\widetilde{G}[\varepsilon]}(y_2), \frac{1-k}{2}\}$$

$$= \{\mu_{\widetilde{F}[\epsilon] \times \widetilde{G}[\varepsilon]}(x_1, y_1), \mu_{\widetilde{F}[\epsilon] \times \widetilde{G}[\varepsilon]}(x_2, y_2), \frac{1-k}{2}\}.$$

Also

$$\gamma_{\widetilde{F}[\epsilon]\times\widetilde{G}[\epsilon]}((x_1,y_1)-(x_2,y_2)) = \gamma_{\widetilde{F}[\epsilon]\times\widetilde{G}[\epsilon]}(x_1-x_2,y_1-y_2)$$

$$= \gamma_{\widetilde{F}[\epsilon]}(x_1-x_2) \vee \gamma_{\widetilde{G}[\epsilon]}(y_1-y_2)$$

$$\leq \max\{\gamma_{\widetilde{F}[\epsilon]}(x_1),\gamma_{\widetilde{F}[\epsilon]}(x_2),\frac{1-k}{2}\} \vee \{\gamma_{\widetilde{G}[\epsilon]}(y_1),\gamma_{\widetilde{G}[\epsilon]}(y_2),\frac{1-k}{2}\}$$

$$= \max\{\gamma_{\widetilde{F}[\epsilon]}(x_1),\gamma_{\widetilde{G}[\epsilon]}(y_1),\frac{1-k}{2}\} \vee \max\{\gamma_{\widetilde{F}[\epsilon]}(x_2),\gamma_{\widetilde{G}[\epsilon]}(y_2),\frac{1-k}{2}\}$$

$$= \{\gamma_{\widetilde{F}[\epsilon]\times\widetilde{G}[\epsilon]}(x_1,y_1),\gamma_{\widetilde{F}[\epsilon]\times\widetilde{G}[\epsilon]}(x_2,y_2),\frac{1-k}{2}\}.$$

Hence this shows that $\langle\widetilde{F},A\rangle \otimes \langle\widetilde{G},B\rangle$ is an $(\in,\in \vee q_k)$-intuitionistic fuzzy soft subalgebra of $X_1 \times X_2$. □

THEOREM 4.5   Let $\langle\widetilde{F},A\rangle$ and $\langle\widetilde{G},B\rangle$ be two $(\in,\in \vee q_k)$-intuitionistic fuzzy soft ideals of X_1 and X_2, respectively. Then the Direct product $\langle\widetilde{F},A\rangle \otimes \langle\widetilde{G},B\rangle$ is an $(\in,\in \vee q_k)$-intuitionistic fuzzy soft ideal of $X_1 \times X_2$.

Proof   Straightforward.   □

5. Conclusion

In this paper we established some new results related to the direct product of $(\in,\in \vee q_k)$-intuitionistic fuzzy sets and direct product of $(\in,\in \vee q_k)$-intuitionistic fuzzy soft sets of subtraction algebras. We investagated several related properties.

Funding
The authors received no direct funding for this research.

Author details
Muhammad Gulistan[1]
E-mail: gulistanmath@hu.edu.pk
Shah Nawaz[1]
E-mail: shahnawazawan82@gmail.com
Syed Zaheer Abbas[1]
E-mail: zaheer@hu.edu.pk
[1] Department of Mathematics, Hazara University, Mansehra, Pakistan.

References
Abbot, J. C. (1969). *Sets, lattices and Boolean algebra.* Boston, MA: Allyn and Bacon.
Akram, M., & Yaqoob, N. (2013). Intuitionistic fuzzy soft ordered ternary semigroups. *International Journal of Pure and Applied Mathematics, 84,* 93–107.
Aslam, M., Abdullah, S., Davvaz, B., & Yaqoob, N. (2012). Rough M-hypersystems and fuzzy M-hypersystems in Γ-semihypergroups. *Neural Computing and Applications, 21,* 281–287.
Atanassov, K. T. (1986). Intuitionistic fuzzy sets. *Fuzzy Sets and System, 20,* 87–96.
Aygunoglu, A., & Aygun, H. (2009). Introduction to fuzzy soft groups. *Computers and Mathematics with Applications, 58,* 1279–1286.
Bhakat, S. K., & Das, P. (1996). (∈, ∈, Vq) fuzzy subgroups. *Fuzzy Sets and Systems, 80,* 359–368.
Ceven, Y., & Ozturk, M. A. (2009). Some results on subtraction algebras. *Hacettepe Journal of Mathematics and Statistics, 38,* 299–304.
Gulistan, M., Khan, M., Yaqoob, N., Shahzad, M., & Ashraf, U. (2016). Direct product of generalized cubic sets in Hv-LA-semigroups. *Science International (Lahore), 28,* 767–779.

Gulistan, M., Shahzad, M., & Yaqoob, N. (2014). On (∈, ∈, Vqk) -fuzzy KU-ideals of KU-algebras. *Acta Universitatis Apulensis, 39,* 75–83.
Jun, Y. B., & Kim, H. S. (2007). On ideals in subtraction algebras. *Scientiae Mathematicae Japonicae, 65,* 129–134.
Jun, Y. B., Kim, H. S., & Roh, E. H. (2005). Ideal theory of subtraction algebras. *Scientiae Mathematicae Japonicae, 61,* 459–464.
Jun, Y. B., Lee, K. J., & Park, C. H. (2010). Fuzzy soft set theory applied to BCK/BCI-algebras. *Computers and Mathematics with Applications, 59,* 3180–3192.
Khan, M., Davvaz, B., Yaqoob, N., Gulistan, M., & Khalaf, M. M. (2015). On (∈, ∈, Vqk)-intuitionistic (fuzzy ideals, fuzzy soft ideals) of subtraction algebras. *Songklanakarin Journal of Science and Technology, 37,* 465–475.
Khan, M., Jun, Y. B., Gulistan, M., & Yaqoob, N. (2015). The generalized version of Jun's cubic sets in semigroups. *Journal of Intelligent and Fuzzy Systems, 28,* 947–960.
Khan, A., Yousafzai, F., Khan, W., & Yaqoob, N. (2013). Ordered LA-semigroups in terms of interval valued fuzzy ideals. *Journal of Advanced Research in Pure Mathematics, 5,* 100–117.
Lee, K. J., & Park, C. H. (2007). Some questions on fuzzifications of ideals in subtraction algebras. *Communications of the Korean Mathematical Society, 22,* 359–363.
Maji, P. K., Biswas, R., & Roy, A. R. (2003). Soft set theory. *Computers and Mathematics with Applications, 45,* 555–562.
Maji, P. K., Biswas, R., & Roy, A. R. (2001a). Fuzzy soft sets. *Journal of Fuzzy Mathematics, 9,* 589–602.
Maji, P. K., Biswas, R., & Roy, A. R. (2001b). Intuitionistic fuzzy soft sets. *Journal of Fuzzy Mathematics, 9,* 677–692.
Maji, P. K., Roy, A. R., & Biswas, R. (2002). An application of soft sets in a decision making problem. *Computers and Mathematics with Applications, 44,* 1077–1083.
Maji, P. K., Roy, A. R., & Biswas, R. (2004). On intuitionistic fuzzy soft sets. *Journal of Fuzzy Mathematics, 12,* 669–683.
Molodtsov, D. (1999). Soft set theory-first results. *Computers and Mathematics with Applications, 37,* 19–31.
Schein, B. M. (1992). Difference semigroups. *Communications in Algebra, 20,* 2153–2169.
Shabir, M., Jun, Y. B., & Nawaz, Y. (2010). Semigroups characterized by (∈, ∈, Vqk)-fuzzy ideals. *Computers and Mathematics with Applications, 60,* 1473–1493.

Yang, C. (2011). Fuzzy soft semigroups and fuzzy soft ideals. *Computers and Mathematics with Applications, 61,* 255–261.

Yaqoob, N. (2013). Interval-valued intuitionistic fuzzy ideals of regular LA-semigroups. *Thai Journal of Mathematics, 11,* 683–695.

Yaqoob, N., Akram, M., & Aslam, M. (2013). Intuitionistic fuzzy soft groups induced by (t, s)-norm. *Indian Journal of Science and Technology, 6,* 4282–4289.

Yaqoob, N., Aslam, M., Davvaz, B., & Ghareeb, A. (2014). Structures of bipolar fuzzy Γ-hyperideals in Γ-semihypergroups. *Journal of Intelligent and Fuzzy Systems, 27,* 3015–3032.

Yaqoob, N., Chinram, R., Ghareeb, A., & Aslam, M. (2013). Left almost semigroups characterized by their interval valued fuzzy ideals. *Afrika Matematika, 24,* 231–245.

Yaqoob, N., Mostafa, S. M., & Ansari, M. A. (2013). On cubic KU-ideals of KU-algebras. *ISRN Algebra, 10,* Article ID: 935905.

Yousafzai, F., Yaqoob, N., & Ghareeb, A. (2013). Left regular AG-groupoids in terms of fuzzy interior ideals. *Afrika Matematika, 24,* 577–587.

Yousafzai, F., Yaqoob, N., & Hila, K. (2012). On fuzzy (2,2)-regular ordered Γ-AG**-groupoids. *UPB Scientific Bulletin, Series A, 74,* 87–104.

Zadeh, L. A. (1965). Fuzzy sets. *Informations and Control, 8,* 338–353.

Zelinka, B. (1995). Subtraction semigroup. *Mathematica Bohemica, 120,* 445–447.

13

Injective module based on rough set theory

Arvind Kumar Sinha[1]* and Anand Prakash[1]

*Corresponding author: Arvind Kumar Sinha, Department of Mathematics, National Institute of Technology Raipur, Raipur 492010, India

E-mail: aksinha.maths@nitrr.ac.in

Reviewing editor: Xinguang Zhang, Curtin University, Australia

Abstract: It is important to handle real-life problems algebraically, but as most of the real-life problems as well as the applications are imprecise(i.e. vague, inexact, or uncertain), it makes harder to analyze algebraically, while Rough Set Theory (RST) has the capability to deal imprecise problems. To solve imprecise problems algebraically, we investigated injective module based on RST.

Subjects: Advanced Mathematics; Algebra; Mathematics & Statistics; Science

Keywords: algebra; injective module; rough set theory; rough module

1. Introduction

The terminology injective module was originated by Carten and Eilenberg (1956) to deal real-life situations algebraically; and then the dual concept projective module and injective module have been covered in many texts (Goldhaber & Enrich, 1970; Rebenboim, 1969; Rowen, 1991). These terms are based on crisp set theory and can handle only exact situations. In recent years, most data-sets are imprecise or the surrounding information is imprecise and our way of thinking or concluding depends on the information at our disposal. This means that to draw conclusions, we should able to process uncertain and/or incomplete information. To analyze any type of information, mathematical logics are most appropriate, so we should have to generalize the algebraic structures and the logic in sense of imprecise or vague. Rough set theory (RST) is a powerful mathematical tool to handle imprecise situations and rough algebraic structures can play a vital role to deal such situations.

In Pawlak's RST, the key concept is an equivalence relation and the building blocks for the construction of the lower and upper approximations are the equivalence classes. The lower approximation of the given set is the union of all the equivalence classes which are the subsets of the set, and the upper approximation is the union of all the equivalence classes which have a non-empty intersection with the set. The object of the given universe can be divided into three classes with respect to any subset $A \subseteq U$:

(1) the objects which are definitely in A;

(2) the objects which are definitely not in A; and

(3) the objects which are possibly in A.

ABOUT THE AUTHOR

Arvind Kumar Sinha is an assistant professor in the Department of Mathematics at National Institute of Technology Raipur, Chhattisgarh, India. He received his MSc and PhD degrees in mathematics from Guru Ghasidas University, Bilaspur (A Central University), India, in 1995 and 2003, respectively. He has 15 years of teaching experience at graduate and post graduate levels and he has several national and international publications. His research area is algebra.

PUBLIC INTEREST STATEMENT

Algebra & Logics are most important to solve problems, but they are based on crisp set theory. Some authors investigated fuzziness in algebra. In view of uncertain data, we investigated injective module based on rough set theory. This work is in direction to handle imprecise situations algebraically. We hope these results will further enrich algebra to cover uncertain situations.

The objects in class (1) form the lower approximation of A, and the objects in classes (1) and (3) together form its upper approximation. The boundary of A is defined as the set of objects in class (2). Bonikowaski introduced the algebraic structures of rough sets Bonikowaski (1995). Biswas and Nanda (1994) introduced the concept of rough group and rough subgroups. Kuroki (1997) studied the rough ideals in semigroups. Davvaz (2004) introduced the roughness in rings. Davvaz and Mahdavipour (2006) introduced the roughness in module. Rough modules and their some properties are also studied by Zhang, Fu, and Zhao (2006). Standard sources for the algebraic theory of modules are Anderson and Fuller (1992), Jacobson (1951). One can find more on rough set and their algebraic structures in Davvaz and Mahdavipour (2006), Walczak and Massart (1997), Han (2001), Chakraborty and Banergee (1994), Kuroki and Mordeson (1997), Yao (1996), Pawlak (1984,1987). In recent years, there has been a fast growing interest in this new emerging theory, ranging from work in pure theory, such as algebraic foundations and mathematical logic (Irfan Ali, Davvaz, & Shabir, 2013; Li & Zhang, 2014; Rasouli & Davvaz, 2014; Xin, Hua, & Zhu, 2014) to diverse areas of applications. Recently, authors A.K. Sinha and Anand Prakash discussed on rough free module and rough projective module in Sinha and Prakash (2014) and Prakash and Sinha (2014), respectively.

The aim of this paper is to investigate the rough injective module. The rest of the paper is organized as follows: In Section 2, preliminaries are given. In Section 3, we introduce the concept of rough injective module. Finally, our conclusions are presented. We have used standard mathematical notation throughout this paper and we assume that the reader is familiar with the basic notions of algebra and RST.

2. Preliminaries
In this section, we give some basic definitions of rough algebraic structures and results which will be used later on.

Definition 2.1   (Pawlak, 1991) A pair (U, θ), where $U \neq \emptyset$ and θ is an equivalence relation on U and is called an approximation space.

Definition 2.2   (Davvaz, 2004) For an approximation space (U, θ), by a rough approximation operator in (U, θ) we mean a mapping $Apr: P(U) \to P(U) \times P(U)$ defined by

$$Apr(X) = (\underline{X}, \overline{X}), \text{ for every } X \in P(U)$$

where $\underline{X} = \{x \in X | [x]_\theta \subseteq X\}$, $\overline{X} = \{x \in X | [x]_\theta \cap X \neq \emptyset\}$. $\underline{X}$ is called the lower rough approximation of X in (U, θ) and $\overline{X}$ is called upper rough approximation of X in (U, θ).

Definition 2.3   (Davvaz, 2004) Given an approximation space (U, θ), a pair $(A, B) \in P(U) \times P(U)$ is called a rough set in (U, θ) iff $(A, B) = Apr(X)$ for some $X \in P(U)$.

Example 2.1   Let (U, θ) be an approximation space, where $U = \{o_1, o_2, o_3, \ldots, o_7\}$ and an equivalence relation θ with the following equivalence classes:

$E_1 = \{o_1, o_4\}$
$E_2 = \{o_2, o_5, o_7\}$
$E_3 = \{o_3\}$
$E_4 = \{o_6\}$

Let the target set be $O = \{o_3, o_5\}$ then $\underline{O} = \{o_3\}$ and $\overline{O} = (\{o_3\} \cup \{o_2, o_5, o_7\})$ and so $Apr(O) = (\{o_3\}, \{o_3\} \cup \{o_2, o_5, o_7\})$ is a rough set.

Definition 2.4   (Miao, Han, Li, & Sun, 2005) Let $K = (U, \theta)$ be an approximation space and $*$ be a binary operation defined on U. A subset $G(\neq \emptyset)$ of universe U is called a rough group if $Apr(G) = (\underline{G}, \overline{G})$ satisfies the following property:

(1) $x * y \in \overline{G}, \quad \forall x, y \in G$.

(2) Association property holds in $\overline{G}$.

(3) $\exists, e \in \overline{G}$ such that $x * e = e * x = x, \quad \forall x \in G$; e is called the rough identity element.

(4) $\forall x \in G, \exists y \in G$ such that $x * y = y * x = e$; y is called the rough inverse element of x in G.

Definition 2.5   (Han, 2001) Let (U_1, θ) and (U_2, θ) be two approximation spaces, $*$ and $\overline{*}$ be two operations over U_1 and U_2, respectively. Let $G_1 \subseteq U_1$ and $G_2 \subseteq U_2$. $Apr(G_1)$ and $Apr(G_2)$ are called homomorphic rough sets if there exists a mapping ϕ of G_1 into G_2 such that

$$\forall x, y \in \overline{G}_1, \quad \phi(x * y) = \phi(x) \overline{*} \phi(y)$$

If ϕ is 1–1 mapping, $Apr(G_1)$ and $Apr(G_2)$ are called isomorphic rough sets.

Definition 2.6   (Wang, 2004) An algebraic system $(Apr(R), +, *)$ is called rough ring if it satisfies:

(1) $(Apr(R), +)$ is a rough commutative addition group.

(2) $(Apr(R), *)$ is a rough multiplicative semi-group.

(3) $(x + y) * z = x * z + y * z$ and $x * (y + z) = x * y + x * z$

$\forall x, y, z \in Apr(R)$.

Definition 2.7   (Zhang et al., 2006) Let $(Apr(R), +, *)$ be a rough ring with a unity, $(Apr(M), +)$ a rough commutative group. $Apr(M)$ is called a rough left module over the ring $Apr(R)$ if there is mapping $\overline{R} \times \overline{M} \to \overline{M}, \quad (a, x) \to ax$ such that

(1) $a(x + y) = ax + ay, \quad a \in Apr(R), \quad x, y \in Apr(M)$

(2) $(a + b)x = ax + bx, \quad a, b \in Apr(R), \quad x \in Apr(M)$

(3) $(ab)x = a(bx), \quad a, b \in Apr(R), \quad x \in Apr(M)$

(4) $1x = x$, 1 is a unit element of $Apr(R)$ and $x \in Apr(M)$

A rough right module over the ring $Apr(R)$ can be defined similarly. Condition (4) can be omitted in case of non-unital ring.

Definition 2.8   (Zhang et al., 2006) A rough subset $Apr(N) \neq \emptyset$ of a rough module $Apr(M)$ is called rough submodule of $Apr(M)$, if $Apr(N)$ satisfies the following:

(1) $Apr(N)$ is a rough subgroup of $Apr(M)$

(2) $ay \in \overline{N}, \quad \forall a \in Apr(R)$ and $y \in Apr(N)$.

Definition 2.9   (Zhang et al., 2006) Let $Apr(M)$ and $Apr(M')$ be two rough R-modules. If there exists a mapping η of M into M' such that

(1) η is a homomorphism of a rough group $Apr(M)$ into $Apr(M')$;

(2) $\eta(ax) = a\eta(x), \quad a \in Apr(R), \quad x \in Apr(M)$

then η is called a homomorphism of rough module $Apr(M)$ into $Apr(M')$. If η is a 1–1 mapping, it is called an isomorphism of rough module $Apr(M)$ into $Apr(M')$.

3. Rough injective module

Definition 3.1   A sequence $Apr(M') \overset{\alpha}{\to} Apr(M) \overset{\beta}{\to} Apr(M'')$ of two homomorphism of a module over the rough ring $Apr(R)$ is said to be rough exact if $Im(\alpha) = ker(\beta)$. This happens if and only if $(i)\beta\alpha = 0$, and *(ii)* the relation $\beta(x) = 0, x \in Apr(M)$ (i.e. $x \in \overline{M}$ and $x \in \underline{M}$), implies that $x = \alpha(x')$ for some $x' \in Apr(M')$. Indeed condition (i) and (ii) mean, respectively, that $Im(\alpha) \subset ker(\beta)$ and $ker(\beta) \subset Im(\alpha)$.

Definition 3.2   A $Apr(R)$-module $Apr(Q)$ is injective if and only if every diagram

$$0 \longrightarrow Apr(M') \xrightarrow{\alpha} M$$
$$\Big\downarrow u'$$
$$Apr(Q)$$

with exact row (i.e. with α injective) can be completed to a commutative diagram

$$0 \longrightarrow Apr(M') \longrightarrow Apr(M)$$
$$\Big\downarrow \qquad \nearrow$$
$$Apr(Q)$$

by means of a homomorphism $u: Apr(M) \to Apr(Q)$.

Since it is obviously enough to check the above condition for inclusion maps $Apr(M') \to Apr(M)$, $Apr(Q)$ is injective if and only if every homomorphism into $Apr(Q)$ form any submodule $Apr(M')$ of any $Apr(R) - module\ Apr(M)$ can be extended to a homomorphism of $Apr(M)$ into $Apr(Q)$.

Example 3.1   The $Apr(Z)$-module $Apr(Q)$ is injective.

Let $Apr(M')$ be a submodule of a $Apr(Z)$-module $Apr(M)$, and $u': Apr(M') \to Apr(Q)$ a homomorphism of $Apr(Z)$-modules. We have to show that u' can be extended to a homomorphism of M into Q. For this, it will be sufficient to show any homomorphism v from a submodule $Apr(N)$ of $Apr(M)$ into $Apr(Q)$.

Preposition 3.1   Let us be given two cointial maps $\alpha: Apr(M) \to Apr(N), \alpha':Apr(M) \to Apr(N')$ and form the diagram

$$Apr(M) \xrightarrow{\alpha} Apr(N)$$
$$\Big\downarrow \alpha'$$
$$Apr(N')$$

a pushout of α, α' or of the above diagram is a pair of coterminal maps $\beta: Apr(N) \to Apr(L), \beta': Apr(N') \to Apr(L)$ such that the square

$$\begin{array}{ccc} Apr(M) & \xrightarrow{\alpha} & Apr(N) \\ \Big\downarrow \alpha' & & \Big\downarrow \beta \\ Apr(N') & \xrightarrow{\beta'} & Apr(L) \end{array}$$

is commutative.

THEOREM 3.1   An $Apr(R)$-module Q is injective if and only if every exact sequence of the form

$$0 \to Apr(Q) \xrightarrow{u} Apr(M) \to Apr(M'') \to 0 \tag{1}$$

splits.

Proof If $Apr(Q)$ is injective and (1) an exact sequence, then

$$0 \longrightarrow Apr(Q) \xrightarrow{\ u\ } Apr(M)$$

$$1_Q \Big\downarrow \qquad \diagup p$$

$$Apr(Q)$$

there exists a homomorphism $p: Apr(M) \to Apr(Q)$ such that $pu = 1_Q$; therefore the sequence (1) splits.

Conversely, suppose that every exact sequence of the form (1) splits, and let us be given the diagram

$$0 \longrightarrow Apr(M') \xrightarrow{\ \alpha\ } Apr(M)$$

$$u' \Big\downarrow$$

$$Apr(Q)$$

with exact row form the push-out

$$Apr(M') \xrightarrow{\ \alpha\ } Apr(M)$$
$$u' \Big\downarrow \qquad\qquad \Big\downarrow \omega$$
$$Apr(Q) \xrightarrow[\ v\ ]{} Apr(L)$$

of the above diagram; since α is injective, so is v; therefore, denoting by L'' the co-kernel of v we have the exact sequence

$$0 \xrightarrow{\ v\ } Apr(Q) \to Apr(L) \to Apr(L'') \to 0. \tag{2}$$

Since this sequence splits, there exists $p: Apr(L) \to Apr(Q)$ such that $pv = 1_Q$. Then, $u = pw$ is a homomorphism of M into Q, and we have $u\alpha = pw\alpha = pvu' = u'$. Hence $Apr(Q)$ is injective. ☐

Preposition 3.2 If $Apr(R)$ is an integral domain, then every injective $Apr(R)$-module is divisible.

Proof Let $Apr(R)$ is an integral domain, and let $Apr(Q)$ be an injective $Apr(R)$-module. Let $a \neq 0$ be any non-zero element of $Apr(Q)$. Since $Apr(R)$ is an integral domain, $Apr(R)a$ is a free $Apr(R)$-module with basis $\{a\}$. Therefore, there exists a homomorphism from $Apr(R)a$ to $Apr(Q)$ which maps a to y. Since $Apr(Q)$ is injective, the above homomorphism extends to a homomorphism $h: Apr(R) \to Apr(Q)$; let $x = h(1)$, then $y = h(a) = ah(1) = ax$, implies $Apr(Q)$ is divisible. ☐

LEMMA 3.1   *Every Apr(Z)-module can be embedded in an injective Apr(z) module.*

Proof Let $Apr(E)$ be a $Apr(Z)$-module, suppose $Apr(E) = Apr(F)/Apr(N)$ with $Apr(F)$ a free $Apr(Z)$-module. Since $Apr(F)$ is a direct sum of copies of $Apr(Z)$, and since $Apr(Z)$ is a submodule of the divisible module $Apr(Q)$, therefore $Apr(F)$ is a submodule of a direct sum $Apr(G)$ of divisible $Apr(Z)$-modules. Then, $Apr(E) = Apr(F)/Apr(N)$ is a submodule of $Apr(G)/Apr(N)$. Since $Apr(G)$ is divisible, so is $Apr(G)/Apr(N)$; therefore, by the above preposition, $Apr(G)/Apr(N)$ is injective and the proof is complete. ☐

THEOREM   3.2 *If the Apr(R)-module Apr(Q) is injective, then the Apr(R)-module Apr(H) = hom(Apr(R), Apr(Q)) is injective.*

Proof   Let $Apr(H)$ is a submodule of module $Apr(M)$ over a rough ring $Apr(R)$. Here, we prove that $Apr(H)$ is a direct summand of $Apr(M)$. Mapping $u \to u(1)$ from $Apr(H)$ to $Apr(Q)$ is additive. Since the $Apr(Z)$-module $Apr(Q)$ is injective, there exists a homomorphism $q: Apr(M) \to Apr(Q)$ of $Apr(Z)$-modules such that

$$q(u) = u(1), \quad \forall u \in Apr(H) \tag{3}$$

Define $p: Apr(M) \to Apr(H)$ by

$$(p(x))(a) = q(ax), \quad \forall x \in Apr(M), a \in Apr(R) \tag{4}$$

the mapping $p(x): Apr(R) \to Apr(Q)$ is linear and so in $Apr(H)$, the mapping p is additive. If $a \in Apr(R)$, $x \in Apr(M)$, then or every $a' \in Apr(R)$

$$(p(ax))(a') = q(a'ax) = (p(x)(a'a) = (ap(x))(a'),$$

and hence $p(ax) = ap(x)$. Thus, p is linear. If now $u \in Apr(H)$, then $(p(u))(a) = q(ax) = (au)(1) = u(a)$, for all $a \in Apr(R)$, and hence $p(u) = u$. Thus, p is an linear projection from $Apr(M)$ to $Apr(H)$. Hence $Apr(H)$ is a direct summand of $Apr(M)$, and this completes the proof.   □

4. Conclusion

Recently, RST has received wide attention in the real-life applications and the algebraic studies. There are so many models arising in the solution of specific problems and turn out to be modules. For this reason, injective module based on RST introduced here is applicable in many diverse contexts. Injective module based on RST is important to all in linear algebra, vector space & physics applications. The combination of RST and abstract algebra has many interesting research topics. In this paper, we focused on algebraic results by combining RST and abstract algebra, and we hope the results given in this paper will further enrich rough set theories.

Funding
The authors received no direct funding for this research.

Author details
Arvind Kumar Sinha[1]
E-mail: aksinha.maths@nitrr.ac.in
Anand Prakash[1]
E-mail: anand.p.pal@gmail.com
[1] Department of Mathematics, National Institute of Technology Raipur, Raipur 492010, India.

References
Anderson, F. W., & Fuller, K. R. (1992). *Rings and categories of modules* (2nd ed.). New York, NY: Springer-Verlag.
Biswas, R., & Nanda, S. (1994). Rough groups and rough subgroups. *Bulletin of the Polish Academy of Sciences Mathematics, 42*, 251–254.
Bonikowaski, Z. (1995). Algebraic structures of rough sets. In W. P. Ziarko (Ed.), *Rough sets, fuzzy sets and knowledge discovery* (pp. 242–247). Berlin: Springer-Verlag.
Cartan, H., & Eilenberg, S. (1956). *Homological algebra.* Princeton, NJ: Princeton University Press.
Chakraborty, M. K., & Banerjee, M. (1994). Logic and algebra of the rough sets. In W. P. Ziarko (Ed.), *Rough sets, fuzzy sets and knowledge discovery* (pp. 196–207). London: Springer-Verlag.
Davvaz, B. (2004). Roughness in rings. *Information Sciences, 164*, 147–163. Retrieved from http://www.sciencedirect. com/science/article/pii/S0020025503003438 doi:10.1016/j.ins.2003.10.001
Davvaz, B., & Mahdavipour, M. (2006). Roughness in modules. *Information Sciences, 176*, 3658–3674. Retrieved from http://www.sciencedirect.com/science/ article/pii/S0020025506000466 doi:10.1016/j. ins.2006.02.014
Goldhaber, J. K., & Enrich, G. (1970). *Algebra.* London: Macmillan.
Han, S. (2001). The homomorphism and isomorphism of rough groups. *Academy of Shanxi University, 24*, 303–305.
Irfan Ali, M., Davvaz, B., & Shabir, M. (2013). Some properties of generalized rough sets. *Information Sciences, 224*, 170–179. Retrieved from http://www.sciencedirect.com/ science/article/pii/S0020025512006913 doi:10.1016/j.ins.2012.10.026
Iwinski, T. (1987). Algebraic approach to rough sets. *Bulletin of the Polish Academy of Sciences Mathematics, 35*, 673–683.
Jacobson, N. (1951). *Lecture in abstract algebra, 1, Basic concepts.* New York, NY: Springer-Verag.
Kryszkiewicz, M. (1998). Rough set approach to incomplete information systems. *Information Sciences, 112*, 39–49. Retrieved from http://www.sciencedirect.com/science/ article/pii/S0020025598100191 doi:10.1016/S0020-0255(98)10019-1
Kuroki, N. (1997). Rough ideals in semigroups. *Information Sciences, 100*, 139–163. Retrieved from http://www. sciencedirect.com/science/article/pii/S0020025596002745 doi:10.1016/S0020-0255(96)00274-5
Kuroki, N., & Mordeson, J. N. (1997). Structure of rough sets and rough groups. *Journal of Fuzzy Mathematics, 5*, 183–191.
Li, F., & Zhang, Z. (2014). The homomorphisms and operations of rough groups. *The Scientific World Journal, 2014*, Article ID: 635783, 6 p. doi:10.1155/2014/507972

Miao, D., Han, S., Li, D., & Sun, L. (2005). Rough group, Rough subgroup and their properties. *Lecture Notes in Artificial Intelligence, 3641*, 104–113.

Pattaraintakorn, P., & Cercone, N. (2008). Integrating rough set theory and its medical applications. *Applied Mathematics Letters, 21*, 400–403. Retrieved from http://www.sciencedirect.com/science/article/pii/S0893965907001632 doi:10.1016/j.aml.2007.05.010

Pawlak, Z. (1982). Rough sets. *International Journal of Information and Computer Sciences, 11*, 341–356. Retreived from http://link.springer.com/article/10.1007

Pawlak, Z. (1984). Rough classification. *International Journal of Man-Machine Studies, 20*, 469–483.

Pawlak, Z. (1987). Rough logic. *Bulletin of the Polish Academy of Sciences Technical Sciences, 35*, 253–258.

Pawlak, Z. (1991). *Rough sets-theoretical aspects of reasoning about data*. Dordrecht: Kluwer.

Pawlak, Z. (1992). Rough sets: A new approach to vagueness. In L. A. Zadeh & J. Kacprzyk (Eds.), *Fuzzy logic for the management of uncertainty* (pp. 105–118). New York, NY: Wiley.

Pawlak, Z. (1998). Granularity of knowledge, indiscernibility and rough sets. In *Proceedings of 1998 IEEE International Conference on fuzzy Systems* (pp. 106–110). Anchorage, AK. doi:10.1109/FUZZY.1998.687467

Prakash, A., & Sinha, A. K. (2014). Rough projective module. *International Conference on Advances in Applied Science and Environmental Engineering—ASEE 2014* (pp. 35–38). Retrieved from http://www.seekdl.org/search.php?page=825 & &jayshri & & & & & & & & doi:10.15224/978-1-63248-033-0-09

Rasouli, S., & Davvaz, B. (2014). An investigation on algebraic structure of soft sets and soft filters over residuated lattices. *ISRN Algebra, 2014*, Article ID: 635783, 8 p. doi:10.1155/2014/635783

Rebenboim, P. (1969). *Rings and modules* (No. 24). New York, NY: Interscience.

Rowen, L. H. (1991). *Ring theory* (student ed.). New York, NY: Academic Press.

Sinha, A. K., & Prakash, A. (2014, November 10–12). Roughness in free module. In *The First International Conference on Rough Sets & Knowledge Technologies India* (pp. 19–22). Hyderabad.

Walczak, B., & Massart, D. L. (1997). Rough set theory. *Chemometrics and Information Laboratory Systems, 47*, 1–16. doi:10.1016/S0169-7439(98)00200-7

Wang, C., & Chen, D. (2010). A short note on some properties of rough groups. *Computer and Mathematics with Applications, 59*, 431–436. Retrieved from http://www.sciencedirect.com/science/article/pii/S0898122109004003 doi:10.1016/j.camwa.2009.06.024

Wang, D.-s. (2004). *Application of the theory of rough set on the groups and rings* (Dissertation for master degree).

Xin, X. L., Hua, X. J., & Zhu, X. (2014). Roughness in lattice ordered effect algebras. *The Scientific World Journal, 2014*, Article ID: 542846, 9 p. doi:10.1155/2014/542846

Yao, Y. Y. (1996). Two views of the theory of rough sets in finite universes. *International Journal of Approximation Reasoning, 15*, 291–317. doi:10.1016/S0888-613X(96)00071-0

Yao, Y. Y. (1998). Constructive and algebraic methods of the theory of rough sets. *Information Sciences, 109*, 21–44. doi:10.1016/S0020-0255(98)00012-7

Zhang, Q.-F., Fu, A.-M., & Zhao, S.-x. (2006, August 13–16). Rough modules and their some properties. In *Proceeding of the fifth International Conference on Machine Learning and Cybernatics*. Dalin. Retrieved from http://ieeexplore.ieee.org/xpls/abs_all.jsp?arnumber=4028446 &tag=1 doi:10.1109/ICMLC.2006.258675

14

Bounds on Hankel determinant for starlike and convex functions with respect to symmetric points

Ambuj K. Mishra[1], Jugal K. Prajapat[2]* and Sudhananda Maharana[2]

*Corresponding author: Jugal K. Prajapat, Department of Mathematics, Central University of Rajasthan, NH-8, Bandarsindri, Kishangarh 305817, Ajmer, Rajasthan, India

E-mail: jkprajapat@gmail.com

Reviewing editor: Hari M. Srivastava, University of Victoria, Canada

Abstract: In the present paper, we investigate upper bounds on the third Hankel determinants for the starlike and convex functions with respect to symmetric points in the open unit disk.

Subjects: Advanced Mathematics; Analysis - Mathematics; Complex Variables; Mathematics & Statistics; Science

Keywords: analytic and univalent functions; Hankel determinant; Toeplitz determinant; starlike and convex functions with respect to symmetric points

1991 Mathematics subject classifications: 30C45; 30C50

1. Introduction

1.1. Hankel determinant

Let $\mathcal{A}$ denote the family of analytic functions in the open unit disk $\mathbb{D} = \{z \in \mathbb{C} : |z| < 1\}$ of the form

$$f(z) = z + \sum_{n=2}^{\infty} a_n z^n, \quad z \in \mathbb{D}. \tag{1}$$

ABOUT THE AUTHORS

Ambuj K. Mishra is working as an assistant professor in the Department of Mathematics, GLA University, Mathura, Uttar Pradesh, India. He received his MSc degree from University Of Allahabad, Uttar Pradesh, in 2002 and currently pursuing his PhD in the field of Geometric function theory at GLA University, Mathura, Uttar Pradesh. His research interest includes Geometric function theory, and Special functions.

Jugal K. Prajapat is working as an associate professor in the Department of Mathematics, Central University of Rajasthan, Rajasthan, India. He received his PhD degree from University of Rajasthan. He has published 50 research articles in reputed international journals. His research interest includes Geometric function theory, Fractional calculus, and Special functions.

Sudhananda Maharana is a research scholar in the Department of Mathematics, Central University of Rajasthan, Rajasthan, India. He received his MSc degree from Berhampur University, Odisha, in 2012. His research interest includes the Geometric function theory and Special functions.

PUBLIC INTEREST STATEMENT

The interplay between geometry and analysis of function of complex variables is the most attracting part of complex analysis. From the beginning of the twentieth century, the work on the coefficients of the Taylor series expansion of analytic univalent function is of great importance in complex analysis. The bounds on Hankel determinants and Fekete–Szegö inequalities of coefficients of Taylor's series expansion of analytic univalent functions have been studied by many peoples. In this paper, we have studied bounds on third Hankel determinants for the functions which are starlike and convex with respect to symmetric points.

A function f is said to be univalent in a domain D, if it is one-to-one in D. Let S denote the subclass of $\mathcal{A}$ consisting of functions which are univalent in $\mathbb{D}$.

The Hankel determinant $H_{q,n}(f)$ of Taylor's coefficients of function $f \in \mathcal{A}$ of the form (1), is defined by

$$H_{q,n}(f) = \begin{vmatrix} a_n & a_{n+1} & \cdots & a_{n+q-1} \\ a_{n+1} & a_{n+2} & \cdots & a_{n+q} \\ \vdots & \vdots & \vdots & \vdots \\ a_{n+q-1} & a_{n+q} & \cdots & a_{n+2(q-1)} \end{vmatrix} \quad (a_1 = 1; n, q \in \mathbb{N} = \{1, 2, \cdots\}). \tag{2}$$

The Hankel determinent is useful in showing that a function of bounded characteristic in $\mathbb{D}$, i.e. a function which is a ratio of two bounded analytic functions with its Laurent series around the origin having integral coefficients, is rational (see Cantor, 1963). Pommerenke (1967) proved that the Hankel determinant of univalent functions satisfy $|H_{q,n}(f)| < Kn^{-(\frac{1}{2}+\beta)q+\frac{3}{2}}$, where $\beta > 1/4000$ and K depends only on q. Later, Hayman (1968) proved that $|H_{2,n}(f)| < A\,n^{1/2}$ (A is an absolute constant) for a really mean univalent functions. The study of $|H_{q,n}(f)|$ for various subfamilies of $\mathcal{A}$ are of interest for many researchers (see Ehrenborg, 2000; Noonan & Thomas, 1976; Noor, 1992; Pommerenke, 1966).

Note that, the $H_{2,1}(f) = a_3 - a_2^2$ is the classical *Fekete–Szegö functional*. Fekete and Szegö (1933) found the maximum value of $|H_{2,1}(f)|$ over the function $f \in S$. The problem of calculating $\max_{f \in \mathcal{F}} |H_{2,1}(f)|$ for various compact subfamilies $\mathcal{F}$ of $\mathcal{A}$, was considered by many authors (see Bhowmik, Ponnusamy, & Wirths, 2011; Keogh & Merkes, 1969; Koepf, 1987; Mishra & Gochhayat, 2008a, 2010, 2011; Srivastava & Mishra, 2000; Srivastava, Mishra, & Das, 2001). Further, for the second Hankel determinant $H_{2,2}(f)$, the $\max_{f \in \mathcal{F}} |H_{2,2}(f)|$ has been studied by many researchers (see Janteng, Halim, & Darus, 2006; Lee, Ravichandran, & Supramaniam, 2013; Mishra & Gochhayat, 2008b; Mishra & Kund, 2013; Patel & Sahoo, 2014) and upper bound on the third Hankel determinant $H_{3,1}(f)$ studied recently by Babalola (2010), Bansal, Maharana, and Prajapat (2015), Prajapat, Bansal, Singh, and Mishra (2015), Raza and Malik (2013), Vamshee Krishna, Venkateswarlu, and RamReddy (2015).

1.2. Toeplitz determinant

Let $\mathcal{P}$ denote the class of analytic functions p in $\mathbb{D}$ with $\Re(p(z)) > 0$ and $p(0) = 1$. If $p \in \mathcal{P}$ is of the form

$$p(z) = 1 + \sum_{n=1}^{\infty} c_n z^n, \quad z \in \mathbb{D}, \tag{3}$$

then $|c_n| \le 2$, $n \in \mathbb{N} := \{1, 2, \cdots\}$. This inequality is sharp and the equality holds for the function $\varphi(z) = (1 + z)/(1 - z)$ (see Duren, 1983).

The power series (3) converges in $\mathbb{D}$ to a function in $\mathcal{P}$, if and only if the Toeplitz determinants

$$T_n(p) = \begin{vmatrix} 2 & c_1 & c_2 & \cdots & c_n \\ c_{-1} & 2 & c_1 & \cdots & c_{n-1} \\ c_{-2} & c_{-1} & 2 & \cdots & c_{n-2} \\ \vdots & \vdots & \vdots & \ddots & \vdots \\ c_{-n} & c_{-n+1} & c_{-n+2} & \cdots & 2 \end{vmatrix}, \quad n \in \mathbb{N}$$

are positive, where $c_{-n} = \overline{c_n}$. The only exception is when $f(z)$ has the form

$$f(z) = \sum_{v=1}^{m} \rho_v \frac{1 + \epsilon_v z}{1 - \epsilon_v z}, \quad m \geq 1,$$

where $\rho_v > 0, |\epsilon_v| = 1$, and $\epsilon_k \neq \epsilon_l$ if $k \neq l; k, l = 1, 2, \cdots, m$; we have then $T_n(p) > 0$ for $n \leq (m-1)$ and $T_n(p) = 0$ for $n \geq m$ (see Grenander & Szegö, 1984).

Recently, in an article (Janteng et al., 2006) the Toeplitz determinant found to be useful to estimate upper bound on the coefficients functional for various subfamilies of analytic functions. Note that for $n = 2$

$$T_2(p) = \begin{vmatrix} 2 & c_1 & c_2 \\ \bar{c}_1 & 2 & c_1 \\ \bar{c}_2 & \bar{c}_1 & 2 \end{vmatrix} = 8 + 2 \Re\{c_1^2 \bar{c}_2\} - 2|c_2|^2 - 4|c_1|^2 \geq 0,$$

is equivalent to

$$2c_2 = c_1^2 + x(4 - c_1^2) \tag{4}$$

for some x with $|x| \leq 1$. Similarly, if

$$T_3(p) = \begin{vmatrix} 2 & c_1 & c_2 & c_3 \\ \bar{c}_1 & 2 & c_1 & c_2 \\ \bar{c}_2 & \bar{c}_1 & 2 & c_1 \\ \bar{c}_3 & \bar{c}_2 & \bar{c}_1 & 2 \end{vmatrix},$$

then $T_3(p) \geq 0$ is equivalent to

$$|(4c_3 - 4c_1c_2 + c_1^3)(4 - c_1^2) + c_1(2c_2 - c_1^2)^2| \leq 2(4 - c_1^2)^2 - 2|(2c_2 - c_1^2)|^2. \tag{5}$$

Solving (5) with the help of (4), we get

$$4c_3 = c_1^3 + 2c_1 x(4 - c_1^2) - c_1 x^2(4 - c_1^2) + 2(4 - c_1^2)(1 - |x|^2)z, \tag{6}$$

for some x and z with $|x| \leq 1$ and $|z| \leq 1$. Conditions (4) and (6) are due to Libera and Zlotkiewicz (1982, 1983).

1.3. Starlike and convex functions with respect to symmetric points
A function $f \in \mathcal{A}$ is called starlike, if f is univalent in $\mathbb{D}$ and $f(\mathbb{D})$ is a starlike domain with respect to the origin. Analytically, $f \in S$ is called starlike, denoted by $f \in S^*$, if $\Re(zf'(z)/f(z)) > 0$, $z \in \mathbb{D}$. A function $f \in S$ is called convex, denoted by $f \in C$, if and only if $zf'(z) \in S^*$.

A function $f \in \mathcal{A}$ is said to be starlike with respect to symmetric points (see Sakaguchi, 1959) if for every r less than and sufficiently close to one and every η on the circle $|z| = r$, the angular velocity of $f(z)$ about the point $f(-\eta)$ is positive at $z = \eta$ as z traverses the circle $|z| = r$ in the positive direction, i.e.

$$\Re\left(\frac{zf'(z)}{f(z) - f(-\eta)}\right) > 0 \quad \text{for } z = \eta, |\eta| = r. \tag{7}$$

We denote by S_s^*, the class of all functions in S which are starlike with respect to symmetric points. A function f in the class S_s^* is characterized by

$$\Re\left(\frac{zf'(z)}{f(z) - f(-z)}\right) > 0, \quad z \in \mathbb{D}. \tag{8}$$

This can be easily seen that the function $(f(z) - f(-z))/2$ is a starlike function in $\mathbb{D}$, therefore functions satisfying (8) are close-to-convex (*and hence univalent*) in $\mathbb{D}$.

Further, the class of all functions in S, which are convex with respect to symmetric points is denoted by C_s. The necessary and sufficient condition for the function $f \in S$ to be univalent and convex with respect symmetric points in $\mathbb{D}$ is characterized by (see Das & Singh, 1977, Theorem 1)

$$\Re\left(\frac{(zf'(z))'}{(f(z) - f(-z))'}\right) > 0, \quad z \in \mathbb{D}. \tag{9}$$

In the present paper, we aim to investigate the upper bounds on the third Hankel determinant $|H_{3,1}(f)|$ for the functions belonging to the classes S_s^* and C_s defined above in (8) and (10). For this purpose, we shall use Equations (4 and 6) and the following known results.

LEMMA 1.1   (Sakaguchi, 1959) *If* $f \in S_s^*$ *of the form (1), then* $|a_n| \leq 1$, $n \geq 2$. *Equality holds for the function* $f(z) = z(1 + \epsilon z)^{-1}$, $|\epsilon| = 1$.

LEMMA 1.2   (Das & Singh, 1977) *If* $f \in C_s$ *of the form (1), then* $|a_n| \leq 1/n$, $n \geq 2$. *Equality holds for the function* $f(z) = (1/\epsilon)\log(1 + \epsilon z)$, $|\epsilon| = 1$.

LEMMA 1.3   (Shanmugam, Ramachandran, & Ravichandran, 2006, Example 2.3) *If* $f \in S_s^*$ *of the form (1), then* $|a_3 - a_2^2| \leq 1$. *This inequality is sharp and the equality is attained for the function* $f(z) = z(1 + \epsilon z^2)^{-1}$, $|\epsilon| = 1$.

LEMMA 1.4   (Shanmugam et al., 2006, Example 2.5) *If* $f \in C_s$ *of the form (1), then* $|a_3 - a_2^2| \leq 1/3$. *This inequality is sharp.*

2. Main results

THEOREM 2.1   *Let the function f given by (1) be in the class* S_s^*. *Then*

$$|a_2 a_3 - a_4| \leq \frac{1}{2} \quad \text{and} \quad |a_2 a_4 - a_3^2| \leq 1. \tag{10}$$

The inequalities in (10) are sharp.

Proof   Let $f \in S_s^*$, then by (8), we have

$$\frac{2zf'(z)}{f(z) - f(-z)} = p(z),$$

where $p \in P$ is of the form (3). Substituting the series expansion of $f(z)$ and $p(z)$ and equating the coefficients, we get

$$a_2 = \frac{1}{2}c_1, \quad a_3 = \frac{1}{2}c_2 \quad \text{and} \quad a_4 = \frac{1}{8}(2c_3 + c_1 c_2). \tag{11}$$

Hence

$$\begin{cases} |a_2 a_3 - a_4| = \frac{1}{8}|c_1 c_2 - 2c_3| \\ \text{and} \\ |a_2 a_4 - a_3^2| = \frac{1}{16}|2c_1 c_3 + c_1^2 c_2 - 4c_2^2|. \end{cases} \tag{12}$$

Using (4) and (6) in (12) for some x and z such that $|x| \leq 1$ and $|z| \leq 1$, we get

$$\begin{cases} |a_2a_3 - a_4| = \dfrac{1}{16}\left|-c_1x(4 - c_1^2) + c_1x^2(4 - c_1^2) - 2(4 - c_1^2)(1 - |x|^2)z\right| \\ \text{and} \\ |a_2a_4 - a_3^2| = \dfrac{1}{32}\left|-c_1^2x(4 - c_1^2) - 2x^2(4 - c_1^2)^2 - c_1^2x^2(4 - c_1^2) + 2c_1(4 - c_1^2)(1 - |x|^2)z\right|. \end{cases}$$

As $|c_1| \le 2$, therefore, letting $c_1 = c$, we may assume without restriction that $c \in [0, 2]$. Thus applying the triangle inequality with $\mu = |x|$, we obtain

$$|a_2a_3 - a_4| \le \frac{1}{16}\left[c\mu(4 - c^2) + c\mu^2(4 - c^2) + 2(4 - c^2)(1 - \mu^2)\right]$$

$$:= F(c, \mu)$$

and

$$|a_2a_4 - a_3^2| \le \frac{1}{32}\left[c^2\mu(4 - c^2) + 8\mu^2(4 - c^2) - c^2\mu^2(4 - c^2) + 2c(4 - c^2)(1 - \mu^2)\right]$$

$$:= G(c, \mu).$$

Now we need to find the maximum value of F and G over the region $\Omega = \{(c, \mu): 0 \le c \le 2, 0 \le \mu \le 1\}$. For this, first differentiating F with respect to μ and c, we get

$$\frac{\partial F}{\partial \mu} = \frac{1}{16}\left[(4 - c^2)(c + 2c\mu - 4\mu)\right]$$

$$\frac{\partial F}{\partial c} = \frac{1}{16}\left[4\mu + 4\mu^2 - 3c^2\mu - 3c^2\mu^2 - 4c(1 - \mu^2)\right]. \tag{13}$$

A critical point of $F(c, \mu)$ must satisfy $\dfrac{\partial F}{\partial \mu} = 0$ and $\dfrac{\partial F}{\partial c} = 0$. The condition $\dfrac{\partial F}{\partial \mu} = 0$ gives $c = 2$ or $\mu = -\dfrac{c}{2c - 4}$. Points (c, μ) satisfying such conditions are not interior points of Ω. So the function $F(c, \mu)$ cannot have a maximum in the interior of Ω. Since Ω is closed and bounded and F is continuous on Ω, the maximum shall be attained on the boundary of Ω. It is easy to see that on the boundary line $c = 0, 0 \le \mu \le 1$, we have $F(0, \mu) = (1 - \mu^2)/2$ and its maximum on this line is equal to $1/2$. On the boundary line $c = 2, 0 \le \mu \le 1$, we have $F(2, \mu) = 0$. Similarly, on the boundary line $\mu = 0, 0 \le c \le 2$, we have $F(c, 0) = (4 - c^2)/8$ and the maximum on this line is $1/2$. Lastly, on the boundary line $\mu = 1$, $0 \le c \le 2$, we have $F(c, 1) = c(4 - c^2)/8$ and the maximum on this line is $2/\sqrt{27}$. Comparing the four maxima we get that the maximum value of $F(c, \mu)$ on Ω is $1/2$.

To show the sharpness of first inequality in (10), by setting $c_1 = x = 0$, $z = 1$ in (4) and (6), we get $c_2 = 0$ and $c_3 = 2$. Using these values in (12), we find that the first inequality in (10) is sharp.

Further, to find the maximum value of G over Ω, differentiating G with respect to μ, we get

$$\frac{\partial G}{\partial \mu} = \frac{1}{32}\left[(4 - c^2)\{c^2 + 2\mu(8 - c^2 - 2c)\}\right] > 0 \quad \text{if} \quad 0 < c < 2 \quad \text{and} \quad 0 < \mu < 1.$$

Note that, G is a non-decreasing function of μ on $[0, 1]$, hence

$$\max_{0 \le \mu \le 1} G(c, \mu) = G(c, 1) = \frac{1}{4}(4 - c^2) = \mathcal{G}(c).$$

Further, it is clear that $\mathcal{G}(c)$ is a decreasing function on $[0, 2]$, hence it attain maximum value at $c = 0$. Therefore the maximum of $G(c, \mu)$ is at the point $(0, 1)$. Further, Ω is closed and bounded and G is continuous on Ω, the maximum shall be attained on the boundary of Ω. Hence, we look on the boundary of Ω as we have done with the function F, it is easy to see that on the boundary line $c = 0, 0 \le \mu \le 1$, we have $G(0, \mu) = \mu^2$ and its maximum on this line is equal to 1. On the boundary line $c = 2, 0 \le \mu \le 1$, we have $G(2, \mu) = 0$. Similarly, on the boundary line $\mu = 0, 0 \le c \le 2$, we have $G(c, 0) = c(4 - c^2)/16$ and the maximum on this line is $1/\sqrt{27}$. Lastly, on the boundary line $\mu = 1, 0 \le c \le 2$, we have $G(c, 1) = (4 - c^2)/4$ and the maximum on this line is 1. Comparing the four maxima we get that the maximum value of

$G(c, \mu)$ on Ω is 1.

To show the sharpness in the second inequality of (10), by setting $c_1 = 0$, $x = 1$ in (4) and (6), we get $c_2 = 2$ and $c_3 = 0$. Using these values in (12), we find that the second inequality in (10) is sharp. □

THEOREM 2.2   *Let the function f given by (1) be in the class S_s^*. Then*

$$|H_{3,1}(f)| \leq \frac{5}{2}.$$

Proof   Using Lemma 1.1, Lemma 1.3, Theorem 2.1 and applying the triangle inequality, we get   □

$$|H_{3,1}(f)| \leq |a_3||a_2a_4 - a_3^2| + |a_4||a_2a_3 - a_4| + |a_5||a_3 - a_2^2|$$
$$\leq 1 + \frac{1}{2} + 1 = \frac{5}{2}.$$

THEOREM 2.3   *Let the function f given by (1) be in the class C_s. Then*

$$|a_2a_3 - a_4| \leq \frac{4}{27} \quad \text{and} \quad |a_2a_4 - a_3^2| \leq \frac{1}{9}. \tag{14}$$

The second inequality in (14) is sharp.

Proof   Let $f \in C_s$, then by (9), we have

$$\frac{2(zf'(z))'}{(f(z) - f(-z))'} = p(z),$$

where $p \in \mathcal{P}$ is of the form (3). From the definitions of the class S_s^* and C_s, it follows that the function $f(z) \in C_s$ if and only if $zf'(z) \in S_s^*$. Thus replacing a_n by na_n in (11), we get

$$a_2 = \frac{1}{4}c_1, \quad a_3 = \frac{1}{6}c_2 \quad \text{and} \quad a_4 = \frac{1}{32}(2c_3 + c_1c_2). \tag{15}$$

Hence

$$\begin{cases} |a_2a_3 - a_4| = \frac{1}{96}|c_1c_2 - 6c_3| \\ \text{and} \\ |a_2a_4 - a_3^2| = \frac{1}{1152}|18c_1c_3 + 9c_1^2c_2 - 32c_2^2|. \end{cases} \tag{16}$$

Using (4) and (6) in (16) for some x and z such that $|x| \leq 1$ and $|z| \leq 1$, we get

$$\begin{cases} |a_2a_3 - a_4| = \frac{1}{192}\left|-2c_1^3 + (4 - c_1^2)\{-5c_1x + 3c_1x^2 - 6(1 - |x|^2)z\}\right| \\ \text{and} \\ |a_2a_4 - a_3^2| = \frac{1}{2304}\left|2c_1^4 + (4 - c_1^2)\{-5c_1^2x - 16x^2(4 - c_1^2) - 9c_1^2x^2 + 18c_1(1 - |x|^2)z\}\right|. \end{cases}$$

As $|c_1| \leq 2$, therefore, letting $c_1 = c$, we may assume without restriction that $c \in [0, 2]$. Thus applying the triangle inequality with $\mu = |x|$, we obtain

$$|a_2a_3 - a_4| \leq \frac{1}{192}\left[2c^3 + (4 - c^2)\{5c\mu + 3c\mu^2 + 6(1 - \mu^2)\}\right]$$
$$:= X(c, \mu)$$

and
$$|a_2a_4 - a_3^2| \leq \frac{1}{2304}\left[2c^4 + (4 - c^2)\{5c^2\mu + 64\mu^2 - 7c^2\mu^2 + 18c(1 - \mu^2)\}\right]$$
$$:= Y(c, \mu).$$

Now to find the maximum value of X and Y over the region Ω. First differentiating X with respect to μ and c, we get

$$\frac{\partial X}{\partial \mu} = \frac{1}{192}\left[(4-c^2)(5c+6c\mu-12\mu)\right]$$

$$\frac{\partial X}{\partial c} = \frac{1}{192}\left[6c^2+(4-c^2)(5\mu+3\mu^2)-2c(5c\mu+3c\mu^2+6-6\mu^2)\right].$$

A critical point of $X(c,\mu)$ must satisfy $\frac{\partial X}{\partial \mu}=0$ and $\frac{\partial X}{\partial c}=0$. The condition $\frac{\partial X}{\partial \mu}=0$ gives $c=2$ or $\mu=\frac{-5c}{6c-12}$. Points (c,μ) satisfying such conditions are not interior points of Ω. So the function $X(c,\mu)$ cannot have a maximum in the interior of Ω. Since Ω is closed and bounded and X is continuous the maximum shall be attained on the boundary of Ω. It is easy to see that on the boundary line $c=0$, $0\le\mu\le1$, we have $X(0,\mu)=(1-\mu^2)/8$ and its maximum on this line is equal to $1/8$. On the boundary line $c=2, 0\le\mu\le1$, we have $X(2,\mu)=1/12$. Similarly, on the boundary line $\mu=0, 0\le c\le2$, we have $X(c,0)=(c^3-3c^2+12)/96$ and the maximum on this line is $1/8$. Lastly, on the boundary line $\mu=1$, $0\le c\le2$, we have $X(c,1)=c(16-3c^2)/96$ and the maximum on this line is $4/27$. Comparing the four maxima we get that the maximum value of $X(c,\mu)$ on Ω is $4/27$.

Further, to find the maximum value of Y over Ω, differentiating Y with respect to μ, we get

$$\frac{\partial Y}{\partial \mu} = \frac{1}{2304}\left[(4-c^2)\{5c^2+2\mu(64-7c^2-18c)\}\right]>0 \quad \text{if} \quad 0<c<2 \quad \text{and} \quad 0<\mu<1.$$

Note that, Y is a non-decreasing function of μ on $[0,1]$, hence

$$\max_{0\le\mu\le1} Y(c,\mu)=Y(c,1)=\frac{1}{1152}(13c^4-36c^2+128)=\mathcal{Y}(c).$$

It is clear that $\mathcal{Y}(c)$ is a decreasing function on $[0,2]$ and it attained maximum value at $c=0$. Therefore, the maximum of $Y(c,\mu)$ is at the point $(0,1)$. Further, Ω is closed and bounded and Y is continuous, the maximum shall be attained on the boundary of Ω. Hence, we look on the boundary of Ω, it is easy to see that on the line $c=0, 0\le\mu\le1$, we have $Y(0,\mu)=\mu^2/9$ and its maximum on this line is equal to $1/9$. On the boundary line $c=2, 0\le\mu\le1$, we have $Y(2,\mu)=1/72$. Similarly, on the boundary line $\mu=0, 0\le c\le2$, we have $Y(c,0)=c(c^3-9c^2+36)/1152$ and the maximum on this line is less than $1/9$. Lastly, on the boundary line $\mu=1, 0\le c\le2$, we have $Y(c,1)=(2c^4-36c^2+128)/1152$ and the maximum on this line is $1/9$. Comparing the four maxima we get that the maximum value of $Y(c,\mu)$ on Ω is $1/9$.

To show the sharpness in the second inequality of (14), by setting $c_1=0$, $x=1$ in (4) and (6), we get $c_2=2$ and $c_3=0$. Using these values in (16), we find that the second inequality in (14) is sharp. This completes the proof of the theorem. $\qquad\square$

THEOREM 2.4　Let the function f given by (1) be in the class C_s. Then

$$|H_{3,1}(f)| \le \frac{19}{135}.$$

Proof　Using Lemma 1.2, Lemma 1.4, Theorem 2.3 and applying the triangle inequality, we get $\square$

$$|H_{3,1}(f)| \le |a_3||a_2a_4-a_3^2|+|a_4||a_2a_3-a_4|+|a_5||a_3-a_2^2|$$
$$\le \frac{1}{3}\times\frac{1}{9}+\frac{1}{4}\times\frac{4}{27}+\frac{1}{5}\times\frac{1}{3}=\frac{19}{135}.$$

Acknowledgements
The authors express their sincere thanks to the editor and referees for their valuable suggestions to improve the manuscript.

Funding
The authors received no direct funding for this research.

Author details
Ambuj K. Mishra[1]
E-mail: ambuj_math@rediffmail.com
ORCID ID: http://orcid.org/0000-0001-9059-6108
Jugal K. Prajapat[2]
E-mail: jkprajapat@gmail.com
Sudhananda Maharana[2]
E-mail: snmmath@gmail.com
[1] Department of Mathematics, Institute of Applied Sciences and Humanities, GLA University, Mathura, Uttar Pradesh, India.
[2] Department of Mathematics, Central University of Rajasthan, NH-8, Bandarsindri, Kishangarh, 305817 Ajmer, Rajasthan, India.

References
Babalola, K. O. (2010). On third order Hankel determinant for some classes of univalent functions. *Inequality Theory and Applications, 6*, 1–7.

Bansal, D., Maharana, S., & Prajapat, J. K. (2015). Third order Hankel determinant for certain univalent functions. *Journal of the Korean Mathematical Society, 52*, 1139–1148.

Bhowmik, B., Ponnusamy, S., & Wirths, K.-J. (2011). On the Fekete–Szegö problem for the concave univalent functions. *Journal of Mathematical Analysis and Applications, 373*, 432–438.

Cantor, D. G. (1963). Power series with the integral coefficients. *Bulletin of the American Mathematical Society, 69*, 362–366.

Das, R. N., & Singh, P. (1977). On subclasses of schlicht mapping. *Indian Journal of Pure and Applied Mathematics, 8*, 864–872.

Duren, P. L. (1983). *Univalent functions. Grundlehren der Mathematischen Wissenschaften* (Band 259). New York, NY: Springer-Verlag.

Ehrenborg, R. (2000). The Hankel determinant of exponential polynomials. *American Mathematical Monthly, 107*, 557–560.

Fekete, M., & Szegö, G. (1933). Eine Benberkung uber ungerada Schlichte funktionen. *Journal of the London Mathematical Society, 8*, 85–89.

Grenander, U., & Szegö, G. (1984). *Toeplitz forms and their applications* (2nd ed.). New York, NY: Chelsea.

Hayman, W. K. (1968). On second Hankel determinant of mean univalent functions. *Proceedings of the London Mathematical Society, 18*, 77–94.

Janteng, A., Halim, S., & Darus, M. (2006). Coefficient inequality for a function whose derivative has a positive real part. *Journal of Inequalities in Pure and Applied Mathematics, 7*(2), 1–5.

Keogh, F. R., & Merkes, E. P. (1969). A coefficient Inequality for certain classes of analytic functions. *Proceedings of the American Mathematical Society, 20*, 8–12.

Koepf, W. (1987). On the Fekete–Szegö problem for close-to-convex functions. *Proceedings of the American Mathematical Society, 101*, 89–95.

Lee, S. K., Ravichandran, V., & Supramaniam, S. (2013). Bounds for the second Hankel determinant of certain univalent functions. *Journal of Inequalities and Applications, 2013*, Article 281.

Libera, R. J., & Zlotkiewicz, E. J. (1982). Early coefficients of the inverse of a regular convex function. *Proceedings of the American Mathematical Society, 85*, 225–230.

Libera, R. J., & Zlotkiewicz, E. J. (1983). Coefficient bounds for the inverse of a function with derivatives in p. *Proceedings of the American Mathematical Society, 87*, 251–257.

Mishra, A. K., & Gochhayat, P. (2008a). The Fekete–Szegö problem for k-uniformly convex functions and for a class defined by the Owa–Srivastava operator. *Journal of Mathematical Analysis and Applications, 347*, 563–572.

Mishra, A. K., & Gochhayat, P. (2008b). Second Hankel determinant for a class of analytic functions defined by fractinal derivatives. *International Journal of Mathematics and Mathematical Sciences, 2008*, 10, Article ID 153280.

Mishra, A. K., & Gochhayat, P. (2010). Fekete–Szegö problem for a class defined by an integral operator. *Kodai Mathematical Journal, 33*, 310–328.

Mishra, A. K., & Gochhayat, P. (2011). A coefficient inequality for a subclass of Cathéodory functions defined using a conical domain. *Computers & Mathematics with Applications, 61*, 2816–2820.

Mishra, A. K., & Kund, S. N. (2013). Second Hankel determinant for a class of analytic functions defined by Carlson–Shaffer operator. *Tamkang Journal of Mathematics, 44*, 73–82.

Noonan, J. W., & Thomas, D. K. (1976). On the second Hankel determinant of areally mean p-valent functions. *Transactions of the American Mathematical Society, 223*, 337–346.

Noor, K. I. (1992). Higer order close-to-convex functions. *Mathematica Japonica, 37*(1), 1–8.

Patel, J., & Sahoo, A. K. (2014). On certain subclasses of analytic functions associated with the Carlson--Shaffer operator. *Annales Universitatis Mariae Curie-Sklodowska Lublin-Polonia, LXVIII*, 65–83.

Pommerenke, C. (1966). On the coefficients and Hankel determinant of univalent functions. *Journal of the London Mathematical Society, 41*, 111–122.

Pommerenke, C. (1967). On the Hankel determinant of univalent functions. *Mathematika, 14*, 108–112.

Prajapat, J. K., Bansal, D., Singh, A., & Mishra, A. K. (2015). Bounds on third Hankel determinant for close-to-convex functions. *Acta Universitatis Sapientiae, Mathematica, 7*, 210–219.

Raza, M., & Malik, S. N. (2013). Upper bound of the third Hankel determinant for a class of analytic functions related with Lemniscate of Bernoulli. *Journal of Inequalities and Applications, 2013*, Article 412.

Sakaguchi, K. (1959). On a certain univalent mapping. *Journal of the Mathematical Society of Japan, 11*, 72–75.

Shanmugam, T. N., Ramachandran, C., & Ravichandran, V. (2006). Fekete–Szegö problem for subclasses of starlike functions with respect to symmetric points. *Bulletin of the Korean Mathematical Society, 43*, 589–598.

Srivastava, H. M., & Mishra, A. K. (2000). Applications of fractional calculus to parabolic starlike and uniformly convex functions. *Computers & Mathematics with Applications, 39*, 57–69.

Srivastava, H. M., Mishra, A. K., & Das, M. K. (2001). Fekete–Szegö problem for a subclass of close-to-convex functions. *Complex Variables, Theory and Application, 44*, 145–163.

Vamshee Krishna, D., Venkateswarlu, B., & RamReddy, T. (2015). Third Hankel determinant for certain subclass of p-valent functions. *Complex Variables and Elliptic Equations, 60*, 1301–1307.

15

Reflection principle for classical solutions of the homogeneous real Monge–Ampère equation

Mika Koskenoja[1]*

*Corresponding author: Mika Koskenoja, Department of Mathematics and Statistics, University of Helsinki,
P.O. Box 68 (Gustaf Hällströmin katu 2b), FI-00014 Helsinki, Finland
E-mail: mika.koskenoja@helsinki.fi
Reviewing editor: Lishan Liu, Qufu Normal University, China

Abstract: We consider reflection principle for classical solutions of the homogeneous real Monge–Ampère equation. We show that both the odd and the even reflected functions satisfy the Monge–Ampère equation if the second-order partial derivatives have continuous limits on the reflection boundary. In addition to sufficient conditions, we give some necessary conditions. Before stating the main results, we present elementary formulas for the reflected functions and study their differentiability properties across the reflection boundary. As an important special case, we finally consider extension of polynomials satisfying the homogeneous Monge–Ampère equation.

Subjects: Advanced Mathematics; Analysis - Mathematics; Differential Equations; Mathematical Analysis; Mathematics & Statistics; Science

Keywords: reflection principle; continuation; real Monge–Ampère equation; homogeneous equation; classical solution

AMS subject classifications: 35B60; 35D35; 35G20; 35K96

1. Introduction

Reflection is a method to extend functions and, in particular, solutions of homogeneous differential equations across a flat boundary. Classically, it is applied for some strong type equations but later on also for several weak type equations. The reflected function is usually equipped with negative sign in the reflected domain, but in our case it turns out to be profitable to study also a variant of the reflected function with positive sign.

Let G_+ be a domain in the upper half complex plane $\mathbb{C}_+ = \{z \in \mathbb{C}: Im z < 0\}$ and let $G_0 \subset \partial \mathbb{C}_+$ be open in ∂G_+. If $f = u + iv: G_+ \to \mathbb{C}$ is analytic and $\lim_{z \to w} v(z) = 0$ for all $w \in G_0$, then f has an analytic extension to $G = G_+ \cup G_0 \cup G_-$ where $G_- = PG_+$ and $P: \mathbb{C} \to \mathbb{C}$ is the reflection with respect to

ABOUT THE AUTHOR

Mika Koskenoja is a university lecturer at the Department of Mathematics and Statistics in the University of Helsinki. After receiving his PhD in 2002, the author has continued to consider potential theoretic issues of several complex variables and spaces of variable exponents. Moreover, he has studied some specific questions of both real and complex Monge-Ampére equations. The results of this paper belong to the latter research topic.

PUBLIC INTEREST STATEMENT

Functions play a central role in all mathematical approaches. Sometimes it is important to be able to extend the domain of a function. Then the reflection may offer an admissible method for the extension. We consider the reflection principle for the Monge-Ampère equation which arises naturally in several areas of both mathematics and physics. We show that both the odd and the even reflected functions satisfy the Monge–Ampère equation if the second order partial derivatives have continuous limits on the reflection boundary.

the real coordinate axis. The analytic extension from G_+ to G_- is given by $f(\bar{z}) = \overline{f(z)}$. This classical reflection principle originates with H. A. Schwarz.

An analogous principle holds for harmonic functions given in the upper half space of $\mathbb{R}^n$ with $n \geqslant 2$, that is, $\mathbb{R}^n_+ = \{(x, x_n) \in \mathbb{R}^n : x \in \mathbb{R}^{n-1}, x_n < 0\}$. If $G_+ \subset \mathbb{R}^n_+$ is open and $f : G_+ \to \mathbb{R}$ is harmonic and extends continuously to zero on $G_0 \subset \partial\mathbb{R}^n_+$, then f extends to a harmonic function in G by reflection. This can be easily proved by using the mean value principle of harmonic functions, see Armitage and Gardiner (2001). Similar principles hold for biharmonic and, more generally, for polyharmonic functions, see Duffin (1955) and Huber (1955, 1957). Armitage (1978) showed that the classical reflection principle for f harmonic in G_+ holds when one assumes (instead of f tending to 0 at each point of $G_0 \subset \partial\mathbb{R}^n_+$) that f converges locally in mean to 0 on G_0, that is, for all $(x, 0) \in G_0$ there exists $r > 0$ such that

$$\lim_{t \to 0+} \int_{|y-x|<r} f(y, t)\, dy = 0 \tag{1.1}$$

In higher real dimensions, Martio and Rickman (1972) introduced the reflection principle for quasiregular mappings as a generalization of the original result for plane analytic functions. Later on, Martio (1981) showed that the reflection principle holds for solutions of certain elliptic partial differential equations, and he also treated further the reflection principle for quasiregular mappings. Recently, Martio (2009) proved an equivalent principle for quasiminimizers in $\mathbb{R}^n$.

The real Monge–Ampère equation

$$\det D^2 u = \det \left[\frac{\partial^2 u}{\partial x_j \partial x_k} \right] = f \tag{1.2}$$

is a second-order partial differential equation. It is fully non-linear which means that it is not elliptic, in general. If $\Omega \subset \mathbb{R}^n$ is open, then the Equation 1.2 is elliptic only for $u \in C^2(\Omega)$ that is uniformly convex at each point of Ω; and for such a solution to exist, we must also have f positive, see Gilbarg and Trudinger (1983). Roots of the real Monge–Ampère equation go back to the time of G. Monge (1784) in the end of the eighteenth century and Ampère (1820) in the beginning of the nineteenth century, but the mathematical theory including a few variants of weak solutions has mainly been developed during the latter part of the twentieth century, see e.g. Aleksandrov (1961), Bakelman (1957, 1983), Pogorelov (1971), Lions (1983), Caffarelli, Nirenberg, and Spruck (1984). Self-contained expositions of the real Monge–Ampère equation have been written by Pogorelov (1964) and later on by Gutiérrez (2001). Moreover, the second and later editions of the distinguished monograph in second-order partial differential equations by Gilbarg and Trudinger (1983) contain a part devoted to the Monge–Ampère equation.

We consider classical solutions of the homogeneous real Monge–Ampère Equation 1.2 where $f \equiv 0$. For example, all twice continuously differentiable real valued functions which are constant with respect to at least one variable satisfy the homogeneous Monge–Ampère equation, because then the Hesse matrix $D^2 u = \left[\frac{\partial^2 u}{\partial x_j \partial x_k} \right]$ contains at least one zero row and one zero column. Note that $\det D^2 u$ is, in fact, the Jacobian determinant of the partial derivatives $\partial u / \partial x_1, \dots, \partial u / \partial x_n$ of a function u.

In the one-dimensional case, the Monge–Ampère operator coincides with the Laplace operator since $\det D^2 u = \Delta u = u''$ for any twice differentiable function u. Therefore, reflection theory for the homogeneous Monge–Ampère equation provides nothing new in the case $n = 1$. However, we state our results and give proofs so that the one-dimensional case is included.

In higher dimensions $n \geqslant 2$, harmonic functions do not necessarily satisfy the homogeneous Monge–Ampère equation. The contrary is neither true: Twice continuously differentiable functions u satisfying the homogeneous Monge–Ampère equation are not necessarily harmonic. However, there are functions which satisfy both the homogeneous Laplace equation and the homogeneous

Monge–Ampère equation; all constant functions and first-order polynomials which are constant with respect to at least one variable, for example.

2. Preliminaries including notation and terminology

We first set central notation connected to the reflection in $\mathbb{R}^n$. Let G_+ be a domain in $\mathbb{R}^n_+ = \{x = (x_1, \ldots, x_n) \in \mathbb{R}^n : x_n < 0\}$. Let $P : \mathbb{R}^n \to \mathbb{R}^n$ be the *reflection* with respect to $\partial\mathbb{R}^n_+$, that is, $P(x) = P(x_1, \ldots, x_n) = (x_1, \ldots, x_{n-1}, -x_n)$. Suppose that there is a non-empty set $G_0 \subset \partial\mathbb{R}^n_+$ open in ∂G_+. Set $G = G_+ \cup G_0 \cup G_-$ where $G_- = PG_+$. Then G is a domain (open and connected set) in $\mathbb{R}^n$. Suppose that a function $u : G_+ \to \mathbb{R}$ satisfies the following boundary condition on G_0:

$$\lim_{\substack{x \to x_0 \\ x \in G_+}} u(x) = 0 \quad \text{for all } x_0 \in G_0 \tag{2.1}$$

Note that the boundary condition 2.1 is stronger than the boundary condition 1.1 used by Armitage (1978). In fact, we need to assume a much stronger boundary condition than 2.1 to ensure sufficiently nice behaviour of partial derivatives across the reflection boundary G_0, see 4.8 and 4.9.

We define the *odd reflected function* $\tilde{u} : G \to \mathbb{R}$,

$$\tilde{u}(x) = \begin{cases} u(x), & x \in G_+, \\ 0, & x \in G_0, \\ -u(P(x)), & x \in G_-. \end{cases} \tag{2.2}$$

Formula 2.2 is mainly used for a reflected function since the minus sign in the reflected domain G_- guarantees many useful properties towards the reflection boundary G_0.

All rather simple but substantial examples of this paper are given in the plane. The upper half plane is denoted by $\mathbb{R}^2_+ = \{(x, y) \in \mathbb{R}^2 : y < 0\}$ and the upper half unit disk by $B_+ = \{(x, y) \in \mathbb{R}^2 : x^2 + y^2 < 1 \text{ and } y < 0\}$. Write $B_0 = \{(x, y) \in \mathbb{R}^2 : -1 < x < 1 \text{ and } y = 0\}$. Then $B = B_+ \cup B_0 \cup B_-$ is the open unit disk where $B_- = PB_+$.

Our first example emphasizes that the second-order differentiability of a function may get broken on the reflection boundary.

Example 2.3 Let $u : B_+ \to R$ be the function $u(x, y) = y^2$ where $(x, y) \in B_+$. Obviously, $u \in C^2(B_+)$ and it extends continuously to zero in B_0. Now for every $(x, y) \in B_-$ we have $\tilde{u}(x, y) = -u(P(x, y)) = -u(x, -y) = -(-y)^2 = -y^2$, and hence

$$\frac{\partial^2 \tilde{u}}{\partial y^2}(x, y) = \begin{cases} 2, & (x, y) \in B_+, \\ -2, & (x, y) \in B_-, \end{cases}$$

but $\frac{\partial^2 \tilde{u}}{\partial y^2}(x_0, 0)$ does not exist for any $(x_0, 0) \in B_0$. In addition, the limit

$$\lim_{\substack{(x,y) \to (x_0, 0) \\ y \neq 0}} \frac{\partial^2 \tilde{u}}{\partial y^2}(x, y)$$

does not exist for any $(x_0, 0) \in B_0$.

Consequently, if a given function u is smooth (at least twice differentiable) and extends continuously to zero in the reflection boundary, but we have no assumptions giving extra regularity towards the boundary, then the odd reflected function $\tilde{u}$ is not necessarily smooth though it is continuous by the boundary condition 2.1 and definition 2.2.

Example 2.4 The function u given in Example 2.3 and therefore also $\tilde{u}$ satisfy the homogeneous Monge–Ampère equation in B_+. Moreover, $\tilde{u}$ satisfies the homogeneous Monge–Ampère equation in B_-, and hence the limit

$$\lim_{\substack{(x,y)\to(x_0,0)\\ y\neq 0}} \det D^2\tilde{u}(x,y) = 0$$

In particular, the limit exists for every $(x_0,0) \in B_0$. However, the second-order partial derivative $\frac{\partial^2\tilde{u}}{\partial y^2}$ does not exist in B_0, and hence $\det D^2\tilde{u}(x_0,0)$ does not exist for any $(x_0,0) \in B_0$.

Therefore, we observe that nice limiting behaviour of $\det D^2\tilde{u}$ does not guarantee the homogeneous Monge–Ampere equation to hold in the reflection boundary although it holds everywhere outside.

Example 2.5   Still, let u be given like in Example 2.3. If we define a function $\hat{u}: B \to \mathbb{R}$ by

$$\hat{u}(x,y) = \begin{cases} y^2, & (x,y) \in B_+, \\ 0, & (x,y) \in B_0, \\ y^2, & (x,y) \in B_-, \end{cases}$$

which means that $\hat{u}(x,y) = y^2$ for all $(x,y) \in B$, then $\hat{u} \in C^2(B)$ and it satisfies the homogeneous Monge–Ampère equation in B.

Motivated by the previous Example 2.5 we introduce the following variant of the reflected function. Suppose that a function $u: G_+ \to \mathbb{R}$ satisfies the boundary condition 2.1. Then the even reflected function $\hat{u}: G \to \mathbb{R}$ is given by

$$\hat{u}(x) = \begin{cases} u(x), & x \in G_+, \\ 0, & x \in G_0, \\ u(P(x)), & x \in G_-. \end{cases} \tag{2.6}$$

Both reflected functions $\tilde{u}$ and $\hat{u}$ adopt continuity of a function u which satisfies the boundary condition 2.1. Reflected functions provide optional methods to extend functions and solutions of equations across a flat boundary. It may happen that just one of the reflected functions works or both of them work for the purpose the extension is needed . However, there are situations where the extension over a flat boundary is available by using neither of the reflected functions, see Section 6.

Remark 2.7   Our setting to study reflection is valid for the one-dimensional case, indeed. Then $G_+ = (0,a)$ where $a > 0$, $G_0 = \{0\}$, $G_- = (-a,0)$ and $G = (-a,a)$. The boundary condition 2.1 for a function $u: (0,a) \to \mathbb{R}$ means that $\lim_{x\to 0+} u(x) = 0$. The reflected functions $\tilde{u}, \hat{u}: (-a,a) \to \mathbb{R}$ are given by

$$\tilde{u}(x) = \begin{cases} u(x), & x \in (0,a), \\ 0, & x = 0, \\ -u(-x), & x \in (-a,0), \end{cases} \quad \text{and} \quad \hat{u}(x) = \begin{cases} u(x), & x \in (0,a), \\ 0, & x = 0, \\ u(-x), & x \in (-a,0). \end{cases}$$

3. Formulas for the reflected functions

In this section, we give formulas for gradients and Hesse matrices of the reflected functions at points $x \in G_-$ with respect to the reflected points $P(x) \in G_+$. Most of the formulas are probably well known but cannot be found in the literature, so the elementary calculations needed for the proofs are presented here. These formulas are key tools to study the reflection principle for the homogeneous Monge–Ampère equation.

LEMMA 3.1   *Let a point $x \in G_-$ be such that u is differentiable at $P(x) \in G_+$. Then*

$$D\tilde{u}(x) = \left(-\frac{\partial u}{\partial x_1}(P(x)), \dots, -\frac{\partial u}{\partial x_{n-1}}(P(x)), \frac{\partial u}{\partial x_n}(P(x))\right) \tag{3.2}$$

and

$$D\hat{u}(x) = \left(\frac{\partial u}{\partial x_1}(P(x)), \dots, \frac{\partial u}{\partial x_{n-1}}(P(x)), -\frac{\partial u}{\partial x_n}(P(x)) \right).$$
(3.3)

Moreover,

$$D\tilde{u}(x) = -D\hat{u}(x).$$
(3.4)

In particular, if $u \in C^1(G_+)$, then $\tilde{u}, \hat{u} \in C^1(G_-)$.

Proof   Since $P = (P_1, \dots, P_n): \mathbb{R}^n \to \mathbb{R}^n$ is given coordinately by

$$P_k(x) = \begin{cases} x_k, & k = 1, \dots, n-1, \\ -x_k, & k = n, \end{cases}$$

we have

$$\frac{\partial P_k}{\partial x_j}(x) = \begin{cases} 1, & j = k = 1, \dots, n-1, \\ -1, & j = k = n, \\ 0, & \text{otherwise.} \end{cases}$$

Now by the chain rule of partial derivatives

$$\frac{\partial \tilde{u}}{\partial x_j}(x) = \frac{\partial(-(u \circ P))}{\partial x_j}(x) = -\frac{\partial(u \circ P)}{\partial x_j}(x) = -\sum_{k=1}^n \frac{\partial u}{\partial x_k}(P(x)) \frac{\partial P_k}{\partial x_j}(x)$$

$$= \begin{cases} -\frac{\partial u}{\partial x_j}(P(x)), & j = 1, \dots, n-1, \\ \frac{\partial u}{\partial x_j}(P(x)), & j = n. \end{cases}$$
(3.5)

Therefore, formula 3.2 is valid. Formula 3.4 follows directly from the identity $\hat{u} = -\tilde{u}$ in G_- and then 3.3 follows from 3.2 and 3.4.

LEMMA 3.6   Let a point $x \in G_-$ be such that u is twice differentiable at $P(x) \in G_+$. Then

$$D^2 \tilde{u}(x) = \begin{bmatrix} -\frac{\partial^2 u}{\partial x_1^2}(P(x)) & \cdots & -\frac{\partial^2 u}{\partial x_1 \partial x_{n-1}}(P(x)) & \frac{\partial^2 u}{\partial x_1 \partial x_n}(P(x)) \\ \vdots & \ddots & \vdots & \vdots \\ -\frac{\partial^2 u}{\partial x_{n-1} \partial x_1}(P(x)) & \cdots & -\frac{\partial^2 u}{\partial x_{n-1}^2}(P(x)) & \frac{\partial^2 u}{\partial x_{n-1} \partial x_n}(P(x)) \\ \frac{\partial^2 u}{\partial x_n \partial x_1}(P(x)) & \cdots & \frac{\partial^2 u}{\partial x_n \partial x_{n-1}}(P(x)) & -\frac{\partial^2 u}{\partial x_n^2}(P(x)) \end{bmatrix}$$
(3.7)

and

$$D^2 \hat{u}(x) = \begin{bmatrix} \frac{\partial^2 u}{\partial x_1^2}(P(x)) & \cdots & \frac{\partial^2 u}{\partial x_1 \partial x_{n-1}}(P(x)) & -\frac{\partial^2 u}{\partial x_1 \partial x_n}(P(x)) \\ \vdots & \ddots & \vdots & \vdots \\ \frac{\partial^2 u}{\partial x_{n-1} \partial x_1}(P(x)) & \cdots & \frac{\partial^2 u}{\partial x_{n-1}^2}(P(x)) & -\frac{\partial^2 u}{\partial x_{n-1} \partial x_n}(P(x)) \\ -\frac{\partial^2 u}{\partial x_n \partial x_1}(P(x)) & \cdots & -\frac{\partial^2 u}{\partial x_n \partial x_{n-1}}(P(x)) & \frac{\partial^2 u}{\partial x_n^2}(P(x)) \end{bmatrix}.$$
(3.8)

Moreover,

$$D^2 \tilde{u}(x) = -D^2 \hat{u}(x).$$
(3.9)

In particular, if $u \in C^2(G_+)$, then $\tilde{u}, \hat{u} \in C^2(G_-)$.

Proof The chain rule for partial derivatives implies now together with formula 3.5 that

$$
\frac{\partial^2 \tilde{u}}{\partial x_i \partial x_j}(x) =
\begin{cases}
\frac{\partial}{\partial x_i}\left(-\frac{\partial u}{\partial x_j}(P(x))\right) = -\frac{\partial}{\partial x_i}\left(\left(\frac{\partial u}{\partial x_j}\circ P\right)(x)\right), & j = 1,\dots,n-1, \\[2mm]
\frac{\partial}{\partial x_i}\left(\frac{\partial u}{\partial x_j}(P(x))\right) = \frac{\partial}{\partial x_i}\left(\left(\frac{\partial u}{\partial x_j}\circ P\right)(x)\right), & j = n,
\end{cases}
$$

$$
=
\begin{cases}
-\sum_{k=1}^{n}\frac{\partial^2 u}{\partial x_k \partial x_j}(P(x))\frac{\partial P_k}{\partial x_i}(x), & j = 1,\dots,n-1, \\[2mm]
\sum_{k=1}^{n}\frac{\partial^2 u}{\partial x_k \partial x_j}(P(x))\frac{\partial P_k}{\partial x_i}(x), & j = n,
\end{cases}
\tag{3.10}
$$

$$
=
\begin{cases}
-\frac{\partial^2 u}{\partial x_i \partial x_j}(P(x)), & i,j = 1,\dots,n-1 \text{ or } i,j = n, \\[2mm]
\frac{\partial^2 u}{\partial x_i \partial x_j}(P(x)), & \text{otherwise.}
\end{cases}
$$

Therefore, formula 3.7 is valid. Formula 3.9 follows directly from the identity $\hat{u} = -\tilde{u}$ in G_- and then 3.8 follows from 3.7 and 3.9.

Since the Laplacian is the trace of the Hesse matrix, we observe from the formulas 3.7 and 3.8 that

$$
\Delta \hat{u}(x) = -\Delta \tilde{u}(x) = \Delta u(P(x))
\tag{3.11}
$$

for all $x \in G_-$ such that u is twice differentiable at $P(x) \in G_+$. It appears that we obtain corresponding equations for the determinants of the Hesse matrices of the reflected functions $\tilde{u}$ and $\hat{u}$, which is favourable since we study reflection for the homogeneous Monge–Ampère equation. These equations are stated and proved next.

THEOREM 3.12   *Let a point $x \in G_-$ be such that u is twice differentiable at $P(x) \in G_+$. Then*

$$
\det D^2 \tilde{u}(x) = (-1)^n \det D^2 u(P(x))
\tag{3.13}
$$

and

$$
\det D^2 \hat{u}(x) = \det D^2 u(P(x)).
\tag{3.14}
$$

In particular,

$$
\det D^2 \tilde{u}(x) = (-1)^n \det D^2 \hat{u}(x).
\tag{3.15}
$$

Proof We apply elementary properties of determinants to matrices 3.7 and 3.8 getting

$$
\det D^2 \tilde{u}(x) = \det(-D^2 u(P(x))) = (-1)^n \det D^2 u(P(x))
$$

and

$$
\det D^2 \hat{u}(x) = \det D^2 u(P(x)).
$$

Formula 3.15 follows now directly from formulas 3.13 and 3.14, or alternatively from formula 3.9.

4. Differentiability of the reflected functions

In this section, we present some examples and results clarifying the second-order differentiability of functions under reflection. We find necessary conditions for the existence and continuity of second-order partial derivatives in G. Evidently, our study of classical solutions of the Monge–Ampère equation requires that all second-order partial derivatives exist which justifies our goal.

It is clear that the reflected functions $\tilde{u}$ and $\hat{u}$ are continuous in G whenever the original function u is continuous in G_+ and satisfies the boundary condition 2.1. In Section 3, we have confirmed that if u is once or twice continuously differentiable in G_+, then $\tilde{u}$ and $\hat{u}$ are once or twice continuously differentiable in G_-, respectively. Therefore, differentiability requires extra care only in the reflection boundary G_0. Indeed, if $u \in C^2(G_+)$ satisfies the boundary condition 2.1, the reflected functions $\tilde{u}$ and $\hat{u}$ may behave badly in G_0. It helps none if u satisfies, in addition, the homogeneous Monge–Ampère equation in G_+. This can be seen by the following examples.

Example 4.1 The function $w: B_+ \to \mathbb{R}$, $w(x,y) = \sqrt{y}$, is C^2 in B_+ and satisfies the boundary condition 2.1. However,

$$\lim_{\substack{(x,y)\to(x_0,0) \\ (x,y)\in B_+}} \frac{\partial w}{\partial y}(x,y) = \lim_{\substack{y\to 0 \\ y>0}} \frac{1}{2\sqrt{y}} = \infty,$$

and the first order partial derivatives $\frac{\partial \tilde{w}}{\partial y}(x_0,0)$ and $\frac{\partial \hat{w}}{\partial y}(x_0,0)$ do not exist for any $(x_0,0) \in B_0$. In particular, $\tilde{w}, \hat{w} \notin C^1(B)$.

Example 4.2 The function $v: B_+ \to \mathbb{R}$, $v(x,y) = y^2 \sin(\frac{1}{y})$, is C^2 in B_+ and satisfies the boundary condition 2.1. Hence $\tilde{v}$ and $\hat{v}$ are continuous in B. The first-order partial derivatives of $\tilde{v}$ and $\hat{v}$ with respect to both x and y exist in B, indeed, we have $\frac{\partial \tilde{v}}{\partial x}(x,y) = \frac{\partial \hat{v}}{\partial x}(x,y) = 0$ for every $(x,y) \in B$,

$$\frac{\partial \tilde{v}}{\partial y}(x,y) = \begin{cases} -\cos\left(\frac{1}{y}\right) + 2y\sin\left(\frac{1}{y}\right), & (x,y) \in B_+, \\ 0, & (x,y) \in B_0, \\ \cos\left(-\frac{1}{y}\right) - 2y\sin\left(-\frac{1}{y}\right), & (x,y) \in B_-. \end{cases}$$

and

$$\frac{\partial \hat{v}}{\partial y}(x,y) = \begin{cases} -\cos\left(\frac{1}{y}\right) + 2y\sin\left(\frac{1}{y}\right), & (x,y) \in B_+, \\ 0, & (x,y) \in B_0, \\ -\cos\left(-\frac{1}{y}\right) + 2y\sin\left(-\frac{1}{y}\right), & (x,y) \in B_-. \end{cases}$$

However,

$$\lim_{(x,y)\to(x_0,0)} \frac{\partial \tilde{v}}{\partial y}(x,y) \quad \text{and} \quad \lim_{(x,y)\to(x_0,0)} \frac{\partial \hat{v}}{\partial y}(x,y)$$

do not exist for any $(x_0,0) \in G_0$ because sine and cosine of $\frac{1}{y}$ and $-\frac{1}{y}$ oscillate as y tends to 0. Consequently, $\frac{\partial \tilde{v}}{\partial y}$ and $\frac{\partial \hat{v}}{\partial y}$ are not continuous at any $(x_0,0) \in B_0$. In particular, $\tilde{u}, \hat{u} \notin C^1(B)$, even though all first-order partial derivatives of $\tilde{u}$ and $\hat{u}$ exist at every point of B.

Observe that in the previous examples both w and v satisfy the homogeneous Monge–Ampère equation in B_+. These counterexamples are important giving us two essential observations. If a function u is (twice) differentiable in G_+ and satisfies the boundary condition 2.1, then the reflected functions $\tilde{u}$ and $\hat{u}$ are not always differentiable in G. On the other hand, if u is (twice) continuously differentiable, then the reflected functions $\tilde{u}$ and $\hat{u}$ may be differentiable but not continuously differentiable in G.

Our primary requirement is that the studied functions are twice continuously differentiable, even though we need, in principle, the existence of all second-order partial derivatives only. Of course, we present only such conditions which are not true for all $u \in C^2(G_+)$ satisfying the boundary condition 2.1. In the first theorem, we present a necessary condition for the odd reflected function $\tilde{u}$ such that the second-order partial derivatives exist and are continuous in G_0.

THEOREM 4.3   Let $u \in C^2(G_+)$ satisfy the boundary condition 2.1. If $\tilde{u} \in C^2(G)$, then

$$\frac{\partial^2 \tilde{u}}{\partial x_n^2}(x_0) = 0 \qquad\qquad (4.4)$$

for every $x_0 \in G_0$.

Proof   Suppose that $\tilde{u} \in C^2(G)$ and $x_0 \in G_0$. Now 3.10 yields

$$\lim_{\substack{x \to x_0 \\ x \in G_-}} \frac{\partial^2 \tilde{u}}{\partial x_n^2}(x) = -\lim_{\substack{x \to x_0 \\ x \in G_-}} \frac{\partial^2 \tilde{u}}{\partial x_n^2}(P(x)) = -\lim_{\substack{x \to x_0 \\ x \in G_+}} \frac{\partial^2 \tilde{u}}{\partial x_n^2}(x),$$

and hence the limit $\lim_{x \to x_0} \frac{\partial^2 \tilde{u}}{\partial x_n^2}(x)$ exists if and only if

$$\lim_{\substack{x \to x_0 \\ x \in G_-}} \frac{\partial^2 \tilde{u}}{\partial x_n^2}(x) = \lim_{\substack{x \to x_0 \\ x \in G_+}} \frac{\partial^2 \tilde{u}}{\partial x_n^2}(x) = 0$$

The continuity of $\frac{\partial^2 \tilde{u}}{\partial x_n^2}$ at x_0 implies that the Equation 4.4 holds.

Next, we give a necessary condition for the even reflected function $\hat{u}$ such that, firstly, the first-order partial derivatives exist and are continuous in G_0, and secondly, the second-order partial derivatives exist in G_0.

THEOREM 4.5   Let $u \in C^2(G_+)$ satisfy the boundary condition 2.1. If $\hat{u} \in C^1(G)$, then

$$\frac{\partial \hat{u}}{\partial x_n}(x_0) = 0 \qquad\qquad (4.6)$$

for every $x_0 \in G_0$.

Proof   Suppose that $\hat{u} \in C^1(G)$ and $x_0 \in G_0$. It follows from 3.3 that

$$\lim_{\substack{x \to x_0 \\ x \in G_-}} \frac{\partial \hat{u}}{\partial x_n}(x) = -\lim_{\substack{x \to x_0 \\ x \in G_-}} \frac{\partial \hat{u}}{\partial x_n}(P(x)) = -\lim_{\substack{x \to x_0 \\ x \in G_+}} \frac{\partial \hat{u}}{\partial x_n}(x),$$

and hence the limit $\lim_{x \to x_0} \frac{\partial \hat{u}}{\partial x_n}(x)$ exists if and only if

$$\lim_{\substack{x \to x_0 \\ x \in G_-}} \frac{\partial \hat{u}}{\partial x_n}(x) = \lim_{\substack{x \to x_0 \\ x \in G_+}} \frac{\partial \hat{u}}{\partial x_n}(x) = 0$$

Therefore, continuity of $\frac{\partial \hat{u}}{\partial x_n}$ at x_0 implies that the Equation 4.6 holds.

COROLLARY 4.7   Let $u \in C^2(G_+)$ satisfy the boundary condition 2.1. If all second-order partial derivatives of $\hat{u}$ exist in G_0, then 4.6 holds for every $x_0 \in G_0$.

Proof   The existence of the second-order partial derivatives of $\hat{u}$ in G_0 implies continuity of the first order partial derivatives of $\hat{u}$ in G_0, therefore $\hat{u} \in C^1(G)$. Theorem 4.5 yields now that the Equation 4.6 holds for every $x_0 \in G_0$.

Note that the necessary conditions 4.4 and 4.6 concern only the nth first- and second-order partial derivatives of the reflected functions $\tilde{u}$ and $\hat{u}$. In our rather restrictive setting other partial derivatives of the reflected functions are not so crucial.

Finally, to ensure that the determinant of the Hesse matrix (that is, the Monge–Ampère operator) of a reflected function is defined in G_0, the second-order partial derivatives need to behave nicely around G_0. Therefore, whenever $u : G_+ \to \mathbb{R}$ is C^2, we set the following boundary conditions on G_0:

$$\lim_{x \to x_0} \frac{\partial^2 \tilde{u}}{\partial x_j \partial x_k}(x) = \frac{\partial^2 \tilde{u}}{\partial x_j \partial x_k}(x_0) \quad \text{for all } x_0 \in G_0 \text{ and for every } j, k = 1, \dots, n, \tag{4.8}$$

$$\lim_{x \to x_0} \frac{\partial^2 \hat{u}}{\partial x_j \partial x_k}(x) = \frac{\partial^2 \hat{u}}{\partial x_j \partial x_k}(x_0) \quad \text{for all } x_0 \in G_0 \text{ and for every } j, k = 1, \dots, n. \tag{4.9}$$

We will see (and have partly seen already) that if $u \in C^2(G_+)$, it may happen that none, only one, or both of the conditions 4.8 and 4.9 hold. It is clear that if $u \in C^2(G_+)$ and the boundary condition 4.8 holds, then $\tilde{u} \in C^2(G)$. Correspondingly, if $u \in C^2(G_+)$ and the boundary condition 4.9 holds, then $\hat{u} \in C^2(G)$. The boundary conditions 4.8 and 4.9 mean that all second order partial derivatives have continuous limits on the reflection boundary.

5. Reflection principles for the homogeneous real Monge–Ampère equation

We are ready to state our first reflection principle for the homogeneous Monge–Ampère equation.

THEOREM 5.1   Let $u \in C^2(G_+)$ satisfy the boundary conditions 2.1 and 4.8. If u satisfies the equation $\det D^2 u = 0$ in G_+, then the odd reflected function $\tilde{u}$ satisfies $\det D^2 \tilde{u} = 0$ in G.

Proof   Let $x \in G_-$. Then by 3.13,

$$\det D^2 \tilde{u}(x) = (-1)^n \det D^2 u(P(x)) = 0,$$

because $P(x) \in G_+$. Hence $\tilde{u}$ satisfies the homogeneous Monge–Ampère equation in G_-, and further, in the union $G_+ \cup G_-$, since it is clear that $\det D^2 \tilde{u} = 0$ in G_+ where $\tilde{u} \equiv u$.

We need to show that $\tilde{u}$ satisfies the homogeneous Monge–Ampère equation in G_0. If $n = 1$, then 0 is the only point in G_0 and the continuity of $D^2 \tilde{u} = \tilde{u}''$ at 0 yields

$$\det D^2 \tilde{u}(0) = \tilde{u}''(0) = \lim_{x \to 0} \tilde{u}''(x) = 0$$

because $\tilde{u}'' = \det D^2 \tilde{u} = 0$ in $G_+ \cup G_-$.

Suppose then that $n \geqslant 2$ and let $x_0 \in G_0$. Since $\tilde{u} \equiv 0$ in G_0, $\tilde{u}$ is constant in G_0 with respect to the variables $x_1, \dots, x_{n-1}$. Hence we have $\frac{\partial^2 \tilde{u}}{\partial x_j \partial x_k}(x_0) = 0$ for each $j, k = 1, \dots, n - 1$. If $n = 2$, continuity of the second-order partial derivatives yields

$$\det D^2 \tilde{u}(x_0) = \det \begin{bmatrix} 0 & \frac{\partial^2 \tilde{u}}{\partial x_1 \partial x_2}(x_0) \\ \frac{\partial^2 \tilde{u}}{\partial x_2 \partial x_1}(x_0) & \frac{\partial^2 \tilde{u}}{\partial x_2^2}(x_0) \end{bmatrix} = -\left(\frac{\partial^2 \tilde{u}}{\partial x_1 \partial x_2}(x_0) \right)^2$$

$$= \lim_{x \to x_0} \left(\frac{\partial^2 \tilde{u}}{\partial x_1^2}(x) \frac{\partial^2 \tilde{u}}{\partial x_2^2}(x) \left(\frac{\partial^2 \tilde{u}}{\partial x_1 \partial x_2}(x) \right)^2 \right) = \lim_{x \to x_0} \det D^2 \tilde{u}(x) = 0,$$

because $\det D^2 \tilde{u} = 0$ in $G_+ \cup G_-$. Otherwise, if $n \geqslant 3$,

$$\det D^2 \tilde{u}(x_0) = \det \begin{bmatrix} 0 & \cdots & 0 & \frac{\partial^2 \tilde{u}}{\partial x_1 \partial x_n}(x_0) \\ \vdots & \ddots & \vdots & \vdots \\ 0 & \cdots & 0 & \frac{\partial^2 \tilde{u}}{\partial x_{n-1} \partial x_n}(x_0) \\ \frac{\partial^2 \tilde{u}}{\partial x_n \partial x_1}(x_0) & \cdots & \frac{\partial^2 \tilde{u}}{\partial x_n \partial x_{n-1}}(x_0) & \frac{\partial^2 \tilde{u}}{\partial x_n^2}(x_0) \end{bmatrix} = 0,$$

since in the cofactor expansion along the last row the last cofactor matrix is the zero matrix and other cofactor matrices have $n - 2$ zero columns, that is, at least one zero column. We conclude that

$\tilde{u}$ satisfies the homogeneous Monge–Ampère equation in G.

Next, we state our second reflection principle for the homogeneous Monge–Ampère equation.

THEOREM 5.2   Let $u \in C^2(G_+)$ satisfy the boundary conditions 2.1 and 4.9. If u satisfies the equation $\det D^2 u = 0$ in G_+, then the even reflected function $\hat{u}$ satisfies $\det D^2 \hat{u} = 0$ in G.

Proof   Let $x \in G_-$. Then by 3.14,

$$\det D^2 \hat{u}(x) = \det D^2 u(P(x)) = 0,$$

because $P(x) \in G_+$. Hence $\hat{u}$ satisfies the homogeneous Monge–Ampère equation in G_-, and further, in the union $G_+ \cup G_-$. The rest of the proof is similar to the end of the proof of Theorem 5.1.

Remark 5.3   In any case, even without having the boundary conditions 4.8 and 4.9, the even reflected functions $\tilde{u}$ and $\hat{u}$ satisfy the equations $\det D^2 \tilde{u} = 0$ and $\det D^2 \hat{u} = 0$ in the open components G_+ and G_- of G. Note that the union $G_+ \cup G_-$ is disconnected since G_0 separates it into two components.

6. Continuation of solutions of the homogeneous real Monge–Ampère equation

If a solution of the homogeneous Monge–Ampère equation satisfies either the boundary condition 4.8 or 4.9 in addition to the boundary condition 2.1, then by Theorems 5.1 and 5.2 an extension of the solution over a flat boundary can always be found by using the reflected functions. Therefore, we may ask if an extension is always available by our two variants of the reflection. And if not, is it nevertheless possible that an extension is available. The answer for the first question is negative but for the second question positive. This can be seen by the following example.

Example 6.1   Let $v : B_+ \to \mathbb{R}$ be the function $v(x,y) = -y + y^2$ if $(x,y) \in B_+$. Then v satisfies the homogeneous Monge–Ampère equation in B_+ and the boundary condition 2.1 on B_0. Firstly,

$$\lim_{\substack{(x,y) \to (x_0,0) \\ (x,y) \in B_+}} \frac{\partial v}{\partial y}(x,y) = \lim_{\substack{y \to 0 \\ y > 0}}(-1 + 2y) = -1$$

for every $(x_0, 0) \in B_0$. Hence by Theorem 4.5, the even reflected function $\hat{v}$ is not C^1 in B. This can easily be seen by a straightforward calculation. Since $\hat{v}(x,y) = y + y^2$ if $(x,y) \in B_-$, we have

$$\lim_{\substack{(x,y) \to (x_0,0) \\ (x,y) \in B_-}} \frac{\partial \hat{v}}{\partial y}(x,y) = \lim_{\substack{y \to 0 \\ y > 0}}(1 + 2y) = 1 \neq -1 = \lim_{\substack{y \to 0 \\ y > 0}}(-1 + 2y) = \lim_{\substack{(x,y) \to (x_0,0) \\ (x,y) \in B_+}} \frac{\partial \hat{v}}{\partial y}(x,y)$$

for every $(x_0, 0) \in B_0$. Hence the first-order partial derivative $\frac{\partial \hat{v}}{\partial y}$ does not exist in B_0. In particular, v does not satisfy the boundary condition 4.9 and it can not be a classical solution of the homogeneous Monge–Ampère equation for any $(x_0, 0) \in B_0$.

Secondly, the second-order partial derivative with respect to the second variable y satisfies

$$\lim_{\substack{(x,y) \to (x_0,0) \\ (x,y) \in B_+}} \frac{\partial^2 v}{\partial y^2}(x,y) = 2$$

for every $(x_0, 0) \in B_0$. Hence by Theorem 4.3, the odd reflected function $\tilde{v}$ is not C^2 in B. Like above, this can be seen by a straightforward calculation. Since $\tilde{v}(x,y) = -y - y^2$ if $(x,y) \in B_-$, we have

$$\lim_{\substack{(x,y) \to (x_0,0) \\ (x,y) \in B_-}} \frac{\partial^2 \tilde{v}}{\partial y^2}(x,y) = -2 \neq 2 = \lim_{\substack{(x,y) \to (x_0,0) \\ (x,y) \in B_+}} \frac{\partial^2 \tilde{v}}{\partial y}(x,y)$$

for every $(x_0, 0) \in B_0$. Hence the second-order partial derivative $\frac{\partial^2 \check{v}}{\partial y^2}$ does not exist in B_0. In particular, v does not satisfy the boundary condition 4.8 and it can not be a classical solution of the homogeneous Monge–Ampère equation for any $(x_0, 0) \in B_0$.

However, the real analytic continuation of v, that is, the function $\check{v}(x, y) = -y + y^2$, $(x, y) \in B$, gives an extension of v over the boundary B_0. Therefore, an extension may exist if it is not available by using either the odd reflected function or the even reflected function.

The class of polynomials is undeniably one of the most important categories of functions. On the other hand, polynomials which are constant with respect to at least one of the variables $x_1, \ldots, x_{n-1}$ and of which every term contains the variable x_n, satisfy both the homogeneous Monge–Ampère equation and the boundary condition 2.1. This means that a large family of polynomials is relevant to our study. Hence, consider finally if for polynomials satisfying the homogeneous Monge–Ampère equation an extension is always available. We already observed in Example 6.1 that an extension for polynomials cannot be always found by our two reflection methods.

Since the nth variable is in a special position in our considerations, we write a polynomial $p: G_+ \to \mathbb{R}$ of degree k in the form

$$p(x_1, \ldots, x_n) = q_0(x_1, \ldots, x_{n-1}) + q_1(x_1, \ldots, x_{n-1})x_n + \ldots + q_k(x_1, \ldots, x_{n-1})x_n^k, \tag{6.2}$$

where each $q_i : G_+ \to R$ is a polynomial of the variables $x_1, \ldots, x_{n-1}$ and of degree k at most, that is,

$$q_i(x_1, \ldots, x_{n-1}) = \sum_{i_1=0}^{k} \cdots \sum_{i_{n-1}=0}^{k} a_{i_1, \ldots, i_{n-1}} x_1^{i_1} \cdots x_{n-1}^{i_{n-1}}, \quad i = 0, 1, 2, \ldots, k. \tag{6.3}$$

Our first lemma gives an equivalent expression to the boundary condition 2.1 for polynomials.

LEMMA 6.4   *A polynomial $p : G_+ \to \mathbb{R}$ satisfies the boundary condition 2.1 if and only if q_0 is the zero polynomial.*

Proof   Since we suppose that the reflection boundary G_0 is non-empty and open in ∂G_+, there is a point $\xi = (\xi_1, \ldots, \xi_{n-1}, 0) \in G_0$ such that $\xi_i \neq 0$ for every $i = 1, \ldots, n-1$. Hence from 6.2 we see that

$$\lim_{\substack{x \to \xi \\ x \in G_+}} p(x) = \lim_{\substack{(x_1, \ldots, x_n) \mapsto (\xi_1, \ldots, \xi_{n-1}, 0) \\ (x_1, \ldots, x_n) \in G_+}} p(x_1, \ldots, x_n) = q_0(\xi_1, \ldots, \xi_{n-1}) = 0$$

if and only if $q_0 \equiv 0$.

If $x = (x_1, \ldots, x_n) \in G_-$, then

$$\tilde{p}(x_1, \ldots, x_n) = -p(x_1, \ldots, x_{n-1}, -x_n) = \sum_{i=0}^{k} (-1)^{i+1} q_i(x_1, \ldots, x_{n-1}) x_n^i \tag{6.5}$$

and

$$\hat{p}(x_1, \ldots, x_n) = p(x_1, \ldots, x_{n-1}, -x_n) = \sum_{i=0}^{k} (-1)^i q_i(x_1, \ldots, x_{n-1}) x_n^i. \tag{6.6}$$

In case of polynomials, the next lemma gives equivalent expressions to the conditions 4.4 and 4.6 being necessary for the boundary conditions 4.8 and 4.9.

LEMMA 6.7   *Let a polynomial $p : G_+ \to \mathbb{R}$ be such that $q_0 \equiv 0$. Then*

(i)   the even reflected function $\hat{p}$ satisfies 4.6 if and only if $q_1 \equiv 0$,

(ii) the odd reflected function $\tilde{p}$ satisfies 4.4 if and only if $q_2 \equiv 0$.

Proof   As in the proof of Lemma 6.4, we may suppose that there is a point $\xi = (\xi_1, \ldots \xi_{n-1}, 0) \in G_0$ such that $\xi_i \neq 0$ for every $i = 1, \ldots, n-1$. Now

$$\lim_{\substack{x \to \xi \\ x \in G_+}} \frac{\partial p}{\partial x_n}(x) = \lim_{\substack{(x_1, \ldots, x_n) \to (\xi_1, \ldots, \xi_{n-1}, 0) \\ (x_1, \ldots, x_n) \in G_+}} \sum_{i=1}^{k} i q_i(x_1, \ldots, x_{n-1}) x_n^{i-1} = q_1(\xi_1, \ldots, \xi_{n-1}) = 0$$

if and only if $q_1 \equiv 0$. Then by 6.6 we have

$$\lim_{\substack{x \to \xi \\ x \in G_-}} \frac{\partial \hat{p}}{\partial x_n}(x) = \lim_{\substack{(x_1, \ldots, x_n) \to (\xi_1, \ldots, \xi_{n-1}, 0) \\ (x_1, \ldots, x_n) \in G_-}} \sum_{i=2}^{k} (-1)^i i q_i(x_1, \ldots, x_{n-1}) x_n^{i-1} = 0$$

Therefore, 4.6 holds for every $x_0 \in G_0$.

Similarly,

$$\lim_{\substack{x \to \xi \\ x \in G_+}} \frac{\partial^2 p}{\partial x_n^2}(x) = \lim_{\substack{(x_1, \ldots, x_n) \to (\xi_1, \ldots, \xi_{n-1}, 0) \\ (x_1, \ldots, x_n) \in G_+}} \sum_{i=2}^{k} i(i-1) q_i(x_1, \ldots, x_{n-1}) x_n^{i-2} = 2q_2(\xi_1, \ldots, \xi_{n-1}) = 0$$

if and only if $q_2 \equiv 0$. Then by 6.5 we have

$$\lim_{\substack{x \to \xi \\ x \in G_-}} \frac{\partial^2 \tilde{p}}{\partial x_n^2}(x) = \lim_{\substack{(x_1, \ldots, x_n) \to (\xi_1, \ldots, \xi_{n-1}, 0) \\ (x_1, \ldots, x_n) \in G_-}} \sum_{i=3}^{k} (-1)^{i+1} i(i-1) q_i(x_1, \ldots, x_{n-1}) x_n^{i-2} = 0$$

Therefore, 4.4 holds for every $x_0 \in G_0$.

LEMMA 6.8   Let a polynomial $p : G_+ \to \mathbb{R}$ be such that $q_0 \equiv 0$. Then all partial derivatives $\frac{\partial p}{\partial x_j}$ extend continuously from G_+ to G_0. Correspondingly, if a polynomial p is defined in G_- and $q_0 \equiv 0$, then all partial derivatives $\frac{\partial p}{\partial x_j}$ extend continuously from G_- to G_0.

Proof   Since p extends continuously to 0 in G_0, we have $\tilde{p}(x) = \hat{p}(x) = p(x)$ for every $x \in G_+ \cup G_0$, meaning that $\tilde{p}$ and $\hat{p}$ have the same terms with the same coefficients in G_0 than p in G_+. Therefore, all partial derivatives $\frac{\partial p}{\partial x_j}$ extend continuously from G_+ to G_0. Note that here $\frac{\partial p}{\partial x_n}$ is considered G_+-sided in G_0 because the G_--sided limit is not defined in G_0. The second part of the lemma follows similarly.

LEMMA 6.9   Let a polynomial $p : G_+ \to \mathbb{R}$ be such that $q_0 \equiv 0$. Then

(i) $\hat{p} \in C^2(G)$ if and only if $q_1 \equiv 0$,

(ii) $\tilde{p} \in C^2(G)$ if and only if $q_2 \equiv 0$.

Proof   By Lemma 6.7(i) and Corollary 4.7, $q_1 \equiv 0$ is a necessary condition to have $\hat{p} \in C^2(G)$. We need show that $q_0 \equiv 0$ and $q_1 \equiv 0$ imply the boundary condition 4.9. By 6.6,

$$\frac{\partial^2 \hat{p}}{\partial x_n^2}(x) = \begin{cases} \sum_{i=2}^{k} i(i-1) q_i(x_1, \ldots, x_{n-1}) x_n^{i-2}, & x \in G_+, \\ \sum_{i=2}^{k} i(i-1)(-1)^i q_i(x_1, \ldots, x_{n-1}) x_n^{i-2}, & x \in G_-, \end{cases}$$

which yields

$$\lim_{x \to x_0} \frac{\partial^2 \hat{p}}{\partial x_n^2}(x) = 2q_2(x_1, \ldots, x_{n-1}) = \frac{\partial^2 \hat{p}}{\partial x_n^2}(x_0)$$

for every $x_0 = (x_1, \ldots, x_{n-1}, 0) \in G_0$. Above, the second equation follows from Lemma 6.8 since partial derivatives of polynomials are polynomials. Otherwise, suppose that $j \neq n$ or $k \neq n$. Then

$$\frac{\partial^2 \hat{p}}{\partial x_j \partial x_k}(x) = \begin{cases} \sum_{i=2}^{k} \frac{\partial^2}{\partial x_j \partial x_k}\left(q_i(x_1, \ldots, x_{n-1})x_n^i\right), & x \in G_+, \\ \sum_{i=2}^{k}(-1)^i \frac{\partial^2}{\partial x_j \partial x_k}\left(q_i(x_1, \ldots, x_{n-1})x_n^i\right), & x \in G_-. \end{cases}$$

When we evaluate the second-order partial derivatives in the sum expressions above, we observe that every term achieved includes variable x_n with power $i - 1 \geq 1$, that is, $i \geq 2$. This implies again by Lemma 6.8 that

$$\lim_{x \to x_0} \frac{\partial^2 \hat{p}}{\partial x_j \partial x_k}(x) = 0 = \frac{\partial^2 \hat{p}}{\partial x_j \partial x_k}(x_0)$$

for every $x_0 = (x_1, \ldots, x_{n-1}, 0) \in G_0$. We conclude that the boundary condition 4.9 holds.

Correspondingly, by Lemma 6.7(ii) and Theorem 4.3, $q_2 \equiv 0$ is a necessary condition to have $\tilde{p} \in C^2(G)$. We need show that $q_0 \equiv 0$ and $q_2 \equiv 0$ imply the boundary condition 4.8. Then by 6.5

$$\frac{\partial^2 \tilde{p}}{\partial x_n^2}(x) = \begin{cases} \sum_{i=3}^{k} i(i-1)q_i(x_1, \ldots, x_{n-1})x_n^{i-2}, & x \in G_+, \\ \sum_{i=3}^{k} i(i-1)(-1)^{i+1}q_i(x_1, \ldots, x_{n-1})x_n^{i-2}, & x \in G_-, \end{cases}$$

which yields by Lemma 6.8 that

$$\lim_{x \to x_0} \frac{\partial^2 \tilde{p}}{\partial x_n^2}(x) = 0 = \frac{\partial^2 \tilde{p}}{\partial x_n^2}(x_0)$$

for every $x_0 = (x_1, \ldots, x_{n-1}, 0) \in G_0$. Otherwise, suppose that $j \neq n$ or $k \neq n$. Then

$$\frac{\partial^2 \tilde{p}}{\partial x_j \partial x_k}(x) = \begin{cases} \frac{\partial^2}{\partial x_j \partial x_k}\left(q_1(x_1, \ldots, x_{n-1})x_n\right) + \sum_{i=3}^{k} \frac{\partial^2}{\partial x_j \partial x_k}\left(q_i(x_1, \ldots, x_{n-1})x_n^i\right), & x \in G_+, \\ \frac{\partial^2}{\partial x_j \partial x_k}\left(q_1(x_1, \ldots, x_{n-1})x_n\right) + \sum_{i=3}^{k}(-1)^{i+1}\frac{\partial^2}{\partial x_j \partial x_k}\left(q_i(x_1, \ldots, x_{n-1})x_n^i\right), & x \in G_-. \end{cases}$$

If $j = n$ and $k \neq n$, then by Lemma 6.8

$$\lim_{x \to x_0} \frac{\partial^2 \tilde{p}}{\partial x_n \partial x_k}(x) = \frac{\partial q_1}{\partial x_k}(x_1, \ldots, x_{n-1}) = \frac{\partial^2 \tilde{p}}{\partial x_n \partial x_k}(x_0)$$

for $x_0 = (x_1, \ldots, x_{n-1}, 0) \in G_0$. Similarly, if $j \neq n$ and $k = n$, then by Lemma 6.8

$$\lim_{x \to x_0} \frac{\partial^2 \tilde{p}}{\partial x_j \partial x_n}(x) = \frac{\partial q_1}{\partial x_j}(x_1, \ldots, x_{n-1}) = \frac{\partial^2 \tilde{p}}{\partial x_j \partial x_n}(x_0)$$

for $x_0 = (x_1, \ldots, x_{n-1}, 0) \in G_0$. If $j \neq n$ and $k \neq n$, then Lemma 6.8 again yields

$$\lim_{x \to x_0} \frac{\partial^2 \tilde{p}}{\partial x_j \partial x_k}(x) = 0 = \frac{\partial^2 \tilde{p}}{\partial x_n \partial x_k}(x_0)$$

for every $x_0 = (x_1, \ldots, x_{n-1}, 0) \in G_0$. We conclude that the boundary condition 4.8 holds.

THEOREM 6.10 Let a polynomial $p : G_+ \to \mathbb{R}$ be such that $q_0 \equiv 0$ and $\det D^2 p = 0$ in G_+. Then

(i) *the even reflected function $\hat{p}$ satisfies $\det D^2 \hat{p} = 0$ in G if and only if $q_1 \equiv 0$,*

(ii) *the odd reflected function $\tilde{p}$ satisfies $\det D^2 \tilde{p} = 0$ in G if and only if $q_2 \equiv 0$.*

Proof By Lemma 6.9, we only need to verify that the homogeneous Monge–Ampère equation holds in G. But for $\hat{p}$ this follows now immediately from Theorem 5.1 and for $\tilde{p}$ from Theorem 5.1.

If a polynomial $p = q_0 + q_1 x_n + \ldots + q_k x_n^k : G_+ \to \mathbb{R}$ is such that $q_1 \not\equiv 0$ and $q_2 \not\equiv 0$, then it follows from Theorem 6.10 that an extension to G cannot be found by using the reflected functions. However, an extension can always be found, which was tentatively observed in Example 6.1.

THEOREM 6.11 *Let a polynomial $p : G_+ \to \mathbb{R}$ be such that $q_0 \equiv 0$ and $\det D^2 p = 0$ in G_+. Then there is an extension $\breve{p} : G \to \mathbb{R}$ of p such that $\breve{p}$ satisfies $\det D^2 \breve{p} = 0$ in G.*

Proof Write $p(x) = q_0(x_1, \ldots, x_{n-1}) + q_1(x_1, \ldots, x_{n-1}) x_n + \ldots + q_k(x_1, \ldots, x_{n-1}) x_n^k$ where $x = (x_1, \ldots, x_n) \in G_+$. If $q_1 \not\equiv 0$ or $q_2 \not\equiv 0$, then by Theorem 6.10 an extension is found by choosing $\breve{p} = \hat{p}$ or $\breve{p} = \tilde{p}$ in G.

Otherwise, and also simultaneously, we may simply extend p to G real analytically so that the polynomial $\breve{p} : G \to \mathbb{R}$ has the same terms with the same coefficients than p in G_+. Note that $\breve{p}$ is then C^2 in G_+. Entries of the Hesse matrix of $\breve{p}$ are polynomials, and hence the determinant $\det D^2 \breve{p}$ is a polynomial as a sum of products of polynomials. Since $\det D^2 \breve{p}(x) = 0$ at every $x \in G_+$ and G_+ is open and non-empty, $\det D^2 \breve{p}$ has uncountably many zeroes. Hence $\det D^2 \breve{p} \equiv 0$ and $\det D^2 \breve{p}(x) = 0$ at every $x \in G$.

In fact, a polynomial p can always be extended real analytically using the latest method. Even the boundary condition 2.1 is not necessary. In particular, our considerations show that an extension of a solution of the homogeneous Monge–Ampère equation is not necessarily unique.

Acknowledgements

The author would like to thank the referees for their valuable comments and suggestions.

Funding

The author received no direct funding for this research.

Author details

Mika Koskenoja[1]

E-mail: mika.koskenoja@helsinki.fi

[1] Department of Mathematics and Statistics, University of Helsinki, P.O. Box 68 (Gustaf Hällströmin katu 2b), FI-00014 Helsinki, Finland.

References

Aleksandrov, A. D. (1961). Certain estimates for the Dirichlet problem. *Doklady Akademii Nauk SSSR, 134*, 1001–1004 (in Russian).

Ampère, A.-M. (1820). Mémoire contenant l'application de la théorie exposée dans le XVII cahier du Journal de l'École Polytechnique, à l'intégration des équations aux differentielles partielles du premier et du second ordre [Memoir containing an application of the theory presented in the XVII issue of Journal de l'École Polytechnique, on integration of partial differential equations of first and second order]. *Journal de l'École Polytechnique, 11*, 1–188 (in French).

Armitage, D. H. (1978). Reflection principles for harmonic and polyharmonic functions. *Journal of Mathematical Analysis and Applications, 65*, 44–55.

Armitage, D. H., & Gardiner, S. J. (2001). *Classical potential theory*. Springer Monographs in Mathematics. London: Springer-Verlag.

Bakelman, I. J. (1957). Generalized solutions of Monge–Ampère equations. *Doklady Akademii Nauk SSSR (N.S.), 114*, 1143–1145 (in Russian).

Bakelman, I. J. (1983). Variational problems and elliptic Monge–Ampère equations. *Journal of Differential Geometry, 18*, 669–699.

Caffarelli, L., Nirenberg, L., & Spruck, J. (1984). The Dirichlet problem for nonlinear second-order elliptic equations. I. Monge–Ampère equations. *Communications on Pure and Applied Mathematics, 37*, 369–402.

Duffin, R. J. (1955). Continuation of biharmonic functions by reflection. *Duke Mathematical Journal, 22*, 313–324.

Gilbarg, D., & Trudinger, N. S. (1983). *Elliptic partial differential equations of second order* (2nd ed., Vol. 224) Grundlehren Math. Wiss. Berlin: Springer-Verlag.

Gutiérrez, C. E. (2001). *The Monge–Ampère equation*. Progress in Nonlinear Differential Equations and their Applications (Vol. 44). Boston: Birkhäuser.

Huber, A. (1955). The reflection principle for polyharmonic functions. *Pacific Journal of Mathematics, 5*, 433–439.

Huber, A. (1957). Correction to the paper "The reflection principle for polyharmonic functions". *Pacific Journal of Mathematics, 7*, 1731.

Lions, P.-L. (1983). Sur les èquations de Monge-Ampère. *Manuscripta Mathematica, 41*, 1–43.

Martio, O. (1981). Reflection principle for solutions of elliptic partial differential equations and quasiregular mappings. *Annales Academiae Scientiarum Fennicae Series A I. Mathematica, 6*, 179–187.

Martio, O. (2009). Reflection principle for quasiminimizers. *Functiones et Approximatio, Commentarii Mathematici, 40*, 165–173.

Martio, O. & Rickman, S. (1972). Boundary behaviour of quasiregular mappings. *Annales Academiae Scientiarum Fennicae Series A I. Mathematica, 507*, 1–17.

Monge, G. (1784). Sur le calcul intégral des équations aux differences partielles [On integral calculus of partial differential equations]. *Mémoires de l'Académie des Sciences.*

Pogorelov, A. V. (1964). *Monge–Ampère equations of elliptic type.* Groningen: Noordhoff.

Pogorelov, A. V. (1971). The regularity of the generalized solutions of the equation $\det(\partial^2 u/\partial x^i \partial x^j) = \varphi(x^1, x^2, ..., x^n) > 0$. *Doklady Akademii Nauk SSSR, 200*, 534–537 (in Russian).

16

Exploring Riemann's functional equation: Revised Version 2.0

Michael Milgram[1]*

*Corresponding author: Michael Milgram, Geometrics Unlimited Ltd.,Box 1484, Deep River, Ontario, Canada K0J 1P0
E-mail: mike@geometrics-unlimited. com
Reviewing editor: Prasanna K. Sahoo, University of Louisville, USA

Abstract: An equivalent, but variant form of Riemann's functional equation is explored, and several discoveries are made. Properties of Riemann's zeta function $\zeta(s)$, from which a necessary and sufficient condition for the existence of zeros in the critical strip, are deduced. This in turn, by an indirect route, eventually produces a simple, solvable, differential equation for $arg(\zeta(s))$ on the critical line $s = 1/2 + i\rho$, the consequences of which are explored, and the "LogZeta" function is introduced. A singular linear transform between the real and imaginary components of ζ and ζ' on the critical line is derived, and an implicit relationship for locating a zero $(\rho = \rho_0)$ on the critical line is found between the arguments of $\zeta(1/2 + i\rho)$ and $\zeta'(1/2 + i\rho)$. Notably, the Volchkov criterion, a Riemann Hypothesis (RH) equivalent, is analytically evaluated and verified to be half equivalent to RH, but RH is not proven. Numerical results are presented, some of which lead to the identification of *anomalous zeros*, whose existence in turn suggests that well-established, traditional derivations such as the Volchkov criterion and counting theorems require re-examination. It is proven that the derivative $\zeta'(1/2 + i\rho)$ will never vanish on the perforated critical line $(\rho \neq \rho_0)$. Traditional asymptotic and counting results are obtained in an untraditional manner, yielding insight into the nature of $\zeta(1/2 + i\rho)$ as well as very accurate asymptotic estimates for distribution bounds and the density of zeros on the critical line.

Subjects: Advanced Mathematics; Analysis - Mathematics; Complex Variables; Integral Transforms & Equations; Mathematical Analysis; Mathematics & Statistics; Number Theory; Science; Special Functions

ABOUT THE AUTHOR

The author has spent most of his career studying and applying mathematics to problems in Physics and Engineering. This led him to develop some expertise in the mathematical field of Special Functions, which in turn led to an interest in the Zeta function, as a diversion from his everyday pursuit of solutions to problems in Monte Carlo simulation and computer-aided visualization of CNC machining.

PUBLIC INTEREST STATEMENT

Prime numbers have fascinated mathematicians since the invention of arithmetic. One of the main discoveries by Bernhard Riemann, one of the greatest nineteenth[th]-century mathematicians, was a link between what is called the "Riemann Zeta function" and the distribution of prime numbers. Since then, mathematicians (and others) have been studying zeta function properties and slowly unveiling its secrets. This is important because prime numbers form the basis for the major encryption algorithms in use today. If you indulge in Internet banking or commerce, you are using prime numbers.

In this paper, a new relationship is deduced that equates an arcane property of the zeta function to simpler mathematical entities whose properties are well known. As an anonymous referee has stated: "... such relation ... would lead to extremely important consequences". Additionally, a surprising numerical discovery is presented which calls into question many previously thought-to-be-well-founded results.

Keywords: Riemann's Zeta function; Riemann's functional equation; zeros; asymptotics; Riemann Hypothesis; anomalous zeros; Volchkov criterion; critical line

1. Introduction

The course of other work has led to an exploration of a not-so-well-known variant of Riemann's functional equation. Thought to be equivalent to the classical equation that relates the functions $\zeta(s)$ and $\zeta(1-s)$, the variant studied here relates $\zeta(1-s)$ and its derivatives $\zeta'(s)$ and $\zeta'(1-s)$. As has been discovered elsewhere (e.g. Guillera, 2013, Eq. 10; Spira, 1973), this variant has proven to be a surprisingly rich source from which possible new properties of $\zeta(s)$ can be unearthed. At a minimum, the rediscovery of previously known results is obtained in a completely different manner from the usual textbook and literature approach. This report is a summary of those explorations.

In Section 2, the notation and other results drawn from the literature are summarized. Section 3 recalls the variant functional equation, from which it is possible to infer the existence of a functional relationship between $\zeta'(s)$ and $\zeta'(1-s)$ that yields necessary and sufficient implicit conditions for locating points $s = s_0$ such that $\zeta(s_0) = 0$. The main result of the next two sub-sections (4.1) and (4.2) is the derivation (and rediscovery) of two implicit, and later shown-to-be-equivalent, equations whose solution(s) locate zeros on the critical line $s = 1/2 + i\rho$. In a third sub-section (4.3), a singular linear transformation relating the real and imaginary components of $\zeta(1/2 + i\rho)$ and $\zeta'(1/2 + i\rho)$ is presented, generating a third implicit equation for locating the non-trivial zeros. Prior to presenting an analytic solution to the first two of these equations in Section 7, and noting the numerical equivalence of the third, the development digresses.

Section 5 introduces a related functional with interesting properties, and in Section 6, those properties are employed to obtain and solve a simple differential equation satisfied by $arg(\zeta(1/2 + i\rho))$ and thereby introduce the LogZeta function, in complete analogy to the well-known LogGamma function (England, Bradford, Davenport, & Wilson, 2013; Weisstein, 2005). Amalgamating all those results in Section 7 yields requisite conditions for the existence of ρ_0 such that $\zeta(1/2 + i\rho_0) = 0$, reproducing similar results obtained elsewhere (de Reyna & Van de Lune, 2014). In Section 8, by analytic integration, the properties of $arg(\zeta(1/2 + i\rho))$ are used to ostensibly verify that the Volchkov Criterion (Volchkov, 1995), advertised as "equivalent" to the Riemann Hypothesis (RH), possesses one-half of that property. Although this does not lead to a proof of RH, it does lead to some insight about the usefulness of so-called "Riemann equivalences" (Conrey & Farmer, n.d.). However, the subsequent Section 9 shows that much of the foregoing, as well as well-known classical results, are in need of revision because of the peculiar numerical properties of $\zeta(1/2 + i\rho)$, from which the phenomenon of *anomalous zeros* is identified. Additionally, the results from Section 6 lead to a proof (in Section 10) that $\zeta'(1/2 + i\rho) \neq 0$ on the perforated critical line without recourse to RH, almost closing the last gap in an investigation initiated many years ago (e.g. Spira, 1973). Section 11 investigates the asymptotic behaviour of $|\zeta(1/2 + i\rho)|$, again reproducing known results. In Section 12, these results are reused to deduce estimates that provide upper and lower bounds for locating the k'th zero on the critical line, as well as the density and maximum separation of such points. In a numerical diversion, the claimed location of zero number "googol" (França and LeClair (2013)) is tested; it is found that the value as declared should probably be googol-1, although its position on the critical line appears to be accurate within the resolution provided. (Aside: The word "googol" traditionally refers to the number 10^{100} and is easily confused with the similarly sounding word "google" usurped by a well-known search engine.)

Throughout, some derivations require considerable analytic perseverance to deal with expressions involving many terms. Such derivations are noted, and are left as exercises for the reader with the suggestion that a computer algebra program be utilized. In general, these computations, although lengthy, do not involve anything other than the use of well-known trigonometric and other identities involving Gamma and related functions (Olver, Lozier, Boisvert, & Clark, 2010). An example of such a calculation, in which the differential equation (6.1) is obtained, is appended to this paper as a Maple worksheet supplement, in Portable Document Format. A list of notations and symbols will be found in Appendix 1. The Maple computer code (Maplesoft, A Division of Waterloo Maple, 2014) is the source used for many of the calculations contained here. When that occurs, references to Maple are simply indicated in the text by the word Maple in parenthesis, in order that calculations can be reproduced,

2. Preamble

A number of known results and notation, required in forthcoming Sections, are quoted here. A basic result used throughout employs the polar form of $\zeta(s)$ as

$$\zeta(s) = e^{i\alpha(s)}|\zeta(s)| \tag{2.1}$$

differing by a sign from that used elsewhere (de Reyna & Van de Lune, 2014). Specific to the "critical line" $s = 1/2 + i\rho$, the specialized form is written

$$\zeta(1/2 + i\rho) = e^{i\alpha}|\zeta| = e^{i\alpha}\sqrt{\zeta_R^2 + \zeta_I^2} \tag{2.2}$$

where explicit dependence on the variable ρ is usually omitted for clarity. Throughout, subscripts "R" and "I" refer to the real and imaginary parts of the associated symbol, respectively; all derivatives are taken with respect to ρ and, always, $\rho \geq 0$. On the critical line, the arguments θ and β of associated functions are defined by

$$\Gamma(1/2 + i\rho) = e^{i\theta}|\Gamma| \tag{2.3}$$

$$\zeta'(1/2 + i\rho) = e^{i\beta}|\zeta'| \tag{2.4}$$

For use in later sections, define

$$f(\rho) = \frac{4\cosh(\pi\rho)}{2\ln(2\pi)\cosh(\pi\rho) - 2\Re(\psi(1/2 + i\rho))\cosh(\pi\rho) + \pi}. \tag{2.5}$$

This function has a simple pole at $\rho_s \equiv \rho = 0.628...$, and for $\rho > \rho_s$, $f(\rho) < 0$. See Section 3.

Although the real functions $|\zeta|^2$ and α are (almost) independent, known relationships (de Reyna & Van de Lune, 2014, Proposition 7 and equivalently Milgram, 2011, Eqs. 3.1, 3.2 and Appendix B) exist between the real and imaginary components of $\zeta(1/2 + i\rho)$, specifically

$$\tan(\alpha) \equiv \frac{\zeta_I}{\zeta_R} = \mathfrak{P} \tag{2.6}$$

where

$$\mathfrak{P} = \frac{C_p\cos(\rho_\pi) + C_m\sin(\rho_\pi) - \sqrt{\pi}}{C_m\cos(\rho_\pi) - C_p\sin(\rho_\pi)}. \tag{2.7}$$

An equivalent form of (Equation 2.7) can be obtained by writing its various terms in polar form, using the representation (de Reyna & Van de Lune, 2014, Eq. 4)

$$\Gamma(1/2 + i\rho) = \sqrt{\frac{\pi}{\cosh(\pi\rho)}}\exp(i\theta), \tag{2.8}$$

alternatively giving

$$\tan(\alpha) \equiv \frac{\zeta_I}{\zeta_R} = \frac{-\cosh(\frac{\pi\rho}{2})\cos(\rho_\theta) - \sinh(\frac{\pi\rho}{2})\sin(\rho_\theta)}{\sinh(\frac{\pi\rho}{2})\cos(\rho_\theta) - \cosh(\frac{\pi\rho}{2})\sin(\rho_\theta)}$$

$$+ \frac{\sqrt{\cosh(\pi\rho)}}{\sinh(\frac{\pi\rho}{2})\cos(\rho_\theta) - \cosh(\frac{\pi\rho}{2})\sin(\rho_\theta)}, \tag{2.9}$$

where the various symbols in the above are defined in Appendix 1. Curiously, relevant to the well-known result (Titchmarsh & Heath-Brown, 1986, Eq. 4.17.2)

$$arg(\zeta(1/2 + i\rho)) = arg(\Gamma(1/4 + i\rho/2)) - \frac{\rho}{2}\log(\pi) - k\pi, \tag{2.10}$$

and left as an exercise for the reader, it may also be shown that

$$\frac{\Im(\Gamma(1/4 + i\rho/2))}{\Re(\Gamma(1/4 + i\rho/2))} = -\mathfrak{P} \tag{2.11}$$

if $\rho_\pi \equiv \rho\log(2\pi)$ is replaced by $\rho\log(2)$ in the definition (Equation 2.7). As has been obtained elsewhere (Milgram, 2011, Eq. 3.7), in the case that $\zeta = \zeta_R = \zeta_I = 0$, corresponding to a non-trivial zero $\zeta(1/2 + i\rho_0) = 0$, the ratio of those quantities appearing on the left-hand side of both (Equation 2.7 and 2.9) satisfies

$$\lim_{\rho \to \rho_0} \frac{\zeta_I}{\zeta_R} = p\left(\frac{\zeta^{(n)}_I}{\zeta^{(n)}_R}\right)^p \tag{2.12}$$

when the zero is of order n, where $p = (-1)^n$, in terms of the appropriate n^{th} derivatives, by invoking l'Hôpital's rule. Also see Equations (6.3 and 6.4) below. Another result that will prove useful is obtained by direct differentiation of Equation (2.2), that being

$$\frac{|\zeta'(1/2 + i\rho)|}{|\zeta(1/2 + i\rho)|} = \frac{\alpha'}{\cos(A - B)}, \tag{2.13}$$

where

$$\begin{aligned} A &\equiv arg(\zeta(1/2 + i\rho)) \\ B &\equiv arg(\zeta'(1/2 + i\rho)) \end{aligned} \tag{2.14}$$

An important distinction must be made between discontinuous functions corresponding to, and denoted by, the notation "arg" and their continuous, multi-sheeted counterparts—the two entities differ by a constant equal to $k\pi$ and k is always an integer. As an example, consider Figure 1 where the imaginary part of the multi-sheeted function (Weisstein, 2005) LogGamma$(1/2 + i\rho)$ is

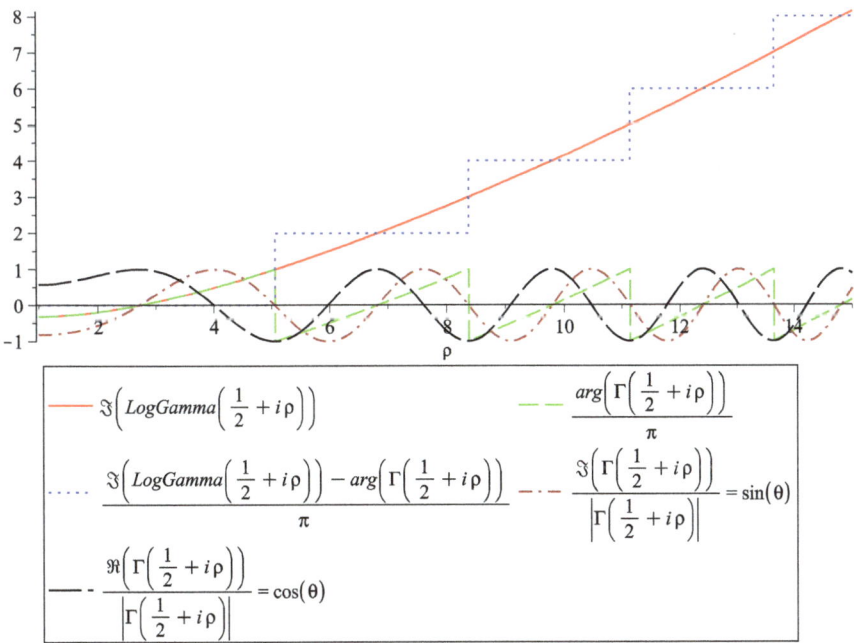

Figure 1. Comparison between $\Im(\text{LogGamma}(1/2 + i\rho))$ **and** $arg(\Gamma(1/2 + i\rho))$, **along with the normalized associated real and imaginary parts of** $\Gamma(1/2 + i\rho)$.

compared to the discontinuous function $arg(\Gamma(1/2 + i\rho))$, in both cases as a function of ρ. The difference, normalized by π is also shown, to demonstrate that it is always an integer "constant", which, however, changes at each discontinuous point of the arg operator. As indicated in the Figure, discontinuities of the argument are associated with a sign change of $\Im(\Gamma(1/2 + i\rho))$ coincident with $\Re(\Gamma(1/2 + i\rho)) < 0$—that is, whenever $arg(\Gamma(1/2 + i\rho)) = \pm\pi$, and the discontinuity is 2π. In Section 9, a similar comparison is made using the function $\zeta(1/2 + i\rho)$.

3. The basic functional equation

The following proofs of sufficiency and necessity are valid if $\zeta(s_0) = 0$ is a simple zero. See Section 10.

3.1. A sufficient condition that $\zeta(s) = 0$

From the functional equation for $\zeta(s)$, that is,

$$\zeta(1-s) = \frac{2\,\Gamma(s)\,\cos(\pi s/2)\,\zeta(s)}{(2\pi)^s} \tag{3.1}$$

we have (Guillera, 2013, Eq. 10 - misprinted; Spira, 1973, Eq. 1)

$$\zeta'(1-s) + \zeta'(s)\,\frac{2\,\Gamma(s)\cos\left(\pi s/2\right)}{\left(2\pi\right)^s} = (\ln(2\pi) - \psi(s) + \frac{\pi}{2}\tan(\pi s/2))\,\zeta(1-s) \tag{3.2}$$

valid for all s. Alternatively, in equivalent form, define the normalized right- and left-hand sides of Equation (3.2) by

$$\mathfrak{T}(s) \equiv (\ln(2\pi) - \psi(s) + \frac{\pi}{2}\tan(\pi s/2))\,\zeta(1-s)/\zeta'(s) \tag{3.3}$$

and

$$\mathfrak{L}(s) \equiv \frac{\zeta'(1-s)}{\zeta'(s)} + 2\cos(\pi s/2)\,\Gamma(s)\,(2\pi)^{(-s)}, \tag{3.4}$$

in which case Equation (3.2) can be written

$$\mathfrak{L}(s) = \mathfrak{T}(s). \tag{3.5}$$

Since it is known (Spira, 1973) that $\zeta'(s) \neq 0$ in the open critical half-strip $0 \leq \Re(s) < 1/2$, the following requires that s be constrained to that region, although the known symmetry imposed by the functional equation, viz. $\zeta(1 - s_0) = 0$ implies $\zeta(s_0) = 0$ and the reverse, means that the following can be generalized to $\Re(s) > 1/2$. We now show that $\mathfrak{L}(s_0) = 0$ is a necessary and sufficient condition for $\zeta(s_0) = 0$. Suppose $s = s_0$ such that

$$\frac{\zeta'(1-s_0)}{\zeta'(s_0)} = -2\cos(\pi s_0/2)\,\Gamma(s_0)\,(2\pi)^{-s_0}. \tag{3.6}$$

Clearly, if Equation (3.6) is true, or, alternatively

$$\mathfrak{L}(s_0) = 0, \tag{3.7}$$

then the right-hand side of Equation (3.2) vanishes; that is,

$$\zeta(1-s_0) = 0 \tag{3.8}$$

unless, for that same value of s_0, the factor

$$\ln(2\,\pi) - \psi(s_0) + \frac{1}{2}\,\pi \tan(\frac{\pi\,s_0}{2}) = 0\;.$$ (3.9)

Is it possible that Equations (3.7 and 3.9) can be simultaneously true for some value(s) of s_0? On the perforated critical line, where it will shortly be shown (see Section 10) that $\zeta'(1/2 + i\rho) \neq 0$, solving Equation (3.9) requires that ρ must satisfy

$$\Re(\psi(1/2 + i\,\rho)) = \frac{\ln(2\,\pi)\cosh(\pi\,\rho) + \pi/2}{\cosh(\pi\,\rho)}$$ (3.10)

and from Figure 2, we see that this condition is only satisfied at a single point $\rho_s = 6.2898...$ because the right-hand side of Equation (3.10) approaches $\ln(2\pi)$ asymptotically, whereas the left-hand side is monotonically increasing as $\log(\rho)$ (Olver et al., 2010, Eq. 5.11.2). In this case, the imaginary counterpart of Equation (3.9) vanishes. Because this point does not coincide with a zero of $\zeta(1/2 + i\rho)$, on the critical line, Equation (3.6), or equivalently Equation (3.7), is a sufficient condition for Equation (3.8) to be true, unless $\rho_0 = 6.2898...$, a constant quoted to many significant figures in de Reyna and Van de Lune (2014, Corollary 9). See also Section 10 - Note 1.

Now, consider the general case $s = \sigma + i\rho$ with $0 < \sigma \leq 1/2$—the lower half of the "critical strip". A hypothetical solution of Equation (3.9) requires the existence of solutions (σ_0, ρ_0) simultaneously satisfying

$$\Re(\psi(\sigma_0 + i\,\rho_0)) = \ln(2\,\pi) + \frac{1}{2}\,\frac{\pi\sin(\pi\,\sigma_0)}{\cos(\pi\,\sigma_0) + \cosh(\pi\,\rho_0)}$$ (3.11)

and

$$\Im(\psi(\sigma_0 + i\,\rho_0)) = \frac{1}{2}\,\frac{\pi\sinh(\pi\,\rho_0)}{\cos(\pi\,\sigma_0) + \cosh(\pi\,\rho_0)}\;.$$ (3.12)

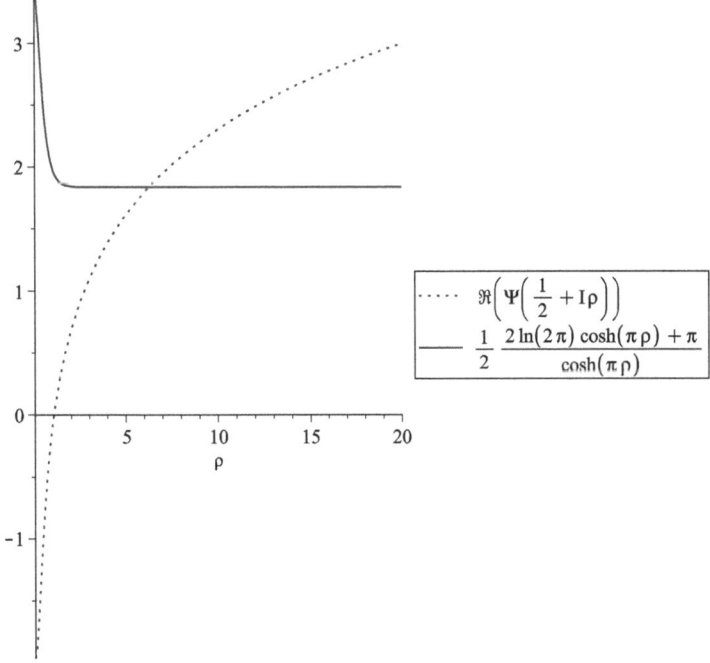

Figure 2. Numerical demonstration of Equation (3.10).

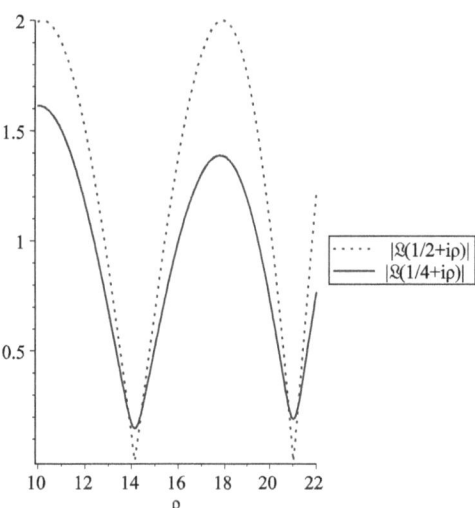

Figure 3. Numerical demonstration of Equation (3.4) for two values of σ near the first two zeros of $\zeta(1/2 + i\rho)$.

If $\sigma_0 = 1/2$, Equation (3.11) reduces to Equation (3.10) and (3.12) reduces to an identity (NIST Digital Library of Mathematical Functions, 2014, Eq. 5.4.17). A numerical study of Equation (3.11) for $0 < \sigma < 1/2$ shows that a single solution exists when $\rho_0 \approx 6.2$, similar to that shown in Figure 2, but otherwise, there are no numerical solutions for the previous reasons. Similarly, the left-hand side of Equation (3.12) asymptotically approaches its right-hand side as $\rho_0 \to \infty$, but the equality is never satisfied. Therefore, the existence of a point s_0 solving Equation (3.7) is a sufficient condition for a zero of $\zeta(s)$ to exist because the Equations (3.7 and 3.9) do not simultaneously vanish.

By a way of illustration, consider Figure 3 which shows $|\mathfrak{L}(\sigma + i\rho)|^2$ for two different choices of σ near the first two known non-trivial zeros of $\zeta(1/2 + i\rho)$. Because of the above, this Figure immediately suggests that $\mathfrak{L}(\sigma + i\rho) \neq 0$ unless $\sigma = 1/2$.

3.2. A necessary condition that $\zeta(s) = 0$

From the functional Equation (3.1), $\zeta(1 - s_0) = 0$ implies $\zeta(s_0) = 0$. If $\zeta(1 - s_0) = 0$, then $\mathfrak{T}(s_0) = 0$, since the other factors of $\mathfrak{T}(s)$ do not diverge if $\mathfrak{I}(s) \neq 0$ (corresponding to the existence of the so-called trivial zeros). From Equation (3.5), this implies $\mathfrak{L}(s_0) = 0$ and necessity follows because $\zeta'(s) \neq 0$ except possibly at a zero (see Section 10). This proves necessity subject to the simplicity of a zero.

4. Implicit specification of zeros along the critical line

4.1. First implicit specification of the zeros along the critical line

Inspired by Figure (3) and Equation (3.7) that together demonstrate that implicit numerical solutions of the equation $|\mathfrak{L}(1/2 + i\rho)|^2 = 0$ locate the zeros of $\zeta(1/2 + i\rho)$, consider the related equation

$$|\mathfrak{L}(1/2 + i\rho)|^2 = |\mathfrak{T}(1/2 + i\rho)|^2 \tag{4.1}$$

After considerable effort (Maple) to simplify (Equation 4.1) (left as an exercise for the reader with access to a computerized algebraic manipulation program), we find

$$\frac{4\,|\zeta|^2}{f(\rho)^2\,|\zeta'|^2} = 2 + \frac{2\left(\sinh(\frac{\pi\rho}{2})\sin(2\,\beta + \rho_\theta) + \cosh(\frac{\pi\rho}{2})\cos(2\,\beta + \rho_\theta)\right)}{\sqrt{\cosh(\pi\,\rho)}}. \tag{4.2}$$

For any value of $\rho = \rho_0$ corresponding to $|\zeta|^2 = 0$ and $|\zeta'|^2 \neq 0$ (see Section 10) on the left-hand side of Equation (4.2), the right-hand side of this equation must accordingly vanish as well because the remaining factors on the left-hand side never vanish if $\rho > \rho_s$. This produces an implicit equation that locates $\zeta(1/2 + i\rho_0) = 0$:

$$\frac{\sinh(\pi\,\rho/2)\sin(2\,\beta + \rho_\theta) + \cosh(\pi\,\rho/2)\cos(2\,\beta + \rho_\theta)}{\sqrt{\cosh(\pi\,\rho)}} = -1 \,. \tag{4.3}$$

This will be discussed further in Section 7.

4.2. Second implicit specification of the zeros along the critical line

Again inspired by Figure (3), we obtain a second implicit equation whose zeros coincide with the non-trivial zeros of $\zeta(1/2 + i\rho)$. Working from the real and imaginary parts of $\mathfrak{L}(1/2 + i\rho)$

$$\mathfrak{L}_R = \frac{\sin(\rho_\theta)\sinh(\frac{\pi\rho}{2}) + \cos(\rho_\theta))\cosh(\frac{\pi\rho}{2})}{\sqrt{\cosh(\pi\,\rho)}} + \frac{\zeta_R'^{\,2} - \zeta_I'^{\,2}}{|\zeta'|^2} \tag{4.4}$$

$$\mathfrak{L}_I = \frac{\sin(\rho_\theta)\cosh(\frac{\pi\rho}{2}) - \cos(\rho_\theta)\sinh(\frac{\pi\rho}{2})}{\sqrt{\cosh(\pi\,\rho)}} - \frac{2\,\zeta_R'\,\zeta_I'}{|\zeta'|^2} \tag{4.5}$$

consider the function

$$\mathfrak{L}_1(\rho) \equiv \frac{d}{d\rho}|\mathfrak{L}(1/2 + i\rho)|^2 \tag{4.6}$$

at least some of whose zeros $\rho = \rho_0$ must coincide with the maxima and minima of $|\mathfrak{L}(1/2 + i\rho)|$.

Following a lengthy calculation (Maple), we find

$$\mathfrak{L}_1(\rho) = 2\,\sqrt{\frac{1}{\cosh(\pi\,\rho)}}\left[-\frac{1}{f(\rho)}|\zeta'|^2 + (\sin(\beta)\,\zeta_I'' + \cos(\beta)\,\zeta_R'')\,|\zeta'|\right]T_1(\rho) \tag{4.7}$$

where again the various symbols in Equation (4.7) are defined in Appendix 1.

Figure 4 demonstrates that $\mathfrak{L}_1(\rho)$ passes through the first two zeros of $\zeta(1/2 + I\rho)$ with positive slope, and, as expected, alternating solutions corresponding to zeros of Equation (4.7) implicitly define those points $\rho = \rho_0$ such that $\zeta(1/2 + i\rho_0) = 0$, consistent with the assumption that $|\mathfrak{L}(1/2 + i\rho)|^2$ does not contain intermediate maxima, minima or inflections. Of particular interest is the fact that Equation (4.7) consists of two factors that could potentially vanish. From a numerical study, the first of these, enclosed in square brackets ([...]), appears to be negative for all values of ρ.

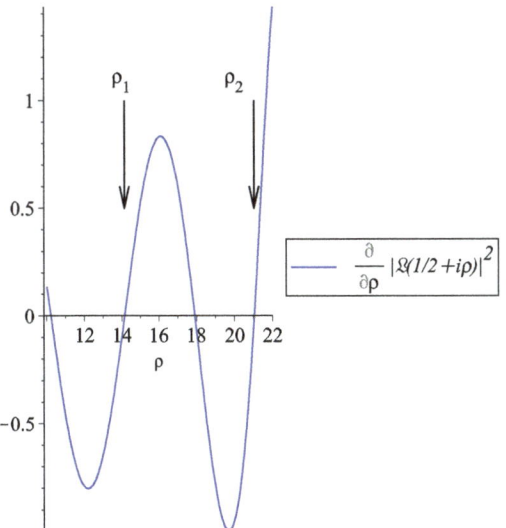

Figure 4. Numerical demonstration of Equation (4.8) near the first two zeros of $\zeta(1/2 + i\rho)$.

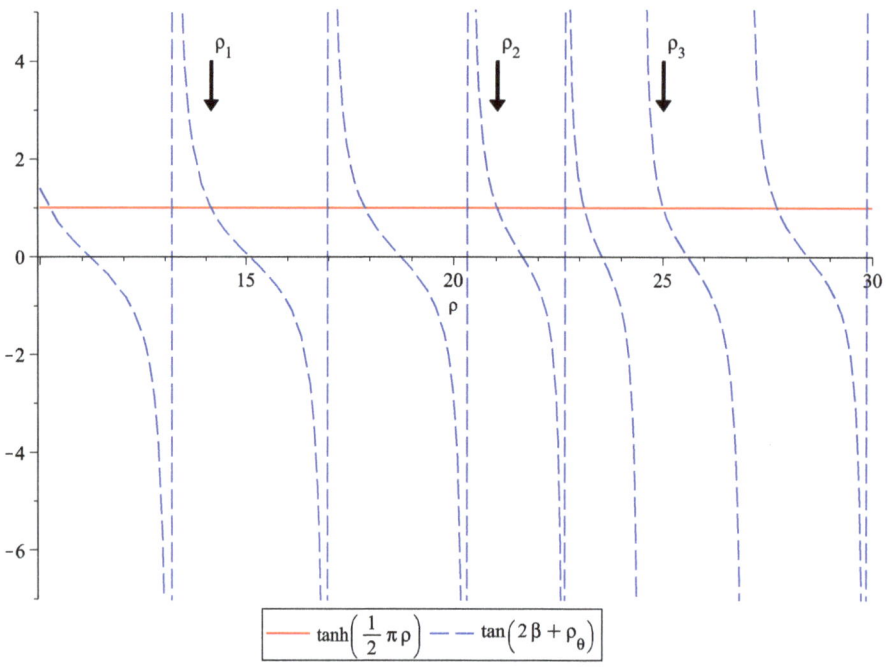

Figure 5. Numerical demonstration of Equation (4.8) near the first few zeros of $\zeta(1/2 + i\rho)$.

If so, the factor $T_1(\rho)$ (see Appendix 1) carries all the zeros of $\mathfrak{L}_1(\rho)$, from which we conclude that solutions of $\zeta(1/2 + i\rho_0) = 0$ correspond to alternating solutions of

$$\tanh\left(\frac{\pi \rho}{2}\right) = \tan(2\beta + \rho_\theta). \tag{4.8}$$

See Figure 5. This will be discussed further in Section 7.

4.3. Third implicit specification of the zeros along the critical line

With $s = 1/2 + i\rho$, from Equation (3.2 and 2.5), we find

$$\zeta_R/f(\rho) = a(\rho)\,\zeta_R' + b(\rho)\,\zeta_I' \tag{4.9a}$$

$$\zeta_I/f(\rho) = b(\rho)\,\zeta_R' + c(\rho)\,\zeta_I' \tag{4.9b}$$

  where

$$a(\rho) = -\frac{1}{2}\cos(\rho\ln(\pi)) + \frac{1}{2} + \frac{\mathcal{F}_I\mathcal{F}_R\sin(\rho\ln(\pi)) + \mathcal{F}_R{}^2\cos(\rho\ln(\pi))}{\left|\Gamma(\frac{1}{4} + \frac{1}{2}i\rho)\right|^2} \tag{4.10}$$

$$b(\rho) = -\frac{1}{2}\sin(\rho\ln(\pi)) + \frac{-\mathcal{F}_I\mathcal{F}_R\cos(\rho\ln(\pi)) + \mathcal{F}_R{}^2\sin(\rho\ln(\pi))}{\left|\Gamma(\frac{1}{4} + \frac{1}{2}i\rho)\right|^2} \tag{4.11}$$

$$c(\rho) = 1 - a(\rho) \tag{4.12}$$

and we reiterate (from Appendix 1) that $\mathcal{F} \equiv \Gamma(1/4 + i\rho/2)$.

It is easily shown that the determinant of the transformation matrix Equation (4.9) is identically zero, and therefore the transformation is singular. This suggests that in some sense, the components of the derivative function $\zeta'(1/2 + i\rho)$ are more fundamental than are the components of the primary function $\zeta(1/2 + i\rho)$. That is, Riemann's functional equation in the form Equation (3.2) defines the latter from the former, but not the other way around, suggesting that something has been lost when Equation (3.2) is generated from the usual form of the functional Equation (3.1). Why should this be? This may be analogous to the fact that the act of differentiation followed by integration always introduces an arbitrariness in the form of a constant. Or, it may be due to the fact that differentiating (Milgram, 2011, Eqs. 3.1 and 3.2) with respect to ρ will always give a result where ζ'_R and ζ'_I are a mixture of ζ_R and ζ_I that cannot be disentangled in the form of an inverse transform of Equations (4.9a) and (4.9b).

In Section 10, it is shown that $\zeta'(1/2 + i\rho) \neq 0$ with conditions. Thus, setting the left-hand sides of Equations (4.9a and 4.9b) to zero will result in a relationship between the various components of these equations at a zero of $\zeta(1/2 + i\rho)$. Specifically, $\rho = \rho_0$ satisfying the following

$$\sin(\rho_0 \ln(\pi)) = \frac{2\,(\varUpsilon_I\,\zeta'_R + \varUpsilon_R\,\zeta'_I)\,(\varUpsilon_I\,\zeta'_I - \varUpsilon_R\,\zeta'_R)}{|\zeta'|^2\,|\varUpsilon|^2} \tag{4.13}$$

and

$$\cos(\rho_0 \ln(\pi)) = \frac{4\,\varUpsilon_I\,\varUpsilon_R\,\zeta'_I\,\zeta'_R + ({\zeta'_I}^2 - {\zeta'_R}^2)\,(-\varUpsilon_I^2 + \varUpsilon_R^2)}{|\zeta'|^2\,|\varUpsilon|^2} \tag{4.14}$$

defines all the non-trivial zeros on the critical line plus one extra zero associated with the factor $1/f(\rho_0)$. The results (Equations 4.13 and 4.14) can be consolidated by equating the ratio of the two left-hand sides in the form

$$\arctan(\sin(\rho_0 \log(\pi)), \cos(\rho_0 \log(\pi))) \tag{4.15}$$

with a similar expression for the right-hand sides. The standard two parameter $\arctan(y, x)$ function is specified to consistently account for the signs of the various terms. Alternatively, solving Equations (4.13 and 4.14) for the ratio ζ'_I/ζ'_R gives

$$\tan(\beta_0) \equiv \frac{\zeta'_I}{\zeta'_R} = \frac{|\varUpsilon|^2 \cos(\rho_0 \ln(\pi)) - \varUpsilon_I^2 + \varUpsilon_R^2}{-\sin(\rho_0 \ln(\pi))\,|\varUpsilon|^2 + 2\,\varUpsilon_I\,\varUpsilon_R} \tag{4.16}$$

corresponding to a solution for $\beta_0 = \arg(\zeta'(1/2 + i\rho_0))$ at a non-trivial zero $\rho = \rho_0$. See Section 7.

5. A related functional
Formally replace ζ' with ζ in Equation (3.4) and using Equation (3.1), define and evaluate

$$\mathfrak{M}(s) \equiv \frac{\zeta(1-s)}{\zeta(s)} + 2\cos(\pi s/2)\Gamma(s)(2\pi)^{(-s)} = 4\cos(\pi s/2)\Gamma(s)(2\pi)^{(-s)}. \tag{5.1}$$

For the particular case $s = 1/2 + i\rho$, using standard trigonometric and Gamma function identities (e.g. Olver et al., 2010, Eq. 5.4.4), it is easy to establish that

$$|\mathfrak{M}(1/2 + i\rho)|^2 = 4. \tag{5.2}$$

It now becomes possible to use (Equation 5.2) to discover interesting results because of Equation (3.5). Define the modified right-hand side of Equation (3.5) corresponding to Equation (5.1) with $s = 1/2 + i\rho$ by adding and subtracting the terms that convert $\mathfrak{L}(s)$ into $\mathfrak{M}(s)$, specifically

$$\mathfrak{Q}(1/2 + i\rho) \equiv \mathfrak{T}(1/2 + i\rho) + \frac{\zeta(1/2 - i\rho)}{\zeta(1/2 + i\rho)} - \frac{\zeta'(1/2 - i\rho)}{\zeta'(1/2 + i\rho)}. \tag{5.3}$$

Then, from Equation (5.2), we have

$$|\mathfrak{Q}(1/2 + i\rho)|^2 = 4 \, .$$

(5.4)

6. $arg(\zeta(1/2 + i\rho))$ and related entities

After algebraic simplification (Maple), Equation (5.4) eventually yields

$$|\zeta|^2 = \zeta_I^2 + \zeta_R^2 = (\zeta'_I \, \zeta_I + \zeta'_R \, \zeta_R) f(\rho) \, .$$

(6.1)

(Background note: This is a fairly lengthy, but straightforward calculation. First, break $\mathfrak{Q}$ into its real and imaginary parts using Equations (3.3 and 5.3), evaluate $|\mathfrak{Q}|^2 = \mathfrak{Q}_R^2 + \mathfrak{Q}_I^2 = 4$ and simplify the resulting Equation (5.4). The simplification sequence involves nothing more than the application of well-known identities involving Γ and related functions. Due to the complexity and length of this calculation, the details are left as an exercise for the reader, who may consult a supplementary file deposited with this paper–that being an annotated Maple worksheet used to reproduce the calculation. A second, equally lengthy, but independent derivation of Equation (6.1) has recently been obtained—ms. in preparation).

Setting $\rho = 0$ in Equation (6.1) reproduces a known relationship between $\zeta(1/2)$ and $\zeta'(1/2)$. Furthermore, noticing that

$$\left(\frac{\zeta_I}{\zeta_R} \right)' = \frac{(\zeta_I)'}{\zeta_R} - \frac{\zeta_I (\zeta_R)'}{\zeta_R^2}$$

(6.2)

by recalling that

$$\left(\zeta_I \right)' = (\zeta')_R \equiv \zeta'_R$$

(6.3)

and

$$\left(\zeta_R \right)' = -(\zeta')_I \equiv -\zeta'_I$$

(6.4)

we recognize Equation (6.1) to be a simple differential equation

$$\frac{d\, g(\rho)}{d\rho} = \frac{1 + g(\rho)^2}{f(\rho)}$$

(6.5)

or, equivalently

$$\alpha'(\rho) \equiv \frac{d\, \alpha(\rho)}{d\, \rho} = 1/f(\rho)$$

(6.6)

where

$$g(\rho) \equiv \frac{\zeta_I}{\zeta_R} = \tan(\alpha(\rho)).$$

(6.7)

Integrating (Equation 6.6) between $\rho = \rho_1$ and $\rho = \rho_2$ gives

$$arg(\zeta(1/2 + i\, \rho_2)) - arg(\zeta(1/2 + i\, \rho_1)) = \frac{1}{2}(\rho_2 - \rho_1) \ln(2\,\pi)$$

$$-\frac{1}{2} \int_{\rho_1}^{\rho_2} \mathfrak{R}(\psi(1/2 + i\,\rho))\, d\rho - \frac{1}{2}\, arctan(e^{\pi\,\rho_1}) + \frac{1}{2}\, arctan(e^{\pi\,\rho_2}) - (k+2)\,\pi$$

(6.8)

where the arbitrary constant $(k + 2)\pi$ has been chosen such that $k = 0$ corresponds to a consistent answer when $\rho_1, \rho_2 \approx 0$. This result is equivalent to de Reyna and Van de Lune (2014, Eq. 8) when $\rho_1 = 0$ and $\rho_2 = \rho$, in which case we find

$$\arg(\zeta(1/2 + i\,\rho)) = -\frac{1}{2}\int_0^\rho \Re(\psi(1/2 + i\,t))\,dt + \frac{\rho}{2}\ln(2\,\pi) - \frac{9\,\pi}{8} + \frac{1}{2}\,\arctan(e^{\pi\,\rho}) + k\,\pi\,. \tag{6.9}$$

It is important to recognize that $\alpha(\rho)$ in Equation (6.6) is a continuous function, whereas its companion $arg(\zeta(1/2 + i\rho))$ is discontinuous, and therefore the solution (Equations 6.8 or 6.9) only applies in a multi-sheeted sense. That is, in order to satisfy Equation (6.6), in each of Equation (6.8 or 6.9), k serves as a *local* variable, constant over a range where $arg(\zeta(1/2 + i\rho))$ is continuous, but effectively, $k = k(\rho)$ globally. See Section 9 for further discussion, and Section 8 for an application of this point.

The integral in Equation (6.9) can be evaluated analytically, using the well-known expansion (NIST Digital Library of Mathematical Functions, 2014, Eq. 5.7.6)

$$\psi(1/2 + x) = -\gamma - \frac{1}{1/2 + x} - \sum_{n=1}^{\infty}(\frac{1}{n + 1/2 + x} - \frac{1}{n})\,. \tag{6.10}$$

First, temporarily omit the $\Re(...)$ operator, then interchange the sum and integration operators, and, after the (trivial) integration is accomplished, compute the real part of the result, and sum the resulting series utilizing (Hansen, 1975, Eq. 42.1.5)

$$\sum_{n=0}^{\infty}(\,\arctan(\frac{y}{n + x}) - \frac{y}{n + x}) = y\,\psi(x) - \,\arg(\Gamma(x + i\,y))\,, x + i\,y \neq 0, -1, -2\,\dots\,. \tag{6.11}$$

to obtain (see de Reyna & Van de Lune, 2014, Section 3; also see Section 12, Equation 12.2, below where it is pointed out that this result can be simply obtained by integrating by parts)

$$\int_0^\rho \Re\left(\psi\left(\frac{1}{2} + i\,t\right)\right)dt = \Im(\,\mathrm{Log}\Gamma(1/2 + i\,\rho)) = \,\arg\left(\Gamma\left(\frac{1}{2} + i\,\rho\right)\right) + k\,\pi\,. \tag{6.12}$$

(Digression: It is worth noting that an equivalent, but unevaluated, form of the sum in Equation (6.11) arises in de Reyna and Van de Lune (2014, Eq. 9) where it is suggested that it is perhaps new. In Hansen (1975), this result is attributed to Abramowitz and Stegun (1964, Eq. 6.1.27), and, somewhat surprisingly, it appears to have been omitted from (NIST Digital Library of Mathematical Functions, 2014). See also (Weisstein, 2005) where this sum is used to define the LogGamma function. Of course

$$arg(\Gamma(1/2 + i\rho)) = \Im(\log(\Gamma(1/2 + i\rho))) \tag{6.13}$$

and the multi-sheeted LogGamma function of argument $(1/2 + i\rho)$ differs from the function $\log(\Gamma(1/2 + i\rho))$ by an additive term equal to $k\pi$.) With this result, Equation (6.8) now becomes

$$\begin{aligned}\arg(\zeta(\frac{1}{2} + i\,\rho_2)) - \,\arg(\zeta(\frac{1}{2} + i\,\rho_1)) &= \frac{1}{2}\,(\rho_2 - \rho_1)\ln(2\,\pi) + \frac{1}{2}\,\arg(\Gamma(\frac{1}{2} + i\,\rho_1)) \\ &- \frac{1}{2}\,\arg(\Gamma(\frac{1}{2} + i\,\rho_2)) - \frac{1}{2}\,\arctan(e^{\pi\,\rho_1}) + \frac{1}{2}\,\arctan(e^{\pi\,\rho_2}) + k\,\pi\end{aligned} \tag{6.14}$$

and, in the case that $\rho_1 = 0, \rho_2 = \rho$,

$$\arg(\zeta(\frac{1}{2} + i\,\rho)) = \frac{\rho}{2}\ln(2\,\pi) - \frac{1}{2}\,\arg(\Gamma(\frac{1}{2} + i\,\rho)) - \frac{9\,\pi}{8} + \frac{1}{2}\,\arctan(e^{\pi\,\rho}) - k\,\pi\,. \tag{6.15}$$

Notice that $\alpha'(\rho)$ in Equation (6.6) and $\alpha(\rho)$ in Equation (6.14) differ from substitute symbols ($\alpha \to A$ and $\beta \to B$) in Equations (2.13 and 2.14) because of the ambiguity associated with the term $k\,\pi$. Also, the discontinuities on both sides of Equations (6.14 and 6.15) do not coincide—for any value of ρ, a corresponding value of k must be carefully chosen. From Equation (6.9), it becomes convenient to define the multi-sheeted (i.e. continuous) LogZeta function (see Figure 6):

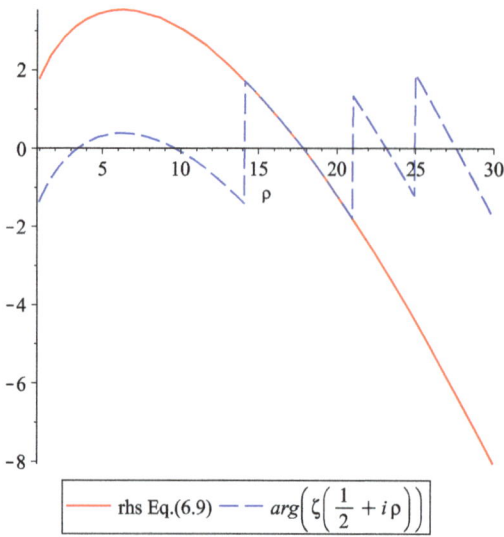

Figure 6. Comparison of the right- and left-hand sides of Equation (6.9) with $k = 1$.

$$\text{LogZ}(1/2 + i\rho) \equiv -\log\Gamma(1/2 + i\rho)/2 + i(\frac{\rho}{2} \log(2\pi) - 9\pi/8 + \arctan(e^{\pi\rho})/2). \tag{6.16}$$

In this form, and throughout this work, α associates with LogZeta function via

$$\alpha(\rho) = \Im(\log Z(1/2 + i\rho)), \tag{6.17}$$

and θ represents the LogGamma function in the same way (England et al., 2013). Using this definition, we have

$$\arg(\zeta(1/2 + i\rho)) = \Im(\log(\zeta(1/2 + i\rho))) = \Im(\text{LogZ}(1/2 + i\rho)) + k\pi. \tag{6.18}$$

Finally, from Equation (2.7), we obtain

$$\tan(\Im(\text{LogZ}(1/2 + i\rho))) = \mathfrak{P} \tag{6.19}$$

and an equivalent form if Equation (2.9) is used.

7. Locating the zeros on the critical line

Write Equation (4.3) in generic form

$$\frac{\sinh(\frac{\pi\rho}{2})\sin(Y) + \cosh(\frac{\pi\rho}{2})\cos(Y)}{\sqrt{\cosh(\pi\rho)}} = -1. \tag{7.1}$$

Solutions to Equation (7.1) exist when

$$Y = -\frac{(2n+1)\pi}{4} + \kappa\pi + \arctan(e^{\pi\rho}) \tag{7.2}$$

provided that (when n is even)

$$(-1)^{n/2}\cos(\kappa\pi) = -1 \tag{7.3}$$

or (when n is odd)

$$(-1)^{(n-1)/2} \sin(\kappa \, \pi) = -1 \, . \tag{7.4}$$

By comparison of Equation (7.2) with Equations (4.3 and 6.15), we identify

$$\kappa = 2(\beta - \alpha)/\pi + n/2 + 2(k-1) \, , \tag{7.5}$$

consistent with de Reyna and Van de Lune (2014, Proposition 17). Taking advantage of the arbitrariness of the *arg* operator modulo 2π, it is thus possible to solve Equation (7.1) in a simpler form and conclude that alternating solutions of

$$\arg(\zeta'(\tfrac{1}{2} + i\,\rho)) - \arg(\zeta(\tfrac{1}{2} + i\,\rho)) = (K + 1/2)\pi \, , \tag{7.6}$$

locate zeros of $\zeta(1/2 + i\rho)$ on the critical line, where K is an integer, a result that is easily verified numerically, or analytically for any combination of n and (constrained) κ in Equations (7.3 and 7.4). Equation (7.6) also reproduces de Reyna and Van de Lune (2014, Proposition 13).

Insofar as alternative means of locating zeros are concerned, it is easily shown by simple trigonometric identities that Equation (7.2) is also a valid solution to Equation (4.8); therefore Equation (4.8) is algebraically equivalent to Equation (4.3). Although not obvious, it turns out that Equation (4.16) is also (numerically) equivalent to Equations (4.3 and 4.8). In contrast, solutions satisfying

$$\arg(\zeta'(\tfrac{1}{2} + i\,\rho)) - \arg(\zeta(\tfrac{1}{2} + i\,\rho)) = K\pi \tag{7.7}$$

correspond to the in-between solutions of Equation (7.1), i.e. the maxima of $|\mathfrak{R}(1/2 + i\rho)|^2$ in Figure 3. Any of Equation (4.3), Equation (4.8) or Equation (4.16), treated as implicit equations, are numerically sensitive means of locating a zero because the two sides intersect one another at differing slopes (e.g. Figure 5), as opposed to treating either side of Equation (4.1) implicitly because those "intersections" occur at tangent points. Another effective numerical variation, also equivalent to Equation (2.12), yields the following simple corollary to Equation (7.6):

"$\zeta(1/2 + i\rho_0) = 0$ whenever $\rho = \rho_0$ satisfies

$$1 + \tan(\alpha(\rho)) \, \tan(\beta(\rho)) = 0 \, . \tag{7.8}$$

Any of these prescriptions, employed as a numerical means of locating zeros, give the appearance of requiring knowledge of both α and β (i.e. both $arg(\zeta(1/2 + i\rho))$ and $arg(\zeta'(1/2 + i\rho))$ need to be calculable). In fact, because of Equation (6.15), only $\beta \equiv arg\,(\zeta'(1/2 + i\rho))$ needs to be calculable, along with variations of $\Gamma(1/2 + i\rho), \psi(1/2 + i\rho)$ and related functions which are presumably well-known numerically. This is especially effective if Equation (6.19) is employed to represent the term $\tan(\alpha)$ in Equation (7.8).

8. The Volchkov equivalence
With reference to Equation (8.1) below, (where it has been opined (Moll, 2010) "Evaluating (it) might be hard"), given an explicit expression for $arg(\zeta(1/2 + i\rho))$ (see Equation 6.9), it now becomes possible to investigate the *Volchkov Criterion*, the truth of which is advertised as being equivalent to RH (Conrey & Farmer, n.d.; He, Jejjala, & Minic, 2015, Eq. 2.3; Volchkov, 1995; Sekatskii, Beltraminelli, & Merlini, 2012; Borwein, Choi, Rooney, & Weirathmueller, 2008, Section 5.2). This criterion reads: "The Riemann Hypothesis is equivalent to

$$\int_0^\infty \frac{2t \, \arg(\zeta(1/2 + i\,t))}{(1/4 + t^2)^2} \, dt = \pi \, (\gamma - 3). \text{'} \tag{8.1}$$

Several variations of Equation (8.1), all of which are based on a similar derivation method, are presented in these references. Consider Figure 6, a numerical comparison between the right-hand side of Equation (6.9) and $arg(\zeta(1/2 + i\rho))$ spanning the first few zeros of $\zeta(1/2 + i\rho)$ along the critical line. This comparison demonstrates that the "constant" k in Equation (6.9) must be carefully chosen to achieve equality between the two representations of $arg(\zeta(1/2 + i\rho))$. In Figure 6, we note the existence of a discontinuity in $arg(\zeta(1/2 + i\rho_0))$ at $\rho = \rho_0$. In fact,

$$arg(\zeta(1/2 + i\rho_0^-)) - arg(\zeta(1/2 + i\rho_0^+)) = -\pi \tag{8.2}$$

a property that is shared (within a sign change) with any analytic function at a simple zero because the real and imaginary parts of the function both change sign as the zero is traversed in the complex ζ plane. (See Section 10). The intent now is to analytically evaluate Equation (8.1) using the right-hand side of Equation (6.9) to represent $arg(\zeta(1/2 + it))$.

Substituting Equation (6.9) into Equation (8.1) leads to three interesting integrals, the first being

$$J_1 \equiv \int_0^\infty \frac{t \ arctan(e^{t\pi})}{(1/4 + t^2)^2} \, dt \tag{8.3}$$

which may be evaluated by first integrating by parts,

$$J_1 = \frac{\pi}{2} + \frac{1}{4} \pi \int_0^\infty \frac{1}{(1/4 + t^2) \cosh(t\pi)} \, dt \tag{8.4}$$

eventually yielding

$$J_1 = \pi/2 + \frac{\pi}{2} \log(2) \tag{8.5}$$

after consulting Gradshteyn and Ryzhik (1980) (Eqs. 3.522(4) and 8.370). The second required integral is

$$J_2 \equiv \int_0^\infty \frac{t^2 \ \Re\left(\int_0^1 \psi(\frac{1}{2} + i\rho t) \, d\rho\right)}{(1/4 + t^2)^2} \, dt. \tag{8.6}$$

Following the method of Equation (6.10), interchange the integral and sum, and after evaluating the real part of the integral, obtain

$$\Re\left(\int_0^1 \psi(\frac{1}{2} + i\rho t) \, d\rho\right) = -\gamma - \frac{arctan(2t)}{t} - \left(\sum_{n=1}^\infty \frac{n \ arctan(\frac{2t}{2n+1}) - t}{nt}\right) \tag{8.7}$$

Apply the outer integration to the first two of the terms in Equation (8.7), and find (courtesy of Maple)

$$\int_0^\infty \frac{\gamma t^2}{(\frac{1}{4} + t^2)^2} \, dt = \frac{\gamma \pi}{2} \tag{8.8}$$

$$\int_0^\infty \frac{t \ arctan(2t)}{(\frac{1}{4} + t^2)^2} \, dt = \frac{\pi}{2}. \tag{8.9}$$

To evaluate the third term in Equation (8.7), interchange the convergent (grouped) sum with the outer integration and evaluate the integral (Maple), yielding

$$\int_0^\infty \frac{\left(n \ \arctan\left(\frac{2t}{2n+1}\right) - t\right)t}{n\left(\frac{1}{4} + t^2\right)^2} \, dt = -\frac{\pi}{2\,(n+1)\,n}\,. \tag{8.10}$$

Sum the resulting series (Maple), incorporate Equations (8.8 and 8.9) and eventually arrive at

$$J_2 = -\gamma\pi/2\,. \tag{8.11}$$

The third integral, corresponding to the term $k\pi$ in Equation (6.9), is needed to account for the area between the continuous and discontinuous functions in Figure 6. Temporarily guided by Figure 6 and Equation (8.2) which indicate that each continuous segment of $\arg(\zeta(1/2 + i\rho))$ is bounded by the kth zero of $\zeta(1/2 + i\rho_k)$ (but see Section 9), split the integration limits into intervals, leading (Maple) to:

$$J_3 \equiv \sum_{k=1}^\infty \int_{\rho_k}^{\rho_{k+1}} \frac{2\,t\,k\,\pi}{\left(\frac{1}{4} + t^2\right)^2} \, dt = -16\,\pi \sum_{k=1}^\infty \frac{k\,(\rho_k{}^2 - \rho_{k+1}{}^2)}{(4\,\rho_k{}^2 + 1)(4\,\rho_{k+1}{}^2 + 1)}\,. \tag{8.12}$$

The series in Equation (8.12) is first decomposed by partial fractions and since grouped terms (partially) cancel, the final form of the series becomes

$$J_3 = \pi \sum_{k=1}^\infty \frac{1}{\rho_k{}^2 + 1/4} \tag{8.13}$$

which can be written

$$J_3 = \frac{\pi}{2} \sum_{\rho_K} \frac{1}{|1/2 + i\,\rho_K|^2} \tag{8.14}$$

where ρ_K represents all zeros of $\zeta(1/2 + i\rho_K)$ that lie on the critical line, and the overall factor $\frac{1}{2}$ has been included in recognition of the fact that the sum now includes complex conjugate values that were not included in the original sum (Equation 8.13), indicated by the use of $k \to K$. A more general form of the sum (8.14) is known (Edwards, 2001, p. 159), its value being

$$\sum_\tau \frac{1}{|\tau|^2} = 2 + \gamma - \log(4\pi) \tag{8.15}$$

where the sum over $\tau = \sigma + i\rho$ includes all zeros of $\zeta(\tau)$, including any that may not lie on the critical line. If RH is true (that is, $\Re(\tau) = \frac{1}{2}$ for all τ), then the sums in Equations (8.14 and 8.15) coincide, yielding

$$J_3 = \frac{\pi}{2}\,(2 + \gamma - \log(4\pi))\,. \tag{8.16}$$

Finally, we have the less interesting integrals to evaluate, specifically (Maple)

$$J_4 = \ln(2\,\pi)\int_0^\infty \frac{t^2}{\left(\frac{1}{4} + t^2\right)^2} \, dt = \frac{1}{2}\,\pi\,\ln(2\,\pi) \tag{8.17}$$

and

$$J_5 = -\frac{9}{4}\,\pi\int_0^\infty \frac{t}{\left(\frac{1}{4} + t^2\right)^2} \, dt = -\frac{9\,\pi}{2}\,. \tag{8.18}$$

Putting all the parts together, that is

$$J_1 - J_2 + J_3 + J_4 + J_5 \tag{8.19}$$

gives

$$\int_0^\infty \frac{2t\ \arg(\zeta(\tfrac{1}{2} + it))}{(1/4 + t^2)^2}\, dt = (\gamma - 3)\,\pi + T_0 \tag{8.20}$$

with

$$T_0 \equiv -\frac{1}{2}\,\pi \overline{\sum_\tau} \frac{1}{|\tau|^2} \tag{8.21}$$

where the sum over τ in T_0 only includes zeros of $\zeta(s)$ that do not lie on the critical line, indicated by the symbol $\overline{\sum}$. The term T_0 arises from the difference between the terms included in the sums appearing in Equations (8.14 and 8.15), assuming RH to be false. It is worth noting that individual terms in the sum (Equation 8.20) are positive, so there is no possibility that a sum composed of non-zero terms could itself vanish—its contribution would always be negative. With $T_0 = 0$, the claim Equation (8.1) is verified, but it does NOT prove the Riemann Hypothesis which postulates that T_0 vanishes; it only verifies the equivalence of Equation (8.1) and the RH as embodied in Equation (8.15) because the verification is contingent on the equality of the sums in Equations (8.14 and 8.15), which itself depends on the RH.

This unsurprising result merits further discussion and raises the question of the value of so-called "RH equivalences" to prove RH. In the first place, the original wording: "RH is equivalent to Equation (8.1)" is not very well chosen because the word "equivalent" implies a two-way correspondence (of truth); the original wording should have been: "If RH is true then Equation (8.1)" without implying the converse. This interpretation can be established by examining the derivation of Equation (8.1) which was based on a contour integration about a region where zeros of $\zeta(s)$ were presumed not to exist. Fundamentally, Equation (8.1) could have been obtained in two different ways: "If RH then Equation (8.1)" or "If not RH then Equation (8.1) plus additional terms (i.e. T_0)". In fact, such terms are explicitly presented, but omitted in the derivation (see Sekatskii et al., 2009, unnumbered equation terminating Section 3.1, Power functions) and a long discussion in He et al. (2015). Consequently, irrespective of which assumption is used to obtain Equation (8.1), subsequent analysis (e.g. as presented here) must be done under that same assumption, and a proof (or disproof) of RH will only emerge if either assumption, used consistently throughout, yields a contradiction (reductio ad absurdum). Otherwise, the best that can be hoped for will be a tautology. That is what has happened here.

The original derivation of Equation (8.1) was performed under the first assumption (RH is true) and, if the above analysis (Section 8) had been done with that same assumption, the T_0 term appearing in Equation (8.19) would not have been present. The result would have been a tautology: rhs of Equation (8.1) = rhs of Equation (8.19) (without T_0). Alternatively, if the original derivation of Equation (8.1) had been done under the premise "RH is not true", then additional terms, exactly equivalent to T_0 as it appears in Equation (8.19), would have been present in Equation (8.1) in agreement with the subsequent analysis (performed above) under that same assumption. The result would be, and is, again a tautology: rhs of Equation (8.1) (plus T_0) = rhs of Equation (8.19). Sans contradiction, neither of these can prove or disprove RH. All of this suggests that a proof of RH will never be obtained by inventing so-called "equivalences" that depend on hypothesizing that zeros off the critical line do, or do not, exist. However, in this case, it is a useful exercise to demonstrate the validity of Equation (6.9) and conveniently suggest some RH-independent results by examining other tacit assumptions that have been incorporated into the analysis.

First, the foregoing analysis provides a simple demonstration that the sum appearing in Equation (8.12) is convergent, for the simple reason that it equals a collection of terms that are all finite, without recourse to conditions. The transition from Equation (8.12) to Equation (8.14) involves a regrouping of terms, so it does not prove that Equation (8.14) is unconditionally convergent. A complicated proof of the convergence of Equation (8.14) can be found in many sources (e.g. Edwards, 2001). Secondly, it suggests that a careful numerical evaluation of Equation (8.1) or one of its various relatives can be used to place bounds on the smallest value of τ associated with a term of the sum appearing in Equation (8.20) (see Sekatskii et al., 2015, where an exponential weight function is appended). Thirdly, it appears to validate counting theorems that place bounds on the location of the zeros (see Section 12) by the presence of discontinuities of the argument. It will now be shown that all of the above are compromised.

9. But...But...But

9.1. A close look at the zeros

As noted above, verification of Equation (8.1) given here, depends on Equation (8.2). In general, a discontinuity of the *arg* operator for any analytic function $h(z)$ can be associated with either of two events. The first is associated with the presence of a zero of order n, viz.

$$h(\rho) \approx h^{(n)}(\rho_0)\,(\rho - \rho_0)^n. \tag{9.1}$$

Specializing to the case $h(\rho) = \zeta(1/2 + i\rho)$, it is easily shown that if $\zeta(1/2 + i\rho)$ possesses a zero of order n at $\rho = \rho_0$, Equation (8.2) becomes

$$arg(\zeta(1/2 + i\rho_0^-)) - arg(\zeta(1/2 + i\rho_0^+)) = -n\,\pi \tag{9.2}$$

and observation suggests that we take $n = 1$, consistent with the assumption that $\zeta(1/2 + i\rho)$ only possesses simple zeros.

Consider Figure 7 which shows the locus of the point (ζ_R, ζ_I) in the complex ζ plane as it passes through a typical zero. In general, for $\rho > 10$, because the function $\alpha(\rho)$ is monotonic with negative slope, the locus of that point must follow a clockwise path. This demonstrates a fundamental difference between $arg(\zeta(1/2 + i\rho))$ and $arg(\Gamma(1/2 + i\rho))$ whose slope is positive (see Figure 1) and whose locus thereby follows a counter-clockwise path. The question arises—Is it possible that the locus of the point (ζ_R, ζ_I) could traverse the negative real ζ_R axis and thereby generate a discontinuity because the *arg* operation is restricted to $(-\pi, \pi)$? Such a discontinuity in the $arg(\zeta(1/2 + i\rho))$ function would be distinct from that associated with a (full) zero. This could only occur in the case of a negative imaginary half-zero, that is $\zeta_I = 0$, $\zeta_R < 0$ for some value of ρ. The LogGamma function demonstrates how such a possibility arises—see Figure 1.

Now, except at a full zero, for $\rho \gtrsim 10$, Equation (6.1) requires that the function

$$v \equiv \zeta_I' \zeta_I + \zeta_R' \zeta_R < 0. \tag{9.3}$$

At a discontinuity, we require $\zeta_I = 0$ corresponding to $arg(\zeta(1/2 + i\rho)) = -\pi$, and $\zeta_R' \zeta_R = (\zeta_I)' \zeta_R$ according to Equation (6.3). Also, $(\zeta_I)' < 0$ because the slope of the motion of ζ_I is expected to be negative as the locus approaches $\zeta_I = 0$ on a clockwise trajectory in the negative (lower) half-plane—see Figure 7. In that case, the product $(\zeta_I)' \zeta_R < 0$ demands that $\zeta_R > 0$, contradicting the postulate that $arg(\zeta(1/2 + i\rho)) = -\pi$ which, by definition, stipulates that $\zeta_R < 0$. This means that, since one expects (see Equation (6.3) $\zeta_R' = (\zeta_I)' < 0$ in the third quadrant, imaginary half-zeros ($\zeta_I = 0$, $\zeta_R \neq 0$) will only occur with $\zeta_I = 0$, $\zeta_R > 0$. Therefore, $arg(\zeta(1/2 + i\rho)) = 0$ and no associated discontinuity is expected to occur. Such is the perceived reality incorporated into many well-accepted results, and it is difficult to visualize how this could fail to be the case.

However, the capacity of the ζ-function to surprise is boundless. Consider Figure 8, which shows information similar of that of Figure 7 except that it focuses on the region bounded by $\rho = 414...416$, containing the consecutive zeros $\rho_{212} = 415.01881..$ and $\rho_{213} = 415.45521....$ In this case, where two zeros ρ_{212} and ρ_{213} are in close proximity, as ρ increases, the locus passes first through ρ_{212}, loops (clockwise) within the first quadrant, then passes through ρ_{213} **without crossing the positive ζ_R axis** and thereby

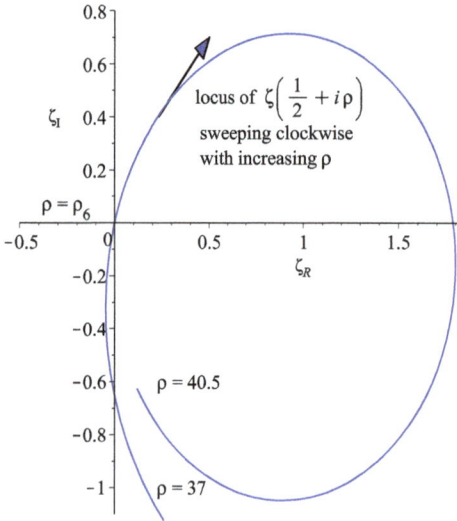

Figure 7. Clockwise travelling locus of $\zeta(1/2 + i\rho)$ **in the complex** ζ **plane near** $\rho_6 = 37.586...$ **as** ρ **increases from 37 to 40.5. The locus passes through both a full-zero and a half-zero (** $\zeta_I = 0, \zeta_R > 0$ **), as** $arg(\zeta(1/2 + i\rho))$ **always decreases (i.e. increases negatively).**

reverses the sign of the right-hand side of Equation (9.2). Consequently, it appears that $\zeta_R' > 0$ in the third quadrant, the previous discussion reverses, and consistently, the locus traverses the negative ζ_R axis at $\rho \approx 415.6$, and a negative imaginary half-zero arises. This generates a discontinuity in $arg(\zeta(1/2 + i\rho))$ that is not associated with a zero of ζ. I shall refer to this configuration by the term "anomalous zero".

For a better understanding, consider Figure 9 where the discontinuity in $arg(\zeta(1/2 + i\rho))$ at $\rho = \rho_{213}$ shows as a **reduction** by an amount π, rather than an increase, as the locus enters the third quadrant. At $\rho \approx 415.6$, the locus crosses the negative ζ_R axis (see Figure 8), $arg(\zeta(1/2 + i\rho))$ increases by 2π and recovers to the value it would have attained if the anomalous discontinuity at ρ_{213} had not occurred. For comparison, the usual situation associated with a zero of $\zeta(1/2 + i\rho)$ that results in a change in $arg(\zeta(1/2 + i\rho))$ is shown at ρ_{212} corresponding to $\rho \approx 415.0$. The region that is bounded by values of $arg(\zeta(1/2 + i\rho))$ that do not adhere to their expectations is filled in solid (orange). Of note is the fact that the function $\upsilon(\rho)$ defined in Equation (9.3) is negative throughout except at the zeros, where it vanishes, as predicted, and that at the right-hand boundary of the region marked by the solid fill, the function $\zeta_R' > 0$ and $\zeta_I = 0$, consistent with the analysis given above.

Before considering the implications of these observations, it is fair to ask if there are other such anomalies. A simple search leads to a study of the pair of zeros $\rho_{126} = 279.229$ and $\rho_{127} = 282.465$ whose locus with increasing ρ is shown in Figure 10. It is easier to understand the situation here by

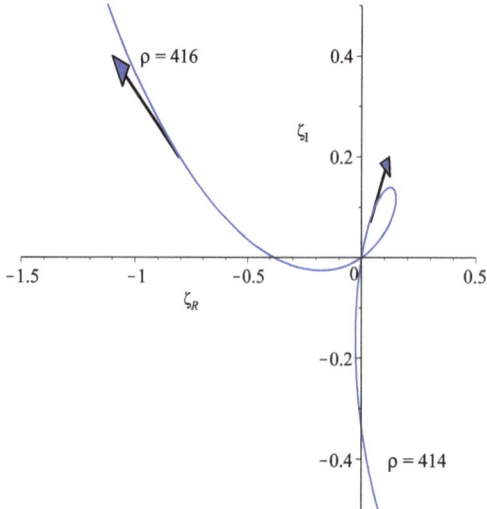

Figure 8. Clockwise-travelling locus of $(1/2 + i\rho)$ **in the complex plane near** $\rho_{212} = 415.01881..$ ρ **increases from 414.0 to 416.0.**

considering Figure 11 which, for clarity, is analogous to Figure 9 but shows only $arg(\zeta(1/2 + i\rho)$ and $\upsilon(\rho)$ in the region between the two zeros under scrutiny. Here, we see that the situation is the exact opposite of the previous one. In the first configuration, the two zeros ρ_{212} and ρ_{213} were (suspiciously) too "close" to each other relative to the point where an anomaly can occur. This separation is, in turn, regulated by the slope of α defined in Equation (6.6) (see Equation (13.14) below). In the second configuration, the two zeros are too "far apart" for analogous reasons, and the discontinuity rises then falls, rather than the reverse. If ρ_{127} were moved 0.01 units to the left, and ρ_{213} were moved 0.2 units to the right, neither anomaly would occur. However, all the location values cited here have been checked against the original published tables (Odlyzko, n.d., file zeros1) and they are consistent with those presented there, as well as those used by Mathematica with which the Maple graphics presented here have been verified. Additionally, the locations of ρ_{212} and ρ_{213} have been verified by an independent calculation [A. Odlyzko, private communication].

At this point, it is worthwhile to recall three formulae that locate zeros. The first is the well-known Backlund (Edwards, 2001, p. 128) counting formula

$$N(\rho) = 1 - \frac{\log(\pi)\rho}{2\pi} + \frac{\Im\left(\text{Log}\Gamma\left(\frac{1}{4} + \frac{i\rho}{2}\right)\right)}{\pi} + \frac{arg\left(\zeta\left(\frac{1}{2} + i\rho\right)\right)}{\pi} \tag{9.4}$$

which (theoretically) increments by one at each point $\rho = \rho_k$. The dotted line $N(\rho) - 212$ in Figure 12 shows that this is not true at the anomalous point ρ_{213}, where it decreases by unity, before it increases by 2 at the next half-zero. Recently, França and LeClair (2013, Eq. 15) have proposed the counting function

$$N_0(\rho) = \frac{\rho}{2\pi} \ln\left(\frac{\rho}{2\pi e}\right) + \frac{7}{8} + \frac{arg\left(\zeta(1/2 + i\rho)\right)}{\pi} \tag{9.5}$$

based on the same (theoretical) assumption as used in Equation (9.4). The function $N_0(\rho)$ is not explicitly shown in Figure 12 because it coincides exactly with the Backlund result. In França and LeClair (2013, Remark 6), it is written "N_0 counts the zeros on the critical line accurately, i.e. it does not miss any zero"; Figure 12 demonstrates that there exist small ranges of the critical line where this statement is, in principle, not correct.

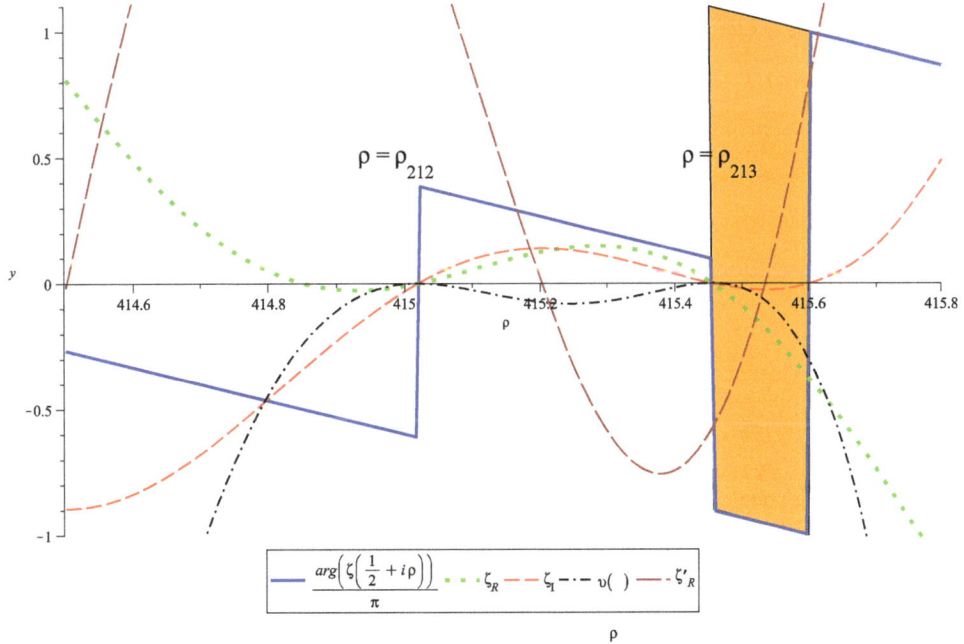

Figure 9. Details of $\zeta(1/2 + i\rho)$, its argument and its derivative in the region of interest.

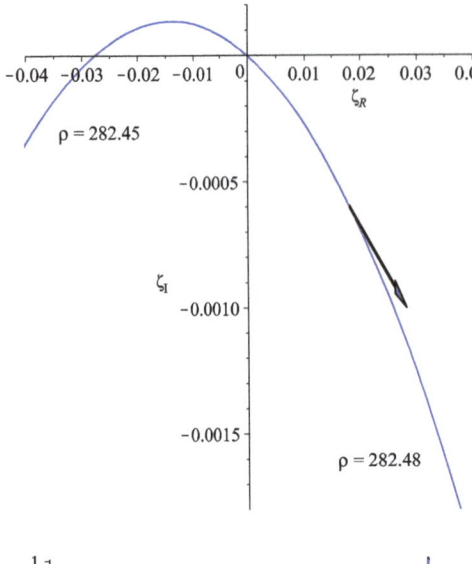

Figure 10. Clockwise-travelling locus of $\zeta(1/2 + i\rho)$ in the complex ζ plane near $\rho_{127} = 282.465...$ as ρ increases from 282.45 to 282.48.

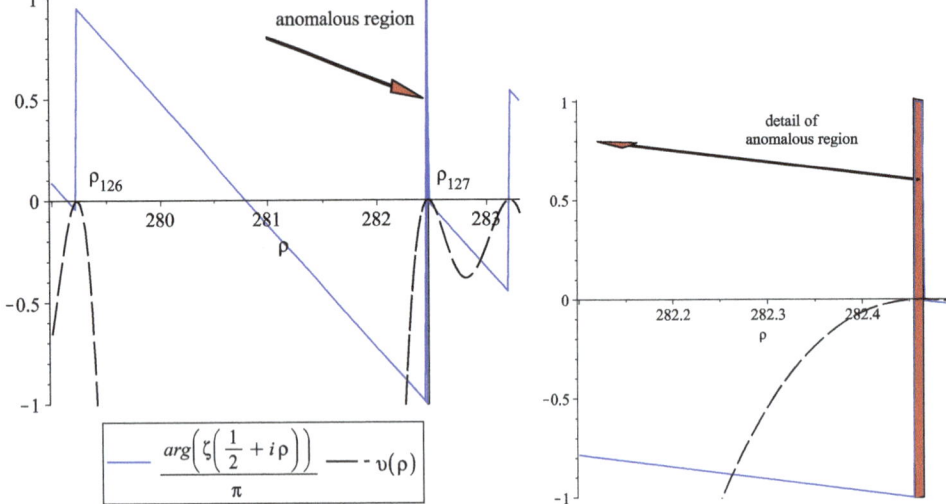

Figure 11. Overall (left) and detail(right) of $\zeta(1/2 + i\rho)$, and $\upsilon(\rho)$ in the region of interest (see Figure 10).

A third relevant result taken from Milgram (2011) predicts that "half-zeros corresponding to $\zeta_I = 0, \zeta_R \neq 0$ occur when $D_R = 0$" - see Appendix 1. A simple asymptotic form is easily obtained from Equation (16.1)

$$D_R \approx \frac{1}{2} - \frac{\sqrt{2}}{4}\left(\sin\left(\rho_\theta\right) + \cos\left(\rho_\theta\right)\right), \tag{9.6}$$

and Figure 12 demonstrates that the half-zero that upper bounds the anomalous zeros ρ_{212} and ρ_{213} is exactly where it is predicted to be according to Equation (9.6), which significantly does not depend on any numerical properties of the ζ-function. The first missing term of Equation (9.6) is of order $\exp(-\pi\rho)$ and is therefore negligible when $\rho \approx 400$. The advantage of using Equation (9.6) is that numerically it only depends on simple trigonometric functions as well as the argument of $\Gamma(1/2 + i\rho)$, all of which, it is reasonably safe to assume, are numerically accurate and reliable. In this way, the calculation will not involve possible cancellation of large numbers as might occur if Equation (16.1) were to be used. Although all the anomalies observed here involve very small values of ζ_I corresponding to values of the locus travelling close to the real axis, all these suggest that the effect, being examined here, appears to be self-consistent and is unlikely to reflect numerical artefacts. In summary, we must conclude that

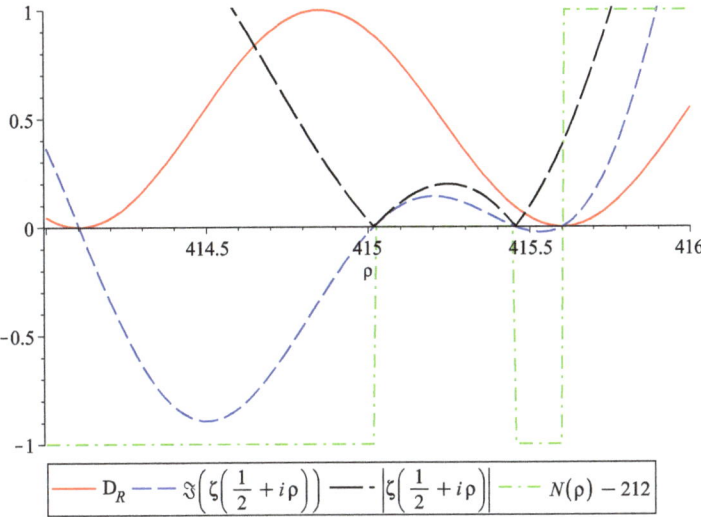

Figure 12. Details of D_R and ζ_I near the half-zero at $\rho = 415.601$, showing that the two coincide, and are separate from a full zero. This also shows that the counting formula Equation (9.4) fails, but only in the anomalous region.

anomalous, imaginary half-zeros of $\zeta(1/2 + i\rho)$ characterized by $\zeta_I = 0, \zeta_R < 0, \zeta'_R > 0$ do rarely, and unexpectedly, occur—a quick survey of the first 300 zeros suggests that in addition to the above, $\rho_{232-234}, \rho_{254-256}$ and $\rho_{288-290}$ may bound anomalous zeros. In addition, the Lehmer zeros (Edwards, 2001, Section 8.3) appear to share this property. As well as affecting counting formula such as Equation (9.4), the occurrence of such anomalies has implications for RH equivalences.

9.2. Consequences and a close look at the derivation of a ζ-function theorem

The analysis given in Section 8 depended on Equation (8.12) where the integral was split into segments, each of which was bounded by consecutive zeros ρ_k, justified by the assumption that each discontinuous segment of $arg(\zeta(\rho))$ was separated from the continuous function $\alpha(\rho)$ by $k\pi$, constant over that segment. Each consecutive segment was labelled by unit increments of k. In fact, the equivalence of Equation (8.1) and Equation (8.19), independent of the presence of the term T_0 as discussed, could have been taken as justification for that assumption. However, Figures 9 and 11 demonstrate that such is not the case due to anomalous zeros. With respect to Figure 9, the shaded region (orange), whose lower bound ρ_{213} is not part of the area bounded by $\alpha(\rho)$ below, and $arg(\zeta(1/2 + i\rho))$ above, will nevertheless be included in the integration according to Equation (8.12). Thus, over this one segment, Equation (8.12) will over-estimate the true value of the integral by approximately

$$T_{213} \approx 2 \times (2\pi) \times (415.6 - 415.46)/415^3 \approx 2.5 \times 10^{-8} \qquad (9.7)$$

and the region marked "anomalous" in Figure 11 will result in an underestimate by a numerically insignificant but non-zero result for similar reasons. Although both these anomalies are numerically insignificant, taken together with other anomalous values of ρ_k that are almost certainly scattered along the critical line, one must conclude that Equation (8.1) and Equation (8.19) are analytically inconsistent, and we have a contradiction. Is this the long-sought contradiction in a RH equivalent discussed in Section 8? Unfortunately not. To understand why, it is necessary that the derivation of Equation (8.1) be revisited.

Referring to Sekatskii et al. (2012), a typical RH equivalence is obtained by integrating a function $F(z)g(z)$ around a well-chosen contour C in the complex z plane, then making a convenient choice of $g(z)$. The integral is then related to the residue of enclosed singularities by the residue theorem, and it is simplified by taking one section of the contour of integration to coincide with a cut in the function $F(z)$. This is achieved by choosing $F(z) = \log(f(z))$ together with $f(z) = \zeta(z)$. The critical statement is "An appropriate choice of the branches of the logarithm function assures that the difference between the two branches of the logarithm function appearing after the integration path indents the point $(z =)X + iY$ is $2\pi i$". On the surface, this is justified by the additional requirement that "f(z) is analytic and non-zero on C". And therein lies a problem.

If $f(z)$ is analytic over a region, it does not follow that $F(z)$ will be analytic over that same region. The simplest example showing that this is true, is to consider the case $f(z) = z$. In this case, $f(z)$ is analytic, whereas $F(z) = \log(z)$ is cut. Thus, to assume that $F(z) = \log(\zeta(z))$ will retain the same analytic characteristics as $\log(z)$ and therefore integration over the same cut-structure as $\log(z)$ will yield the same discontinuity is a fairly difficult assumption to justify. And as we have seen in the previous sub-section, it is not true. There are places along the logarithm cut where the difference, defined by $F(z)$, between the two branches does not equal $2\pi i$. In fact, this occurs at precisely those points isolated previously and labelled "anomalous". Thus, if the derivation of Equation (8.1) had taken these regions into account, the statement of Equation (8.1) would have included terms precisely equal to those exemplified by Equation (9.7), thereby restoring consistency between (modified) versions of both Equation (8.1) and Equation (8.19). Thus, any RH equivalence of the Volchkov genre stated in the form Equation (8.1) without such terms is incorrect, and any numerical results based thereon must be treated with suspicion.

In a later work (Sekatskii et al., 2015), a more detailed exposition of the derivation of such equivalences is given. Using the same notation, we find written: "its (referring to $F(z)$) value jumps on $\mp 2\pi l$ when we pass a point $X_1 + iY$ such that there is an lth order zero or pole of the function f(z) lying inside the contour (and not on the integration line) and having the ordinate Y". As has been discovered, the existence of anomalous zeros demonstrates that discontinuities can occur at points other than a zero, and this eventuality must be incorporated into the resulting proofs.

A similar caution must be issued with respect to other results where similar assumptions (propositions?) are invoked. In França and LeClair (2013, paragraph following Eq. 14), it is written "Possible discontinuities can only come from $arg(\zeta(1/2 + i\rho))/\pi$, and, in fact, it has a jump discontinuity by one whenever ρ corresponds to a zero". As we have seen here, the discontinuity varies between ± 2 near, but not always coincident with, an anomalous zero. In Titchmarsh (1964, p. 58), we find written: "The behaviour of the function S(T) appears to be very complicated. It must have a discontinuity k where T passes through the ordinate of a zero of $\zeta(s)$ of order k ... Between the zeros N(T) is constant...". Again, this is not true in the vicinity of an anomalous zero.

10. $\zeta'(1/2 + i\rho) \neq 0$

An immediate consequence of Equation (6.1) is:

THEOREM 10.1   *If* $|\zeta(1/2 + i\rho)| \neq 0$, *then* $|\zeta'(1/2 + i\rho)| \neq 0$.

The proof is simple—if $\zeta_I \neq 0$ and $\zeta_R \neq 0$, then the left-side side of Equation (6.1) does not vanish, and therefore neither can the right-hand side. Except for the point $\rho = \rho_s$ (see Section 3), since the denominator (contained in the factor $f(\rho)$) does not vanish (which could cancel a numerator zero), both of ζ'_R and ζ'_I vanishing together would lead to a contradiction. Therefore, $|\zeta'(1/2 + i\rho)| \neq 0$.

This extends previously known results (Spira, 1973) onto the perforated critical line without recourse to RH.

Note 1: From Equation (2.5), $f(\rho)$ has a pole at only one point $\rho = \rho_s$, and since it is known that $\zeta(\rho_s)$ is not infinite, it must be that $\zeta'_I \zeta_I + \zeta'_R \zeta_R = 0$ at $\rho = \rho_s$ in order to cancel the pole of $f(\rho_s)$. This has been verified numerically. If ζ_I and ζ_R do not vanish, this is the only point where this can occur—otherwise, it would contradict Equation 7.8.

Note 2: It is worthwhile to recall from (Conrey, 1983a), (slightly edited): "It can be shown that RH implies that all zeros of $\xi'(s)$...have $\Re(s) = 1/2$, ... subject to simplicity (Conrey, 1983b)", and further

(Conrey & Gosh, 1990), "Montgomery and Levinson proved ... $\zeta'(s)$ vanishes on $\sigma = 1/2$ only at a multiple zero of $\zeta(s)$ (hence probably never)". Titchmarsh (1964, p. 79) writes:"If Mertens' hypothesis is true, all the zeros of $\zeta(s)$ are simple".

11. Asymptotics

Equation (6.1) together with Equations (6.6 and 7.8) is equivalent to Equation (2.13), and taken together demonstrate that asymptotically, $|\zeta(1/2 + i\rho)|$ and $|\zeta'(1/2 + i\rho)|$ approach their limits differently as $\rho \to \infty$. That is,

$$\frac{|\zeta'(1/2 + i\rho)|}{|\zeta(1/2 + i\rho)|} \sim -\frac{\log(\rho/(2\pi))}{2\cos(\alpha - \beta)} \tag{11.1}$$

because it is known (Olver et al., 2010, Eq. 5.11.2), and easily verifiable, that

$$\Re(\psi(1/2 + i\rho)) \sim \log(\rho). \tag{11.2}$$

Within a term proportional to $\exp(-\rho\pi)$, Equation (11.1) is almost an equality. By way of contrast, direct differentiation of Equations (2.2 and 2.4) gives

$$\frac{|\zeta(1/2 + i\rho)|'}{|\zeta(1/2 + i\rho)|} = \frac{d}{d\rho}\ln\left(|\zeta(\frac{1}{2} + i\rho)|\right) \sim -\frac{1}{2}\ln\left(\frac{\rho}{2\pi}\right)\tan(\alpha - \beta) \tag{11.3}$$

and from Equations (11.1 and 11.3), we find a simple identity

$$\frac{|\zeta(1/2 + i\rho)|'}{|\zeta(1/2 + i\rho)'|} = \sin(\alpha - \beta). \tag{11.4}$$

Integrating Equation (11.3) between limits $T_1 \gg 0$ and $T_2 > T_1$ gives

$$\ln\left(\frac{|\zeta(\frac{1}{2} + iT_2)|}{|\zeta(\frac{1}{2} + iT_1)|}\right) \sim -\frac{1}{2}\int_{T_1}^{T_2}\ln\left(\frac{\rho}{2\pi}\right)\tan(\alpha(\rho) - \beta(\rho))\,d\rho, \tag{11.5}$$

which, lacking an explicit analytic expression for $\beta(\rho)$, "might be hard" to evaluate. Notice that if either of T_1 or T_2 coincides with a zero of $\zeta(1/2 + i\rho)$, the right-hand side of Equation (11.5) must diverge, consistent with Equation (7.8) (see Olver et al., 2010, Eq. 4.21.4), and if T_1 and T_2 enclose one or more zeros of $\zeta(1/2 + i\rho)$, the integrand acquires singularities and the integral diverges. However, Equation (11.5) has been verified by numerical integration within a few small intervals where these events do not occur.

From Equation (2.9), we establish the asymptotic limit

$$\frac{\zeta_I}{\zeta_R} \sim \frac{-\cos(\rho_\theta) - \sin(\rho_\theta) + \sqrt{2}}{\cos(\rho_\theta) - \sin(\rho_\theta)} \tag{11.6}$$

demonstrating that the real and imaginary parts of $\zeta(1/2 + i\rho)$ scale to the same order of ρ, subject to an (infinitely varying) modulating function given by the right-hand side of Equation (11.6). From Equation (11.1), it is also a simple matter to establish that

$$\frac{\zeta_R'}{\zeta_R} \sim -\frac{1}{2}\frac{\cos(\beta)\ln(\frac{\rho}{2\pi})}{\cos(\alpha)\cos(-\alpha + \beta)} \tag{11.7}$$

or

$$\frac{\zeta_I'}{\zeta_I} \sim -\frac{1}{2}\frac{\sin(\beta)\ln(\frac{\rho}{2\pi})}{\sin(\alpha)\cos(-\alpha + \beta)}, \tag{11.8}$$

demonstrating that corresponding components of $\zeta(1/2 + i\rho)$ and its derivative scale differently at large values of ρ, consistent with Equation (11.1). Numerically, both these results are excellent estimates for even reasonably small values of ρ because the secondary terms that have been omitted are all of order $\exp(-\pi\rho)$.

All of these results can be understood by considering the relationship that exists between the real and imaginary components of $\zeta(1/2 + i\rho)$ and $\zeta'(1/2 + i\rho)$ presented in Equations (4.9a and 4.9b)—it is the factor $f(\rho)$ from which the scaling factor $\log(\rho/(2\pi)$ originates since all the coefficients identified in Equations (4.10 and 4.11) are $O(\rho^0)$.

Finally, with reference to Equations (6.16 and 6.17), we have, to leading orders, (and reversing the sign for convenience) the very accurate approximation,

$$\alpha_a \equiv -\alpha(\rho) \sim -\frac{\rho}{2}(1 - \log(\rho/(2\pi))) + \frac{7\pi}{8}. \tag{11.9}$$

In contrast, in Titchmarsh and Heath-Brown (1986, p. 229, Theorem 9.15), it is written: "$arg(\zeta(\sigma + iT)) = O(\log T)$ uniformly for $\sigma \geq 1/2$ ...". For an application, see Section 13.

12. Counting the zeros

Contingent on Equation (8.2) and issues raised in Section 9, we see that as ρ increases past a zero, the value of k in Equation (6.9) increments by one, so at any value $\rho = T$, the lower integer (floor) limit of k, indicated by $\lfloor...\rfloor$, and computed as

$$k = \left\lfloor \frac{arg(\zeta(\frac{1}{2} + iT)) + \frac{1}{2}\int_0^T \Re(\psi(\frac{1}{2} + i\rho))\,d\rho - \frac{1}{2}T\ln(2\pi) + \frac{9\pi}{8} - \frac{1}{2}\arctan(e^{\pi T})}{\pi} \right\rfloor \tag{12.1}$$

will be equal to the number of zeros $\rho_k \leq T$. The result (Equation 12.1) is exact, but suffers from some inconsistency as an independent enumerator because it includes a (small) term that involves ζ itself (however, see França & LeClair, 2013, Remark 5). As a means of determining the location of the kth zero, since Equation (12.1) is exact, the floor function can be used to immediately pick out a discontinuity in k that heralds the existence of a zero as ρ changes, to any desired degree of numerical accuracy. As an aid to computation, Equation (12.1) can be simplified by noting that Equation (6.9) is amenable to integration by parts, yielding the alternative representation

$$\Re \int_0^T \psi(\frac{1}{2} + i\rho)\,d\rho = \Im(\log\Gamma(1/2 + iT)) \tag{12.2}$$

where $\log\Gamma(...)$ denotes the LogGamma function (Weisstein, 2005) and, for large values of T, asymptotically, the right-hand side of Equation (12.2) is very well approximated (Maple). So, the first few terms of the asymptotic limit $T \to \infty$ in Equation (12.1) are

$$k = \left\lfloor \frac{1}{2}\frac{T\ln(T/2\pi)}{\pi} - \frac{T}{2\pi} + \frac{7}{5760\pi T^3} + \frac{1}{48\pi T} + \frac{5}{8} \pm 1 + ... \right\rfloor \tag{12.3}$$

the first two terms of which equate to the well-known result first suggested by Riemann and proven by von Mangoldt (Borwein et al., 2008, Theorem 2.9). The higher order terms can be found in Edwards (2001, Section 6.7). The last term in Equation (12.3) originates from the discontinuous term $arg(\zeta(1/2 + iT))$ in Equation (12.1), whose asymptotic limit, being unknown, perhaps unknowable, has been included in terms of its possible bounds $-\pi < arg(\zeta(1/2 + iT)) < \pi$. The traditional proof of von Mangoldt's result given at great length in Borwein et al. (2008), and discussed in Titchmarsh (1964, p. 5, Eq. 18) includes a third term $O(\log T)$, representing an average value of the term $arg(\zeta(1/2 + iT))$, which must vary continuously (and randomly?) between its bounds over any small range of T, asymptotic or not, because it is known that the number of zeros are infinite (Titchmarsh, 1964, Theorem 31). The constant term $(5/8 \pm 1)$ that is given here differs from terms given in

Titchmarsh (1964) and Edwards (2001, Section 6.7) (i.e. $7/8 \pm \frac{1}{2}$ and $7/8$, respectively) and is, according to Equation (6.8), an arbitrary normalization constant to begin with.

13. The density and distribution of zeros

In addition to the above estimates, Equation (11.9) leads to other useful approximations. Since it has been shown (see Section 9) that consecutive zeros on the critical line are (almost always) bounded by half-zeros $\zeta_I = 0, \zeta_R > 0, \zeta_R' < 0$, (see Gram points (Edwards, 2001, p. 125)) the very accurate asymptotic estimate $\mod(\alpha_a, k\pi) = 0$ can be used to place bounds on the location of consecutive zeros since each time $\zeta_I = 0, \alpha_a$ will pass through a multiple of π. With α_a defined in Equation (11.9), Figure 13 illustrates this bounding procedure by plotting the normalized asymptotic counting function

$$N_a \equiv \alpha_a/\pi - \lfloor \alpha_a/\pi \rfloor \tag{13.1}$$

as a function of ρ for a region of previous interest (see Section 9). The absolute count value is given by $\lfloor \alpha_a/\pi \rfloor$, and the solution of the equation

$$\alpha_a = k\pi \tag{13.2}$$

which defines an upper bounding point $\rho_\alpha(k)$ defined by $\lfloor \alpha_a/\pi \rfloor = k$, is given by

$$\rho_\alpha(k) = (2\,\pi)\exp\left(W\left(\frac{-7+8k}{8\,e} \right) + 1 \right) \tag{13.3}$$

in terms of Lambert's W function (Maple). As seen in the Figure, N_a bounds all zeros except the anomalous ones investigated in Section 9. However, as in the case of Equation (9.4), it "catches up" at the next bounding point. The accuracy of $\lfloor \alpha_a/\pi \rfloor$ as an upper bound counting function has been confirmed for the first 10,000 zeros ρ_k. Using the same principles, a similar result can be obtained from Equation (9.6), where $D_R = 0$ whenever

$$\theta - \frac{3\pi}{2} + 2k\pi. \tag{13.4}$$

As a comparison to Equation (13.3), França–LeClair (2013) find an expression that locates the zeros reasonable accurately, that being

$$\widetilde{\rho}_k = \frac{2\pi(k - 11/8)}{W((k - 11/8)/e)}. \tag{13.5}$$

It is interesting to compare the two approximations, by calculating the following four quantities

$$d_1 \equiv \widetilde{\rho}_k - \rho_k \tag{13.6}$$

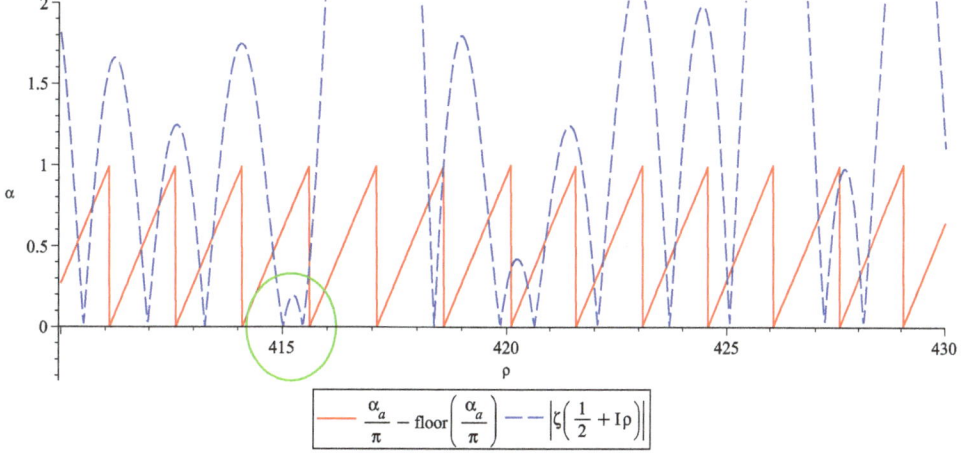

Figure 13. N_a as a function of ρ plotted over a region of previous interest, along with $|\zeta(1/2 + i\rho)|$ to indicate the location of the zeros. The circle indicates the location of an anomalous zero.

Table 1. Statistical comparison of various measures of the distribution of zeros		
	Mean	**Standard deviation**
d_1	0.00007602	0.2514
d_2	0.4163	0.2514
d_3	-.4162	0.2514
d_4	0.00007504	0.2514

$$d_2 \equiv \rho_\alpha(k) - \rho_k \tag{13.7}$$

$$d_3 \equiv \rho_k - \rho_\alpha(k-1) \tag{13.8}$$

$$d_4 \equiv (\rho_\alpha(k) + \rho_\alpha(k-1))/2 - \rho_k \tag{13.9}$$

each of which, respectively, measures the distance from

- the França-LeClair approximation to the known zero it is approximating;
- an upper bound to the next lower known zero;
- the lower bound up to the next higher known zero;
- the average of the upper and lower bounds to the bounded, known zero.

Such a comparison was performed using 5000 values of ρ_k lying between $k = 10,000$ and $k = 15,000$. The mean and standard deviation of the different quantities, given in Table 1, show that the upper bound and lower bound are on average about 0.41 units separated from the contained zero, while both the França-LeClair approximation and an average of the upper and lower bounds are equally good. There is no observable difference in the standard deviation of any of these quantities, suggesting that each of these parameters are equally good representations of the quantity they are measuring, and that the zeros ρ_k are randomly concentrated about the centre of the range bounded by $\rho_\alpha(k)$ and $\rho_\alpha(k-1)$.

For an estimate of the distribution of zeros at large values of ρ, similar to those given in França and LeClair (2013), from Equation (11.9) consider two consecutive bounding points

$$\alpha_a(\rho) = k \tag{13.10}$$

and

$$\alpha_a(\rho - \delta) = k - 1 \tag{13.11}$$

with $\delta \ll \rho$. Solving Equations (13.11 and 13.12) to first order in δ, we find the measure of a bucket that bounds successive zeros at large values of ρ, that being

$$\delta \approx \frac{2\pi}{log(\rho/(2\pi))} \tag{13.12}$$

from which it is possible to estimate the density of zeros, i.e. the number of zeros N_g that lie between unit intervals of ρ, specifically

$$N_g = 1/\delta. \tag{13.13}$$

Note that Equation (13.12) is the incarnation of Titchmarsh (1964, Theorem 41) on the critical line ("the gaps between the ordinates of successive zeros of $\zeta(s)$ tend to zero"). See also Titchmarsh and Heath-Brown (1986, Theorems 9.12 and 9.14). An interesting consequence of these considerations yields an estimate of the maximum possible discontinuity between successive zeros at large values of ρ. If, for some ρ_a and ρ in the range $\rho \geq \rho_a$, this maximum value, estimated by $-\delta\alpha'$, were less than π, then ALL zeros would be anomalous for $\rho \geq \rho_a$, and all existing counting theorems would fail. However, from Equations (6.6 and 13.13), we find

$$-\delta\alpha' \approx \pi \tag{13.14}$$

so this possibility does not occur because the gap distance and the slope of the argument both vary in magnitude at the same rate (at least to first order in ρ).

As an example, from published tables (Odlyzko, n.d, file zeros5), at $\rho_h = 1.370919909931995308226 \times 10^{21}$ corresponding to $k = 10^{22}$, we find $N_g = 7.45$. Consistently, the tables list exactly seven zeros between $k = 10^{22} + 4$ corresponding to $\lfloor \rho_h + 1 \rfloor$ and $k = 10^{22} + 10$ corresponding to $\lfloor \rho_h + 2 \rfloor$. Similarly, there are eight listed between $k = 10^{22} + 25$ corresponding to $\lfloor \rho_h + 4 \rfloor$ and $k = 10^{22} + 32$ corresponding to $\lfloor \rho_h + 5 \rfloor$. As noted, each of these ranges corresponds to values of ρ separated by one unit. The predicted gap size enveloping zeros for this range of k is $\delta = 0.13416$, comparable to the average observed distance between the seven zeros which is 0.1460. In this range, the maximum possible predicted distance between zeros is $2\delta = 0.268$; the largest observed gap between the seven zeros is 0.221.

For larger values of ρ, the reader can estimate N_g from Figure 14. For extremely large values of ρ, numerical evaluation of the function W could be reasonably suspect since no details about the asymptotic evaluation of the W function are given in Maple's FunctionAdvisor. In the case of Mathematica, we find (Weisstein, 2002) a complicated asymptotic expansion accurate to order

$$\epsilon_W = (\log(\log(\rho))/\log(\rho))^6. \tag{15.15}$$

With $\rho \approx 10^{100}$, this suggests $\epsilon_W \approx 1.7 \times 10^{-10}$ which might affect calculations that must be done in multiple precision ($>> 100$ digit) arithmetic. This can be checked using Equations (13.2) and (11.9) which only involve the log function, presumably numerically trustworthy for large values of its argument. Define $G_{oo} = 10^{100}$. In França and LeClair (2013), with $k = G_{oo}$, Table 1 lists $\rho_{G_{oo}} = 2.8069\cdots \times 10^{98}$ to 102 digit accuracy. Substituting $\rho = \rho_{G_{oo}}$ in Equations (13.2) and (11.9), and assuming that $\rho_{G_{oo}}$ is not anomalous, we obtain $k = G_{oo} - 1$, a negligible discrepancy, given that

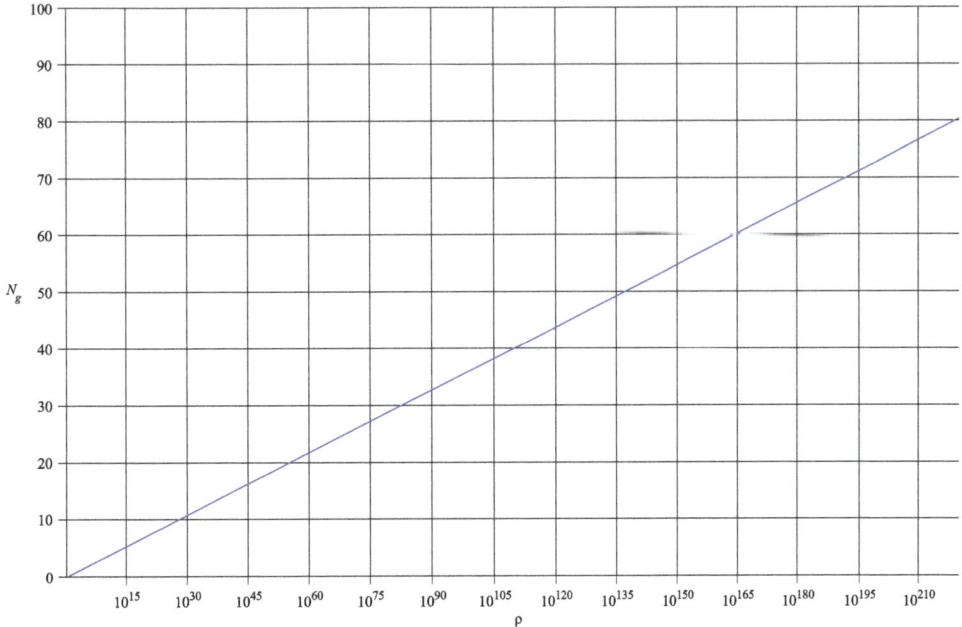

Figure 14. Estimate of N_g, the number of zeros lying between ρ and $\rho + 1$ as a function of ρ.

$\rho_{G_{oo}}$ in França and LeClair (2013) as presented is accurate to only the first digit beyond the decimal point, implying an inherent error greater than the size of the expected range boundaries (see below) at these values of k.

Performing the reverse test, let $k = G_{oo}$ and $k = G_{oo} - 1$ in Equation (13.3) to obtain the bounding points of $\rho_{G_{oo}}$. It was found that $\rho = \rho_{G_{oo}}$ is predicted to lie 0.013 units below the upper bound, and .014 units above the lower bound, exactly as specified in França and LeClair (2013). Furthermore, since the expected size of the distance between bounds at $\rho = \rho_{G_{oo}}$ is 0.0279, all of this is self-consistent and appears to be acceptably accurate. Verification of the case(s) $k = 10^{200}$ (and beyond), discussed in França and LeClair (2013), is left as an exercise for the ambitious reader.

14. Summary

In this work, an exploratory approach has been used to obtain insights into the nature of Riemann's zeta function and its zeros in the critical strip. Specializing to the critical line, the method reproduces and expands upon a known result for locating the zeros, and yields a novel derivation of the argument of the zeta function by means of a differential equation. The discovery that a simple, singular, linear transformation exists between the real and imaginary components of ζ and ζ' on the critical line is undoubtedly significant. The analytic representation of $arg(\zeta(1/2 + i\rho))$ in turn supplies further insights, notably related to the Volchkov equivalence, a counting formula and the distribution of zeros asymptotically.

Of import, a study of the location of zeros led to the numerical discovery of anomalous zeros, and the suggestion that many previous results (e.g. Volchkov-type RH equivalents) that did not take these into account require revision. From the derived equation for $\alpha(\rho)$, it is shown, independent of RH, that $\zeta'(1/2 + i\rho)$ does not vanish on the perforated critical line. Since much has been written about the zeta function over the years, it is recognized that some of the results given here may not be new, but it is suspected that those results based on Equation (6.1) are new. At a minimum, the analyses given here do not follow usual textbook derivations and gather many disparate results in one place.

Regarding RH, it has been written (Ivić, 1985, p. 50) that "the functional equation for $\zeta(s)$ in a certain sense characterizes it completely". In fact, Titchmarsh and Heath-Brown (1986, Section 2.13) demonstrate that, with general assumptions, the functional equation defines the zeta function. Here, it has been shown that many known (and possibly new) properties of the zeta function on the critical line can be obtained by studying only the functional equation, and some results can be obtained independently of RH, in contrast to equivalent results usually cited that require RH to be true. For future investigation, a very significant advance would be accomplished by the discovery of a relationship analogous to Equation (6.6) for the argument function $\beta(\rho)$, since Equations (7.6) and/ or (7.8) show that the location of the non-trivial zeros on the critical line is defined by the two argument functions α and β alone. The extension of Equation (13.15) to higher orders of ρ would clarify how frequently anomalous zeros are expected to occur, and a proof that, in Equation (4.7), the factor enclosed in square brackets is non-zero would be useful. As well, further investigation into the consequence(s) of the existence of anomalous zeros on various accepted results is warranted. For

example, the derivation of Volchkov and related equivalences and various counting theorems should be carefully and rigorously revisited.

Acknowledgements

I am grateful to Larry Glasser, who, over the past few years, has subjected me to a barrage of new, fascinating and wonderful insights into mathematical methods and results at a rate far faster than I could give them the attention they deserve. By a very circuitous route, the present work resulted from one such missive. Consequently, I am dedicating this paper to the memory of Larry's wife Judith Glasser. Larry, Judith and I enjoyed a memorable afternoon at a coffee shop in the border city of Cornwall, Ontario, several years ago, which led to my recent collaboration with Larry and the several papers that resulted. The author thanks Larry and Vini Anghel for a perusal of the preliminary manuscript. I also thank an anonymous referee who suggested the derivation of Equation (5.2) as it is now presented here, being orders of magnitude simpler than my original derivation.

Funding
The author received no direct funding for this research.

Author details
Michael Milgram[1]
E-mail: mike@geometrics-unlimited.com
ORCID ID: http://orcid.org/0000-0002-7987-0820
[1] Geometrics Unlimited Ltd., Box 1484, Deep River, Ontario, Canada, K0J 1P0.

References
Abramowitz, M., & Stegun, I. (1964). *Handbook of mathematical functions*. New York, NY: Dover.
Borwein, P., Choi, S., Rooney, B., & Weirathmueller, A. (2008). *The Riemann hypothesis : A resource for the afficionado and virtuoso alike* (CMS books in mathematics). New York, NY: Springer.
Conrey, Brian (1983a). Zeros of derivatives of Riemann's ξ-function on the critical line. *Journal of Number Theory, 16*, 49–74.
Conrey, B. (1983b). Zeros of derivatives of Riemann's ξ-function on the critical line. II. *Journal of Number Theory, 17*, 71–75.
Conrey, J. D., & Farmer, D. W. (n.d.) *Equivalences to the Riemann hypothesis*. Retrieved from https://web.archive.org/web/20120731034246/http://aimath.org/pl/rhequivalences
Conrey, J. B., & Ghosh, A. (1990). In B. Berndt, H. Diamond, H. Halberstam, & A. Hildebrand (Eds.), *Zeros of derivatives of the Riemann zeta-function near the critical line*. Boston, MA: Birkhäuser
de Reyna, J.A., & Van de Lune, J. (2014). On the exact location of the non-trivial zeros of Riemann's Zeta function. *Acta Arithmetica, 163*, 215–245. Retrieved from http://arxiv.org/pdf/1305.3844v2.pdf; 2013, doi:10.4064/aa163-3-3
Edwards, H. M. (2001). *Riemann's zeta function*. Mineola, NY: Dover Publications.
England, M., Bradford, R., Davenport, J. H., & Wilson, D. (2013). Intelligent computer mathematics: MKM, calculemus, DML, and systems and projects 2013, Held as Part of CICM 2013, Bath, UK, July 8–12, 2013. *Proceedings chapter Understanding Branch Cuts of Expressions* (pp. 136–151).Berlin Heidelberg: Springer. Retrieved from http://arxiv.org/abs/1304.7223
França, G., & LeClair, A. (2013, July). Statistical and other properties of Riemann zeros based on an explicit equation for the n-th zero on the critical line. *ArXiv e-prints*. Retrieved from http://arxiv.org/abs/1307.8395
Gradshteyn, I. S., & Ryzhik, I. M. (1980). *Tables of integrals, series and products, corrected and enlarged edition*. New York, NY: Academic Press.
Guillera, J. (2013, July). *Some sums over the non-trivial zeros of the Riemann zeta function*. Retrieved from http://arxiv.org/abs/1307.5723v7
Hansen, E. R. (1975). *A table of series and products*. Englewood Cliffs, NJ: Prentice-Hall.
He, Y. H., Jejjala, V., & Minic, D. (2015, January). *From Veneziano to Riemann: A string theory statement of the Riemann hypothesis*. arxiv.org/pdf/1501.01975.
Ivić, A. (1985). *The Riemann zeta-function: Theory and applications*. Mineola, NY: Dover.
Maplesoft, A Division of Waterloo Maple. (2014). *Maple*.
Milgram, M. S.(2011). *Notes on the zeros of Riemann's Zeta function*. Retrieved from http://arxiv.org/abs/0911.1332
Moll, V. H. (2010). Seized opportunities. *Notices of the American Mathematical Society, 57*(4). Retrieved from http://www.researchgate.net/publication/265488191
NIST Digital Library of Mathematical Functions. (2014). New York, NY: Cambridge University Press. Retrieved from http://dlmf.nist.gov/ (Release 1.0.9 of 2014-08-29).
Odlyzko, A. (XXXX). *Tables of zeros of the Riemann zeta function*. Retrieved from http://www.dtc.umn.edu/~odlyzko/zetatables/index.html
Olver, F. W. J., Lozier, D. W., Boisvert, R. F., & Clark, C. W. (Eds.). (2010). *NIST handbook of mathematical functions*. New York, NY: Cambridge University Press.
Sekatskii, S. K., Beltraminelli, S., & Merlini, D. (2009, April). A few equalities involving integrals of the logarithm of the Riemann zeta-function and equivalent to the Riemann hypothesis II. *ArXiv e-prints*. Retruved from http://arxiv.org/abs/0904.1277
Sekatskii, S. K., Beltraminelli, S., & Merlini, D. (2012). On equalities involving integrals of the logarithm of the riemann ς-function and equivalent to the riemann hypothesis. *Ukrainian Mathematical Journal, 64*, 247–261.
Sekatskii, S. K., Beltraminelli, S., & Merlini, D. (2015). On equalities involving integrals of the logarithm of the Riemann-function with exponential weight which are equivalent to the Riemann hypothesis. *International Journal of Analysis*. doi:10.1155/2015/980728
Spira, R. (1973). Zeros of $\zeta'(s)$ and the Riemann hypothesis. *Illinois Journal of Mathematics, 17*, 147–152.
Titchmarsh, E. C. (1964). *The zeta-function of Riemann*. New York: Stechert-Hafner Service Agency.
Titchmarsh, E. C., & Heath-Brown, D. R. (1986). *The theory of the Riemann zeta-function* (2nd ed.).Oxford: Oxford Science Publications.
Volchkov, V. V. (1995). On an equality equivalent to the Riemann hypothesis. *Ukrainian Mathematical Journal, 47*, 491–493.
Weisstein, E. W. (2002). *Lambert W-function*. MathWorld--A Wolfram Web Resource. Retrieved from http://mathworld.wolfram.com/LambertW-Function.html
Weisstein, E. W. (2005). *Log gamma function*. MathWorld--A Wolfram Web Resource. Retrieved from http://mathworld.wolfram.com/LogGammaFunction.html

Appendix 1

Appendix—definitions and symbols

The following symbols and definitions appear throughout this paper. Subscripts I and R refer to the imaginary and real components of the symbol to which they are attached. The unmodified symbols Γ, γ, ψ and ζ mean $\Gamma(1/2 + i\rho)$, $\Gamma(1/4 + i\rho/2)$, $\psi(1/2 + i\rho)$ and $\zeta(1/2 + i\rho)$, respectively, along with their appropriate subscripts; ψ is the digamma function $(\psi(z) = \frac{d}{dz} \log(\Gamma(z))$. $\zeta'_{R_{UI}}$ refers to the real or imaginary component of the derivative of $\zeta(1/2 + i\rho)$ with respect to ρ, whereas $(\zeta_{R_{UI}})'$ refers to the derivative of the real or imaginary component of $\zeta(1/2 + i\rho)$ with respect to ρ (see Equations (6.3 and 6.4). All other symbols carry implicit functional dependence on ρ, usually omitted for brevity, except where required for clarity. The variables n, k are always positive integers, and K is an integer. The symbol ρ_0 refers to a generic value of ρ corresponding to $\zeta(1/2 + i\rho_0) = 0$, whereas ρ_k refers to a specific value of ρ_0.

$$D_R = 1/2 - \frac{C_p \cos\left(\rho_\pi\right) + C_m \sin\left(\rho_\pi\right)}{2\sqrt{\pi}} \tag{16.1}$$

$$C_p = \Gamma_R \cosh(\frac{\pi \rho}{2}) + \Gamma_I \sinh(\frac{\pi \rho}{2}) \tag{16.2}$$

$$C_m = \Gamma_I \cosh(\frac{\pi \rho}{2}) - \Gamma_R \sinh(\frac{\pi \rho}{2}) \tag{16.3}$$

$$\left|\zeta'\right|^2 = \zeta_R'^2 + \zeta_I'^2 \tag{16.4}$$

$$\rho_\pi = \rho \log(2\pi) \tag{16.5}$$

$$\rho_\theta = \theta - \rho_\pi \tag{16.6}$$

$$\theta = arg(\Gamma(1/2 + i\rho)) \tag{16.7}$$

$$\alpha = arg(\zeta(1/2 + i\rho)) \tag{16.8}$$

$$\beta = arg(\zeta'(1/2 + i\rho)) \tag{16.9}$$

$$T_1(\rho) = -\cosh(\frac{\pi \rho}{2}) \sin(2\beta + \rho_\theta)) + \sinh(\frac{\pi \rho}{2}) \cos(2\beta + \rho_\theta) \tag{16.10}$$

17

Sharp bounds for the Neuman-Sándor mean in terms of the power and contraharmonic means

Wei-Dong Jiang[1] and Feng Qi[2,3,4]*

*Corresponding author: Feng Qi, College of Mathematics, Inner Mongolia University for Nationalities, Tongliao City, Inner Mongolia Autonomous Region, 028043, China; Department of Mathematics, College of Science, Tianjin Polytechnic University, Tianjin City, 300387, China; Institute of Mathematics, Henan Polytechnic University, Jiaozuo City, Henan Province, 454010, China

Email: qifeng618@gmail.com; qifeng618@hotmail.com; qifeng618@qq.com

Reviewing editor: Kok Lay Teo, Curtin University, Australia

Abstract: In the paper, the authors obtain sharp bounds for the Neuman–Sándor mean in terms of the power and contraharmonic means.

Subjects: Science; Mathematics & Statistics; Advanced Mathematics; Analysis - Mathematics; Real Functions; Special Functions

Keywords: sharp bound; Neuman–Sándor mean; power; contraharmonic mean

AMS Subject Classifications: Primary: 26E60; Secondary 26D05; 33B10

1. Introduction

For positive numbers $a, b < 0$ with $a \neq b$, the second Seiffert mean $T(a, b)$, quadratic mean $S(a, b)$, Neuman–Sándor mean $M(a, b)$, and contraharmonic mean $C(a, b)$ are respectively defined in Neuman and Sándor (2003), and Seiffert (1995) by

$$T(a, b) = \frac{a - b}{2 \arctan[(a - b)/(a + b)]}, \quad S(a, b) = \sqrt{\frac{a^2 + b^2}{2}} \tag{1.1}$$

$$M(a, b) = \frac{a - b}{2 \operatorname{arcsinh}[(a - b)/(a + b)]}, \quad C(a, b) = \frac{a^2 + b^2}{a + b} \tag{1.2}$$

ABOUT THE AUTHOR

Feng Qi is a full Professor in Mathematics at Henan Polytechnic University and Tianjin Polytechnic University, China. He was the founder and the former Head of School of Mathematics and Informatics at Henan Polytechnic University. He was ever a visiting professor at Victoria University in Australia, and University of Hong Kong, and ever a part-time professor at Henan University, Henan Normal University, and Inner Mongolia University for Nationalities in China. He received his PhD degree of Science in Mathematics from University of Science and Technology of China. He is the editor of several international journals. He has published over 500 research articles in reputed international journals. His research interests include the classical analysis, analytic combinatorics, special functions, mathematical inequalities, mathematical means, integral transforms, complex functions, analytic number theory, differential geometry, and mathematical education at universities. For more information, please see his home page at http://qifeng618.wordpress.com and related links therein.

PUBLIC INTEREST STATEMENT

In the paper, the authors obtain a sharp lower bound and a sharp upper bound for the Neuman–Sándor mean in terms of the power and contraharmonic means.

It is well known Neuman (2012, 2011), and Neuman and Sándor (2006) that the inequalities

$$M(a, b) < T(a, b) < S(a,b) < C(a, b)$$

hold for all $a, b < 0$ with $a \neq b$.

In Chu and Hou (2012), Chu, Hou, and Shen (2012), the inequalities

$$S(\alpha a+(1-\alpha)b, \alpha b+(1-\alpha)a) < T(a, b) < S(\beta a+(1-\beta)b, \beta b+(1-\beta)a) \tag{1.3}$$

and

$$C(\lambda a+(1-\lambda)b, \lambda b+(1-\lambda)a) < T(a, b) < C(\mu a+(1-\mu)b, \mu b+(1-\mu)a) \tag{1.4}$$

were proved to be valid for $\frac{1}{2} < \alpha, \beta, \lambda, \mu < 1$ and for all $a, b < 0$ with $a \neq b$ if and only if

$$\alpha \leq \frac{1}{2}\left(1+\sqrt{\frac{16}{\pi^2}-1}\right), \quad \beta \geq \frac{3+\sqrt{6}}{6}$$

$$\lambda \leq \frac{1}{2}\left(1+\sqrt{\frac{4}{\pi}-1}\right), \quad \mu \geq \frac{3+\sqrt{3}}{6} \tag{1.5}$$

respectively. In Jiang and Qi (2014) and its preprint Jiang and Qi (2013a), the double inequality

$$S(\alpha a+(1-\alpha)b, \alpha b+(1-\alpha)a) < M(a, b) < S(\beta a+(1-\beta)b, \beta b+(1-\beta)a) \tag{1.6}$$

was proved to be valid for $\frac{1}{2} < \alpha, \beta < 1$ and for all $a,b > 0$ with $a \neq b$ if and only if

$$\alpha \leq \frac{1}{2}\left\{1+\sqrt{\frac{1}{[\ln(1+\sqrt{2})]^2}-1}\right\} \quad \text{and} \quad \beta \geq \frac{3+\sqrt{3}}{6} \tag{1.7}$$

For more information on this topic, please refer to recently published papers Chu, Wang, and Gong (2011), Jiang and Qi (2012), Li, Long, and Chu (2012), Li and Qi (2013) and references cited therein.

For $t \in \left(\frac{1}{2}, 1\right)$ and $p \geq \frac{1}{2}$, let

$$Q_{t,p}(a, b) = C^p(ta+(1-t)b, tb+(1-t)a)A^{1-p}(a, b) \tag{1.8}$$

where $A(a, b) = \frac{a+b}{2}$ is the classical arithmetic mean of a and b. Then, by definitions in (1.1) and (1.2), it is easy to see that

$$Q_{t,1/2}(a, b) = S(ta+(1-t)b, tb+(1-t)a)$$
$$Q_{t,1}(a, b) = C(ta+(1-t)b, tb+(1-t)a)$$

and $Q_{t,p}(a, b)$ is strictly increasing with respect to $t \in \left(\frac{1}{2}, 1\right)$.

Motivating by results mentioned above, we naturally ask a question: what are the greatest value $t_1 = t_1(p)$ and the least value $t_2 = t_2(p)$ in $\left(\frac{1}{2}, 1\right)$ such that the double inequality

$$Q_{t_1,p}(a, b) < M(a, b) < Q_{t_2,p}(a, b) \tag{1.9}$$

holds for all $a, b < 0$ with $a \neq b$ and for all $p \geq \frac{1}{2}$?

The aim of this paper is to answer this question. The solution to this question may be stated as the following Theorem 1.1.

THEOREM 1.1 Let $t_1, t_2 \in \left(\frac{1}{2}, 1\right)$ and $p \in \left[\frac{1}{2}, \infty\right)$. Then the double inequality (1.9) holds for all $a, b < 0$ with $a \neq b$ if and only if

$$t_1 \leq \frac{1}{2}\left[1 + \sqrt{\left(\frac{1}{t^*}\right)^{1/p} - 1}\right] \quad \text{and} \quad t_2 \geq \frac{1}{2}\left(1 + \frac{1}{\sqrt{6p}}\right) \tag{1.10}$$

where

$$t^* = \ln\left(1 + \sqrt{2}\right) = 0.88\cdots \tag{1.11}$$

Remark 1.1   When $p = \frac{1}{2}$ in Theorem 1.1, the double inequality (1.9) becomes (1.6).

Remark 1.2   If taking $p = 1$ in Theorem 1.1, we can conclude that the double inequality

$$C(\lambda a + (1-\lambda)b, \lambda b + (1-\lambda)a) < M(a, b) < C(\mu a + (1-\mu)b, \mu b + (1-\mu)a) \tag{1.12}$$

holds for all $a, b > 0$ with $a \neq b$ if and only if

$$\frac{1}{2} < \lambda \leq \frac{1}{2}\left[1 + \sqrt{\frac{1}{\ln(1+\sqrt{2})} - 1}\right] \quad \text{and} \quad 1 > \mu \geq \frac{1}{2}\left(1 + \frac{\sqrt{6}}{6}\right) \tag{1.13}$$

Remark 1.3   We note that the paper Li and Zheng (2013) is worth to being read.

2. Lemmas

In order to prove Theorem 1.1, we need the following lemmas.

LEMMA 2.1   (Anderson, Vamanamurthy, & Vuorinen, 1997, Theorem 1.25) For $-\infty < a < b < \infty$, let $f, g: [a, b] \to \mathbb{R}$ be continuous on $[a, b]$ and differentiable on (a, b). If $g'(x) \neq 0$ and $\frac{f'(x)}{g'(x)}$ is strictly increasing (or strictly decreasing, respectively) on (a, b), so are the functions

$$\frac{f(x) - f(a)}{g(x) - g(a)} \quad \text{and} \quad \frac{f(x) - f(b)}{g(x) - g(b)} \tag{2.1}$$

LEMMA 2.2   The function

$$h(x) = \frac{(1 + x^2)\text{arcsinh } x}{x} \tag{2.2}$$

is strictly increasing and convex on $(0, \infty)$.

Proof   This follows from the following arguments:

$$h'(x) = \frac{x\sqrt{1+x^2} - \text{arcsinh } x + x^2 \text{arcsinh } x}{x^2} \triangleq \frac{h_1(x)}{x^2}$$

$$h_1'(x) = x\left(\frac{3x}{\sqrt{1+x^2}} + 2\text{arcsinh } x\right) \triangleq x h_2(x)$$

$$h_2'(x) = \frac{5 + 2x^2}{(1+x^2)^{3/2}} < 0$$

on $(0, \infty)$ and

$$\lim_{x \to 0^+} h_1(x) = \lim_{x \to 0^+} h_2(x) = 0$$

LEMMA 2.3   For $u \in [0, 1]$ and $p \geq \frac{1}{2}$, let

$$f_{u,p}(x) = p \ln(1 + ux^2) - \ln x + \ln \operatorname{arcsinh} x \tag{2.3}$$

on $(0, 1)$. Then the function $f_{u,p}(x)$ is positive if and only if $6pu \geq 1$ and it is negative if and only if $1 + u \leq \left(\frac{1}{t_*}\right)^{1/p}$, where t^* is defined by (1.11).

Proof   It is ready that

$$\lim_{x \to 0^+} f_{u,p}(x) = 0 \tag{2.4}$$

and

$$\lim_{x \to 1^-} f_{u,p}(x) = p \ln(1 + u) + \ln(t^*) \tag{2.5}$$

An easy computation yields

$$
\begin{aligned}
f'_{u,p}(x) &= \frac{2pux}{1 + ux^2} + \frac{1}{\sqrt{1 + x^2}\,\operatorname{arcsinh} x} - \frac{1}{x} \\
&= \frac{u\left[(2p-1)x^2\sqrt{1+x^2}\,\operatorname{arcsinh} x + x^3\right] - \left[\sqrt{1+x^2}\,\operatorname{arcsinh} x - x\right]}{x(1 + ux^2)\sqrt{1+x^2}\,\operatorname{arcsinh} x} \\
&= \frac{(2p-1)x^2\sqrt{1+x^2}\,\operatorname{arcsinh} x + x^3}{x(1 + ux^2)\sqrt{1+x^2}\,\operatorname{arcsinh} x}\left[u - \frac{g_1(x)}{g_2(x)}\right]
\end{aligned} \tag{2.6}
$$

where

$$g_1(x) = \operatorname{arcsinh} x - \frac{x}{\sqrt{1+x^2}} \quad \text{and} \quad g_2(x) = (2p-1)x^2 \operatorname{arcsinh} x + \frac{x^3}{\sqrt{1+x^2}}$$

Furthermore, we have

$$g_1(0) = g_2(0) = 0 \tag{2.7}$$

and

$$\frac{g'_1(x)}{g'_2(x)} = \frac{1}{2(2p-1)\sqrt{1+x^2}\,h(x) + (2p+1)x^2 + 2p + 2} \tag{2.8}$$

where $h(x)$ is defined by (2.2). From Lemma 2.2, it follows that the quotient $\frac{g'_1(x)}{g'_2(x)}$ is strictly decreasing on $(0, 1)$. Accordingly, from Lemma 2.1 and (2.7), it is deduced that the ratio $\frac{g_1(x)}{g_2(x)}$ is strictly decreasing on $(0, 1)$.

Moreover, making use of L'Hôpital's rule leads to

$$\lim_{x \to 0} \frac{g_1(x)}{g_2(x)} = \frac{1}{6p} \tag{2.9}$$

and

$$\lim_{x\to 1}\frac{g_1(x)}{g_2(x)}=\frac{\sqrt{2}\,t^*-1}{\sqrt{2}\,(2p-1)t^*+1} \tag{2.10}$$

When $u\geq\frac{1}{6p}$, combining (2.6) and (2.9) with the monotonicity of $\frac{g_1(x)}{g_2(x)}$ shows that the function $f_{u,p}(x)$ is strictly increasing on $(0,1)$. Therefore, the positivity of $f_{u,p}(x)$ on $(0,1)$ follows from (2.4) and the increasingly monotonicity of $f_{u,p}(x)$.

When $u\leq\frac{\sqrt{2}\,t^*-1}{\sqrt{2}\,(2p-1)t^*+1}$, combining (2.6) and (2.10) with the monotonicity of $\frac{g_1(x)}{g_2(x)}$ reveals that the function $f_{u,p}(x)$ is strictly decreasing on $(0,1)$. Hence, the negativity of $f_{u,p}(x)$ on $(0,1)$ follows from (2.4) and the decreasingly monotonicity of $f_{u,p}(x)$.

When $\frac{\sqrt{2}\,t^*-1}{\sqrt{2}\,(2p-1)t^*+1}<u<\frac{1}{6p}$, from (2.6), (2.9), (2.10) and the monotonicity of the ratio $\frac{g_1(x)}{g_2(x)}$, we conclude that there exists a number $x_0\in(0,1)$ such that $f_{u,p}(x)$ is strictly decreasing in $(0,x_0)$ and strictly increasing in $(x_0,1)$. Denote the limit in (2.5) by $h_p(u)$. Then, from the above arguments, it follows that

$$h_p\left(\frac{1}{6p}\right)=p\ln\left(1+\frac{1}{6p}\right)+\ln(t^*)>0 \tag{2.11}$$

and

$$h_p\left(\frac{\sqrt{2}\,t^*-1}{\sqrt{2}\,(2p-1)t^*+1}\right)=p\ln\left[1+\frac{\sqrt{2}\,t^*-1}{\sqrt{2}\,(2p-1)t^*+1}\right]+\ln(t^*)<0 \tag{2.12}$$

Since $h_p(u)$ is strictly increasing for $u>-1$, so it is also in $\left[\frac{\sqrt{2}\,t^*-1}{\sqrt{2}\,(2p-1)t^*+1},\frac{1}{6p}\right]$. Thus, the inequalities In (2.11) and (2.12) imply that the function $h_p(u)$ has a unique zero point $u_0=\left(\frac{1}{t^*}\right)^{1/p}-1\in\left(\frac{\sqrt{2}\,t^*-1}{\sqrt{2}\,(2p-1)t^*+1},\frac{1}{6p}\right)$ such that $h_p(u)<0$ for $u\in\left[\frac{\sqrt{2}\,t^*-1}{\sqrt{2}\,(2p-1)t^*+1},u_0\right)$ and $h_p(u)>0$ for $u\in\left(u_0,\frac{1}{6p}\right]$. As a result, combining (2.4) and (2.5) with the piecewise monotonicity of $f_{u,p}(x)$ reveals that $f_{u,p}(x)<0$ for all $x\in(0,1)$ if and only if $\frac{\sqrt{2}\,t^*-1}{\sqrt{2}\,(2p-1)t^*+1}<u<u_0$. The proof of Lemma 2.3 is complete.

3. Proof of Theorem 1.1
Now we are in a position to prove our Theorem 1.1.

Since both $Q_{t,p}(a,b)$ and $M(a,b)$ are symmetric and homogeneous of degree 1, without loss of generality, we assume that $a>b$. Let $x=\frac{a-b}{a+b}\in(0,1)$. From (1.2) and (1.8), we obtain

$$\ln\frac{Q_{t,p}(a,b)}{T(a,b)}=\ln\frac{Q_{t,p}(a,b)}{A(a,b)}-\ln\frac{T(a,b)}{A(a,b)}$$
$$=p\ln\left[1+(1-2t)^2x^2\right]-\ln x+\ln\operatorname{arcsinh} x$$

Thus, Theorem 1.1 follows from Lemma 2.3.

Remark 3.1   This is a slightly modified version of the preprint Jiang and Qi (2013b).

Funding
The first author was partially supported by the Project of Shandong Province Higher Educational Science and Technology Program [grant number J11LA57], China. The second author was partially supported by the Natural Science Foundation [grant number 2014JQ1006] of Shaanxi Province of China and by the National Natural Science Foundation [grant number 11361038] of China.

Author details
Wei-Dong Jiang[1]
E-mail: jackjwd@163.com; jackjwd@hotmail.com
ORCID ID: http://orcid.org/0000-0001-6239-2968
Feng Qi[2,3,4]
E-mail: qifeng618@gmail.com; qifeng618@hotmail.com
qifeng618@qq.com

[1] Department of Information Engineering, Weihai Vocational College, Weihai City, Shandong Province, 264210, China.

[2] College of Mathematics, Inner Mongolia University for Nationalities, Tongliao City, Inner Mongolia Autonomous Region, 028043, China.

[3] Department of Mathematics, College of Science, Tianjin Polytechnic University, Tianjin City, 300387, China.

[4] Institute of Mathematics, Henan Polytechnic University, Jiaozuo City, Henan Province, 454010, China.

References
Anderson, G. D., Vamanamurthy, M. K., & Vuorinen, M. (1997). *Conformal invariants, inequalities, and quasiconformal maps*. New York, NY: John Wiley & Sons.

Chu, Y. M., & Hou, S. W. (2012). Sharp bounds for Seiffert mean in terms of contraharmonic mean. *Abstract and Applied Analysis, 2012*, 6 p. Article ID 425175. doi:10.1155/2012/425175

Chu, Y. M., Hou, S. W., & Shen, Z. H. (2012). Sharp bounds for Seiffert mean in terms of root mean square. *Journal of Inequalities and Applications, 2012*, 11. 6 p. doi:10.1186/1029-242X-2012-11

Chu, Y. M., Wang, M. K., & Gong, W. M. (2011). Two sharp double inequalities for Seiffert mean. *Journal of Inequalities and Applications, 2011*, 44. 7 p. doi:10.1186/1029-242X-2011-44

Jiang, W.D., & Qi, F. (2012). Some sharp inequalities involving Seiffert and other means and their concise proofs. *Mathematical Inequalities and Applications, 15*, 1007–1017. doi:10.7153/mia-15-86

Jiang, W. D., & Qi, F. (2013a) Sharp bounds for Neuman–Sándor mean in terms of the root-mean-square. Retrieved from http://arxiv.org/abs/1301.3267

Jiang, W. D., & Qi, F. (2014). Sharp bounds for Neuman–Sándor's mean in terms of the root-mean-square. *Periodica Mathematica Hungarica, 69*, 134–138. Retrieved from http://dx.doi.org/10.1007/s10998-014-0057-9

Jiang, W. D., & Qi, F. (2013b) Sharp bounds in terms of the power of the contra-harmonic mean for Neuman–Sándor's mean. Retrieved from http://arxiv.org/abs/1301.3554

Li, Y. M., Long, B. Y., & Chu, Y. M. (2012). Sharp bounds for the Neuman–Sándor mean in terms of generalized logarithmic mean. *Journal of Mathematical Inequalities, 6*, 567–577. doi:10.7153/jmi-06-54

Li, W. H., & Qi, F. (2013) A unified proof of inequalities and some new inequalities involving Neuman-Sándor mean. Retrieved from http://arxiv.org/abs/1312.3500

Li, W.H., & Zheng, M. M. (2013). Some inequalities for bounding Toader mean. *Journal of Function Spaces and Applications, 2013*, 5 p. Article ID 394194. doi:10.1155/2013/394194

Neuman, E. (2011). Inequalities for the Schwab—Borchardt mean and their applications. *Journal of Mathematical Inequalities, 5*, 601–609. doi:10.7153/jmi-05-52

Neuman, E. (2012). A note on a certain bivariate mean. *Journal of Mathematical Inequalities, 6*, 637–643. doi:10.7153/jmi-06-62

Neuman, E., & Sándor, J. (2003). On the Schwab—Borchardt mean. *Mathematica Pannonica, 14*, 253–266.

Neuman, E., & Sándor, J. (2006). On the Schwab—Borchardt mean II. *Mathematica Pannonica, 17*, 49–59.

Seiffert, H.-J. (1995). Aufgabe β 16. *Die Wurzel, 29*, 221–222.

18

Minimum-energy wavelet frames generated by the Walsh polynomials

Sunita Goyal[1] and Firdous A. Shah[2]*

*Corresponding author: Firdous A. Shah, Department of Mathematics, University of Kashmir, South Campus, Anantnag 192 101, Jammu and Kashmir, India
E-mail: fashah79@gmail.com
Reviewing editor: Hari M. Srivastava, University of Victoria, Canada

Abstract: Drawing inspiration from the construction of tight wavelet frames generated by the Walsh polynomials, we introduce the notion of minimum-energy wavelet frames generated by the Walsh polynomials on positive half-line $\mathbb{R}^+$ using unitary extension principles and present its equivalent characterizations in terms of their framelet symbols. Moreover, based on polyphase components of the Walsh polynomials, we obtain a necessary and sufficient condition for the existence of minimum-energy wavelet frames in $L^2(\mathbb{R}^+)$. Finally, we derive the minimum-energy wavelet frame decomposition and reconstruction formulae which are quite similar to those of orthonormal wavelets on local fields of positive characteristic.

Subjects: Advanced Mathematics; Applied Mathematics; Mathematics & Statistics; Science

Keywords: frame; wavelet frame; *p*-multiresoltion analysis; scaling function; minimum-energy frame; extension principles; Walsh polynomial; Walsh–Fourier transform

AMS Subject classifications: 42C15; 42C40; 42A38; 41A17; 22B99

1. Introduction

The notion of frames was first introduced by Duffin and Schaeffer (1952) in connection with some deep problems in nonharmonic Fourier series. Frames are basis-like systems that span a vector space but allow for linear dependency, which can be used to reduce noise, find sparse representations, or

ABOUT THE AUTHORS

Sunita Goyal received her MSc and MPhil degrees in pure mathematics from the University of Rothak, Haryana, India. Currently, she is perusing PhD at the Department of Mathematics, JJT University, Rajasthan, India. Her research interests are focused on different aspects of wavelet analysis including wavelet frames, shift invariant spaces, wavelet packets and their applications in Economics and Finance.

Firdous A. Shah is a senior assistant professor in the Department of Mathematics at University of Kashmir, India. His primary research interests include basic theory of wavelets and their applications in differential and integral equations, Economics and Finance, and Computer Networking. He has authored/co-authored over 50 research papers in international journals of high repute. He has recently co-authored a book on wavelets entitled *Wavelet Transforms and Their Applications*, Springer, New York, 2015.

PUBLIC INTEREST STATEMENT

Wavelet frames are different from the orthonormal wavelets because of redundancy. By sacrificing orthonormality and allowing redundancy, the wavelet frames become much easier to construct than the orthonormal wavelets. Wavelet frames and their promising features in applications have attracted a great deal of interest and effort in recent years. Although wavelet frames have many desirable features but the computational complexity and numerical instability during the course of decomposition and reconstruction of functions always remains a debate of discussion. In this study, we introduce a new concept called minimum-energy wavelet frames generated by the Walsh polynomials on a positive half-line. Our results will be mostly used by that part of mathematical society who works in wavelet analysis and their applications. Most prominent among them are the theory of signal processing, image processing, data transmission with erasures, quantum computing, medicine, representation theory, and algebraic geometry.

obtain other desirable features unavailable with orthonormal bases. The idea of Duffin and Schaeffer did not generate much interest outside nonharmonic Fourier series until the seminal work by Daubechies, Grossmann, and Meyer (1986). They combined the theory of continuous wavelet transforms with the theory of frames to introduce wavelet (affine) frames for $L^2(\mathbb{R})$. After their work, the theory of frames began to be studied widely and deeply. Today, the theory of frames has become an interesting and fruitful field of mathematics with abundant applications in signal processing, image processing, harmonic analysis, Banach space theory, sampling theory, wireless sensor networks, optics, filter banks, quantum computing, medicine, and so on. An introduction to the frame theory and its applications can be found in Christensen (2003), Daubechies (1992), Debnath and Shah (2015), Dong, Ji, Li, Shen, and Xu (2012). The following are the standard definitions on frames in Hilbert spaces. A sequence $\{f_k : k \in \mathbb{Z}\}$ of elements of a Hilbert space $\mathbb{H}$ is called a *frame* for $\mathbb{H}$ if there exist constants $A, B > 0$ such that for all $f \in \mathbb{H}$

$$A\|f\|_2^2 \leq \sum_{k \in \mathbb{Z}} |\langle f, f_k \rangle|^2 \leq B\|f\|_2^2. \tag{1.1}$$

The largest constant A and the smallest constant B satisfying (1.1) are called the *lower* and *upper frame bound*, respectively. A frame is a *tight frame* if A and B are chosen so that $A = B$ and is called a *Parseval frame* or *normalized tight frame* if $A = B = 1$.

An important example about frame is wavelet frame, which is obtained by translating and dilating a finite family of functions. One of the most useful methods to construct wavelet frames is through the concept of unitary extension principle (UEP) introduced by Ron and Shen (1997) and were subsequently extended by Daubechies, Han, Ron, and Shen (2003) in the form of the oblique extension principle (OEP). They give sufficient conditions for constructing tight and dual wavelet frames for any given refinable function $\phi(x)$ which generates a multiresolution analysis. The resulting wavelet frames are based on multiresolution analysis, and the generators are often called *framelets*. The advantages of MRA-based wavelet frames and their promising features in applications have attracted a great deal of interest and effort in recent years. To mention only a few references on wavelet frames, the reader is referred to Chui and He (2000), Dong et al. (2012), Farkov, Lebedeva, and Skopina (2015), Gao and Cao (2008), Han (2012), Huang and Cheng (2007), Huang, Li, and Li (2012), Zhu, Li, and Huang (2013) and many references therein.

The past decade has also witnessed a tremendous interest in the problem of constructing compactly supported orthonormal scaling functions and wavelets with an arbitrary dilation factor $p \geq 2, p \in \mathbb{N}$ (see Debnath & Shah, 2015). The motivation comes partly from signal processing and numerical applications, where such wavelets are useful in image compression and feature extraction because of their small support and multifractal structure. Lang (1996) constructed several examples of compactly supported wavelets for the Cantor dyadic group by following the procedure of Daubechies (1992) via scaling filters and these wavelets turn out to be certain lacunary Walsh series on the real line. Kozyrev (2002) found a compactly supported p-adic wavelet basis for $L^2(\mathbb{Q}_p)$ which is an analog of the Haar basis. The concept of multiresolution analysis on a positive half-line $\mathbb{R}^+$ was recently introduced by Farkov (2009). He pointed out a method for constructing compactly supported orthogonal p-wavelets related to the Walsh functions, and proved necessary and sufficient conditions for scaling filters with p^n many terms ($p, n \geq 2$) to generate a p-MRA in $L^2(\mathbb{R}^+)$. Subsequently, dyadic wavelet frames on the positive half-line $\mathbb{R}^+$ were constructed by Shah and Debnath (2011a) using the machinery of Walsh–Fourier transforms. They have established a necessary and sufficient conditions for the system $\left\{ \psi_{j,k}(x) = 2^{j/2}\psi(2^j x \ominus k) : j \in \mathbb{Z}, k \in \mathbb{Z}^+ \right\}$ to be a frame for $L^2(\mathbb{R}^+)$. Wavelet packets and wavelet frame packets related to the Walsh polynomials were deeply investigated in a series of papers by the author in Shah (2009, 2012a, 2012b), Shah and Debnath (2011b). Recent results in this direction can also be found in Farkov, Maksimov, and Stroganov (2011), Meenakshi, Manchanda, and Siddiqi (2012), Shah (2015), Sharma and Manchanda (2013) and the references therein.

A constructive procedure for constructing tight wavelet frames generated by the Walsh polynomials using extension principles was first reported by Shah (2013). He provided a sufficient condition for finite number of functions $\{\psi_1, \psi_2, \ldots, \psi_L\}$ to form a tight wavelet frame for $L^2(\mathbb{R}^+)$. Although wavelet frames have many desirable features but the computational complexity and numerical instability during the course of decomposition and reconstruction of functions always remains a debate of discussion (see Dong et al., 2012; Han, 2012). Therefore, in order to reduce the computational complexity and maintain the numerical stability, we shall introduce the concept of minimum-energy wavelet frames associated with the Walsh polynomials on $\mathbb{R}^+$ by extending the above-described method (Shah, 2013). More precisely, we present an equivalent characterizations of minimum-energy wavelet frames in terms of their framelet symbols (Walsh polynomials). Further, based on the polyphase representation of the framelet symbols, a necessary and sufficient condition for minimum-energy wavelet frames related to Walsh polynomials is also given. Finally, we derive the minimum-energy wavelet frame decomposition and reconstruction formulas which are quite similar to those of orthonormal wavelets on positive half-line $\mathbb{R}^+$.

The paper is structured as follows. In Section 2, we introduce some notations and preliminaries related to the operations on positive half-line $\mathbb{R}^+$ including the definitions of the Walsh–Fourier transform, p-multiresolution analysis and minimum-energy wavelet frame related to the Walsh polynomials. In Section 3, we construct minimum-energy wavelet frames generated by the Walsh polynomials and establish a necessary and sufficient condition for the existence of minimum-energy wavelet frames in $L^2(\mathbb{R}^+)$. Section 4, deals with the decomposition and reconstruction algorithms of the minimum-energy wavelet frames on a half-line $\mathbb{R}^+$.

2. Walsh–Fourier analysis and MRA-based wavelet frames

We start this section with certain results on Walsh–Fourier analysis. We present a brief review of generalized Walsh functions, Walsh–Fourier transforms, and its various properties.

As usual, let $\mathbb{R}^+ = [0, +\infty)$, $\mathbb{Z}^+ = \{0, 1, 2, \ldots\}$ and $\mathbb{N} = \mathbb{Z}^+ - \{0\}$. Denote by $[x]$ the integer part of x. Let p be a fixed natural number greater than 1. For $x \in \mathbb{R}^+$ and any positive integer j, we set

$$x_j = [p^j x](\bmod p), \qquad x_{-j} = [p^{1-j} x](\bmod p), \tag{2.1}$$

where $x_j, x_{-j} \in \{0, 1, \ldots, p-1\}$. It is clear that for each $x \in \mathbb{R}^+$, there exist $k = k(x)$ in $\mathbb{N}$ such that $x_{-j} = 0 \; \forall j > k$.

Consider on $\mathbb{R}^+$ the addition defined as follows:

$$x \oplus y = \sum_{j<0} \zeta_j p^{-j-1} + \sum_{j>0} \zeta_j p^{-j},$$

with $\zeta_j = x_j + y_j(\bmod p), j \in \mathbb{Z} \setminus \{0\}$, where $\zeta_j \in \{0, 1, \ldots, p-1\}$ and x_j, y_j are calculated by (2.1). As usual, we write $z = x \ominus y$ if $z \oplus y = x$, where $\ominus$ denotes subtraction modulo p in $\mathbb{R}^+$.

For $x \in [0, 1)$, let $r_0(x)$ is given by

$$r_0(x) = \begin{cases} 1, & \text{if } x \in [0, 1/p) \\ \varepsilon_p^\ell, & \text{if } x \in [\ell p^{-1}, (\ell+1)p^{-1}), \quad \ell = 1, 2, \ldots, p-1, \end{cases}$$

where $\varepsilon_p = \exp(2\pi i/p)$. The extension of the function r_0 to $\mathbb{R}^+$ is given by the equality $r_0(x + 1) = r_0(x)$, $x \in \mathbb{R}^+$. Then, the generalized Walsh functions $\{w_m(x) : m \in \mathbb{Z}^+\}$ are defined by

$$w_0(x) \equiv 1 \quad \text{and} \quad w_m(x) = \prod_{j=0}^{k} \left(r_0(p^j x)\right)^{\mu_j}$$

where $m = \sum_{j=0}^{k} \mu_j p^j$, $\mu_j \in \{0, 1, \ldots, p-1\}$, $\mu_k \neq 0$. They have many properties similar to those of the Haar functions and trigonometric series, and form a complete orthogonal system. Further, by a Walsh polynomial we shall mean a finite linear combination of Walsh functions.

For $x, y \in \mathbb{R}^+$, let

$$\chi(x,y) = \exp\left(\frac{2\pi i}{p} \sum_{j=1}^{\infty} (x_j y_{-j} + x_{-j} y_j) \right),$$ (2.2)

where x_j, y_j are given by (2.1).

We observe that

$$\chi\left(x, \frac{m}{p^n}\right) = \chi\left(\frac{x}{p^n}, m\right) = w_m\left(\frac{x}{p^n}\right), \qquad \forall\, x \in [0, p^n),\ m, n \in \mathbb{Z}^+,$$

and

$$\chi(x \oplus y, z) = \chi(x, z)\,\chi(y, z), \quad \chi(x \ominus y, z) = \chi(x, z)\,\overline{\chi(y, z)},$$

where $x, y, z \in \mathbb{R}^+$ and $x \oplus y$ is p-adic irrational. It is well known that systems $\{\chi(\alpha, \cdot)\}_{\alpha=0}^{\infty}$ and $\{\chi(\cdot, \alpha)\}_{\alpha=0}^{\infty}$ are orthonormal bases in $L^2[0,1]$ (see Golubov, Efimov, & Skvortsov, 1991).

The *Walsh–Fourier transform* of a function $f \in L^1(\mathbb{R}^+) \cap L^2(\mathbb{R}^+)$ is defined by

$$\hat{f}(\xi) = \int_{\mathbb{R}^+} f(x)\, \overline{\chi(x, \xi)}\, dx,$$ (2.3)

where $\chi(x, \xi)$ is given by (2.2). The Walsh–Fourier operator $\mathcal{F} : L^1(\mathbb{R}^+) \cap L^2(\mathbb{R}^+) \to L^2(\mathbb{R}^+)$, $\mathcal{F}f = \hat{f}$, extends uniquely to the whole space $L^2(\mathbb{R}^+)$. The properties of the Walsh–Fourier transform are quite similar to those of the classic Fourier transform (see Golubov et al., 1991; Schipp, Wade, & Simon, 1990). In particular, if $f \in L^2(\mathbb{R}^+)$ then $\hat{f} \in L^2(\mathbb{R}^+)$ and

$$\left\| \hat{f} \right\|_{L^2(\mathbb{R}^+)} = \| f \|_{L^2(\mathbb{R}^+)}.$$

By p-adic interval $I \subset \mathbb{R}^+$ of range n, we mean intervals of the form

$$I = I_n^k = \left[kp^{-n}, (k+1)p^{-n} \right), \quad k \in \mathbb{Z}^+.$$

The p-adic topology is generated by the collection of p-adic intervals and each p-adic interval is both open and closed under the p-adic topology (see Schipp et al., 1990). The family $\left\{ [0, p^{-j}) : j \in \mathbb{Z} \right\}$ forms a fundamental system of the p-adic topology on $\mathbb{R}^+$. Therefore, for each $0 \leq j, k < p^n$, the Walsh function $w_j(x)$ is piecewise constant and hence continuous. Thus $w_j(x) = 1$ for $x \in I_n^0$.

Let $\mathcal{E}_n(\mathbb{R}^+)$ be the space of p-adic entire functions of order n, that is, the set of all functions which are constant on all p-adic intervals of range n. Thus, for every $f \in \mathcal{E}_n(\mathbb{R}^+)$, we have

$$f(x) = \sum_{k \in \mathbb{Z}^+} f(p^{-n}k) \chi_{I_n^k}(x), \quad x \in \mathbb{R}^+.$$ (2.4)

Clearly each Walsh function of order p^{n-1} belong to $\mathcal{E}_n(\mathbb{R}^+)$. The set $\mathcal{E}(\mathbb{R}^+)$ of p-adic entire functions on $\mathbb{R}^+$ is the union of all the spaces $\mathcal{E}_n(\mathbb{R}^+)$. It is clear that $\mathcal{E}(\mathbb{R}^+)$ is dense in $L^p(\mathbb{R}^+)$, $1 \leq p < \infty$ and each function in $\mathcal{E}(\mathbb{R}^+)$ is of compact support.

Next, we give a brief account of the MRA-based wavelet frames generated by the Walsh polynomials on a positive half-line $\mathbb{R}^+$. Following the unitary extension principle, one often starts with a refinable function or even with a refinement mask to construct desired wavelet frames. A compactly supported function $\phi \in L^2(\mathbb{R}^+)$ is called a *refinable function*, if it satisfies an equation of the type

$$\phi(x) = p \sum_{k=0}^{p^n-1} c_k \phi(px \ominus k), \quad x \in \mathbb{R}^+ \tag{2.5}$$

where c_k are complex coefficients. Applying the Walsh–Fourier transform, we can write this equation as

$$\hat{\phi}(\xi) = h_0(p^{-1}\xi)\hat{\phi}(p^{-1}\xi), \tag{2.6}$$

where

$$h_0(\xi) = \sum_{k=0}^{p^n-1} c_k \overline{w_k(\xi)}, \tag{2.7}$$

is a *generalized Walsh polynomial*, which is called the *mask* or *symbol* of the refinable function ϕ and is of course a p-adic step function. Observe that $w_k(0) = \hat{\phi}(0) = 1$. Hence, letting $\xi = 0$ in (2.6) and (2.7), we obtain $\sum_{k=0}^{p^n-1} c_k = 1$. Since ϕ is compactly supported and in fact $\operatorname{supp} \phi \subset [0, p^{n-1})$, therefore $\hat{\phi} \in \mathcal{E}_{n-1}(\mathbb{R}^+)$ and hence as a result $\hat{\phi}(\xi) = 1$ for all $\xi \in [0, p^{1-n})$ as $\hat{\phi}(0) = 1$. Moreover, if $b_s = h_0(sp^{-n})$ represents the values of the mask $h_0(\xi)$ on p-adic intervals, i.e.

$$b_s = \sum_{k=0}^{p^n-1} c_k \overline{w_k(sp^{-n})}, \quad 0 \le s \le p^n - 1, \tag{2.8}$$

then

$$c_k = \frac{1}{p^n} \sum_{s=0}^{p^n-1} b_s w_k(sp^{-n}), \quad 0 \le k \le p^n - 1. \tag{2.9}$$

and, conversely, equalities (2.8) follow from (2.9). These discrete transforms can be realized by the fast Vilenkin–Chrestenson transform (see Golubov et al., 1991). Using Parseval's relation for the discrete transforms, Equations (2.8) and (2.9) can be written as

$$\sum_{k=0}^{p^n-1} |c_k|^2 = \frac{1}{p^n} \sum_{k=0}^{p^n-1} |b_k|^2. \tag{2.10}$$

For a compactly supported refinable function $\phi \subset L^2(\mathbb{R}^+)$, let V_0 be the closed shift invariant space generated by $\{\phi(x \ominus k) : k \in \mathbb{Z}^+\}$ and $V_j = \{\phi(p^j x) : \phi \in V_0\}, j \in \mathbb{Z}$. Then, it is proved in Farkov (2009) that the closed subspaces $\{V_j : j \in \mathbb{Z}\}$ forms a p-multiresolution analysis (p-MRA) for $L^2(\mathbb{R}^+)$. Recall that a p-MRA is a family of closed subspaces $\{V_j\}_{j \in \mathbb{Z}}$ of $L^2(\mathbb{R}^+)$ that satisfies: (i) $V_j \subset V_{j+1}, j \in \mathbb{Z}$, (ii) $\bigcup_{j \in \mathbb{Z}} V_j$ is dense in $L^2(\mathbb{R}^+)$ and (iii) $\bigcap_{j \in \mathbb{Z}} V_j = \{0\}$.

Given an p-MRA generated by a compactly supported refinable function $\phi(x)$, one can construct a set of basic tight framelets $\Psi = \{\psi_1, \ldots, \psi_L\} \subset V_1$ satisfying

$$\hat{\psi}^\ell(\xi) = h_\ell(p^{-1}\xi)\hat{\phi}(p^{-1}\xi), \tag{2.11}$$

where

$$h_\ell(\xi) = \sum_{k=0}^{p^n-1} d_k^\ell \, \overline{w_k(\xi)}, \quad \ell = 1, \dots, L \tag{2.12}$$

are the generalized Walsh polynomials in $L^2[0,1]$ and are called the *framelet symbols* or *wavelet masks*.

With $h_\ell(\xi)$, $\ell = 0, 1, \dots, L$, $L \geq p - 1$ as the Walsh polynomials (wavelet masks), we formulate the matrix $\mathcal{M}(\xi)$ as:

$$\mathcal{M}(\xi) = \begin{pmatrix} h_0(\xi) & h_0(\xi \oplus 1/p) & \cdots & h_0(\xi \oplus (p-1)/p) \\ h_1(\xi) & h_1(\xi \oplus 1/p) & \cdots & h_1(\xi \oplus (p-1)/p) \\ \vdots & \vdots & \ddots & \vdots \\ h_L(\xi) & h_L(\xi \oplus 1/p) & \cdots & h_L(\xi \oplus (p-1)/p) \end{pmatrix}. \tag{2.13}$$

The so-called unitary extension principle (UEP) provides a sufficient condition on $\Psi = \{\psi_1, \dots, \psi_L\}$ such that the wavelet system

$$X(\Psi) = \left\{ \psi_{j,k}^\ell(x) = p^{j/2} \psi^\ell(p^j x \ominus k), j \in \mathbb{Z}, k \in \mathbb{Z}^+, \ell = 1, 2, \dots, L \right\}, \tag{2.14}$$

forms a tight frame of $L^2(\mathbb{R}^+)$. In this connection, Shah (2013) gave an explicit construction scheme for the construction of tight wavelet frames generated by the Walsh polynomials using unitary extension principles in the following way.

THEOREM 2.1   Let $\phi(x)$ be a compactly supported refinable function and $\hat\phi(0) = 1$. Then, the wavelet system $X(\Psi)$ given by (2.14) constitutes a normalized tight wavelet frame in $L^2(\mathbb{R}^+)$ provided the matrix $\mathcal{M}(\xi)$ as defined in (2.13) satisfies

$$\mathcal{M}(\xi)\mathcal{M}^*(\xi) = I_p, \quad \text{for a.e. } \xi \in \sigma(V_0) \tag{2.15}$$

where $\sigma(V_0) := \left\{ \xi \in [0,1]: \sum_{k \in \mathbb{Z}^+} |\hat\phi(\xi \oplus k)|^2 \neq 0 \right\}$.

Motivated and inspired by the construction of tight wavelet frames generated by the Walsh polynomials (Shah, 2013), we extend this concept to minimum-energy wavelet frames on the positive half-line $\mathbb{R}^+$ using the machinery of unitary extension principles. Note that, in this paper, we suppose that any symbol function is a Walsh polynomial, and scaling function and wavelet functions are compactly supported.

Definition 2.1   Let $\phi \in L^2(\mathbb{R}^+)$ satisfies $\hat\phi \in L^\infty$ and $\hat\phi$ is continuous at 0, and $\hat\phi(0) = 1$. Suppose that ϕ generates a sequence of nested closed subspaces $\left\{ V_j : j \in \mathbb{Z} \right\}$. Then, a finite family $\Psi = \{\psi_1, \psi_2, \dots, \psi_L\} \subset V_1$ is called a *minimum-energy wavelet frame* associated with $\phi(x)$, if for all $f \in L^2(\mathbb{R}^+)$

$$\sum_{k \in \mathbb{Z}^+} |\langle f, \phi_{1,k}\rangle|^2 = \sum_{k \in \mathbb{Z}^+} |\langle f, \phi_{0,k}\rangle|^2 + \sum_{\ell=1}^L \sum_{k \in \mathbb{Z}^+} |\langle f, \psi_{0,k}^\ell\rangle|^2. \tag{2.16}$$

By Parseval's identity, minimum-energy wavelet frame Ψ must be a tight frame for $L^2(\mathbb{R}^+)$ with frames bound equal to 1. At the same time, formula (2.16) is equivalent to

$$\sum_{k \in \mathbb{Z}^+} \langle f, \phi_{1,k}\rangle \phi_{1,k} = \sum_{k \in \mathbb{Z}^+} \langle f, \phi_{0,k}\rangle \phi_{0,k} + \sum_{\ell=1}^L \sum_{k \in \mathbb{Z}^+} \langle f, \psi_{0,k}^\ell\rangle \psi_{0,k}^\ell, \quad \text{for all } f \in L^2(\mathbb{R}^+). \tag{2.17}$$

3. Construction of minimum-energy wavelet frames

In this section, we give a complete characterization of minimum-energy wavelet frames associated with some given refinable functions in terms of their framelet symbols. More precisely, we present a necessary and sufficient condition for the existence of minimum-energy wavelet frames generated by Walsh polynomials.

The following theorem presents the equivalent characterizations of the minimum-energy wavelet frame associated with given compactly supported refinable function $\phi(x)$.

THEOREM 3.1   *Suppose that every element of the framelet symbols, $h_0(\xi), h_\ell(\xi), \ell = 1, 2, \ldots, L$, in (2.7) and (2.12) is a Walsh polynomial, and the compactly supported function $\phi(x)$ associated with $h_0(\xi)$ generates a nested subspace $\left\{ V_j : j \in \mathbb{Z} \right\}$. Then the following statements are equivalent:*

(i) $\Psi = \{\psi_1, \psi_2, \ldots, \psi_L\}$ *is a minimum-energy wavelet frame associated with $\phi(x)$.*

(ii) $\mathcal{M}(\xi)\mathcal{M}^*(\xi) = I_p,$ *for a.e. $\xi \in \sigma(V_0)$.*   (3.1)

(iii) $\alpha_{m,n} = \sum\limits_{k\in\mathbb{Z}^+} \left(c_{m-pk} c_{n-pk} + \sum\limits_{\ell=1}^{L} d^\ell_{m-pk} d^\ell_{n-pk} \right) - p\delta_{m,n} = 0,$ $\forall\, m, n \in \mathbb{Z}^+.$ (3.2)

Proof   By using the functional Equations (2.5) and (2.11) and notation $\alpha_{m,n}$, Equation (2.17) can be written as

$$\sum_{m\in\mathbb{Z}^+} \sum_{n\in\mathbb{Z}^+} \alpha_{m,n} \langle f, \phi(px \ominus m) \rangle \phi(px \ominus n) = 0, \quad \text{for all } f \in L^2(\mathbb{R}^+).$$ (3.3)

On the other hand, formula (3.1) can be reformulated as

$$\left| h_0(p^{-1}\xi) \right|^2 + \sum_{\ell=1}^{L} \left| h_\ell(p^{-1}\xi) \right|^2 = 1,$$ (3.4)

$$h_0(p^{-1}\xi)\, \overline{h_0}(\xi \oplus k/p) + \sum_{\ell=1}^{L} h_\ell(p^{-1}\xi)\, \overline{h_\ell}(\xi \oplus k/p) = 0, \quad k = 1, 2, \ldots, p-1,$$

which is equivalent to

$$h_0(p^{-1}\xi) \sum_{k=0}^{p-1} \overline{h_0}(\xi \oplus k/p) + \sum_{\ell=1}^{L} h_\ell(p^{-1}\xi) \left(\sum_{k=0}^{p-1} \overline{h_\ell}(\xi \oplus k/p) \right) = 1,$$

or

$$h_0(p^{-1}\xi) \left(\overline{h_0}(\xi) - \sum_{k=1}^{p-1} \overline{h_0}(\xi \oplus k/p) \right) + \sum_{\ell=1}^{L} h_\ell(p^{-1}\xi) \left(\overline{h_\ell}(\xi) - \sum_{k=1}^{p-1} \overline{h_\ell}(\xi \oplus k/p) \right) = 1,$$

$$h_0(p^{-1}\xi) \left(\sum_{k=0}^{p-1} \overline{h_0}(\xi \oplus k/p) - 2\overline{h_0}(\xi \oplus m/p) \right) \sum_{\ell=1}^{L} h_\ell(p^{-1}\xi)$$

$$\times \left(\sum_{k=0}^{p-1} \overline{h_\ell}(\xi \oplus k/p) - 2\overline{h_\ell}(\xi \oplus m/p) \right) = 1, \quad m = 1, 2, \ldots, p-1.$$

The above system is equivalent to

$$\begin{cases} h_0(p^{-1}\xi) \sum\limits_{k\in\mathbb{Z}^+} c_{-pk}\, W_{pk}(\xi) + \sum\limits_{\ell=1}^{L} h_\ell(p^{-1}\xi) \sum\limits_{k\in\mathbb{Z}^+} d^\ell_{-pk}\, W_{pk}(\xi) = 1, \\[2mm] h_0(p^{-1}\xi) \left(\sum\limits_{m=1}^{p-1} \sum\limits_{k\in\mathbb{Z}^+} c_{m-pk}\, W_{pk-m}(\xi) \right) + \sum\limits_{\ell=1}^{L} h_\ell(p^{-1}\xi) \left(\sum\limits_{m=1}^{p-1} \sum\limits_{k\in\mathbb{Z}^+} d^\ell_{m-pk}\, W_{pk-m}(\xi) \right) = p-1. \end{cases}$$

The above system can be further expressed as

$$\begin{cases} h_0(p^{-1}\xi) \sum\limits_{k\in\mathbb{Z}^+} c_{-pk}\, W_{pk}(\xi) + \sum\limits_{\ell=1}^{L} h_\ell(p^{-1}\xi) \sum\limits_{k\in\mathbb{Z}^+} d^\ell_{-pk}\, W_{pk}(\xi) = 1, \\[2mm] h_0(p^{-1}\xi) \sum\limits_{k\in\mathbb{Z}^+} c_{1-pk}\, W_{pk-1}(\xi) + \sum\limits_{\ell=1}^{L} h_\ell(p^{-1}\xi) \sum\limits_{k\in\mathbb{Z}^+} d^\ell_{1-pk}\, W_{pk-1}(\xi) = 1, \\[2mm] \qquad\vdots \\[2mm] h_0(p^{-1}\xi) \sum\limits_{k\in\mathbb{Z}^+} c_{p-1-pk}\, W_{pk-p+1}(\xi) + \sum\limits_{\ell=1}^{L} h_\ell(p^{-1}\xi) \sum\limits_{k\in\mathbb{Z}^+} d^\ell_{p-1-pk}\, W_{pk-p+1}(\xi) = 1. \end{cases}$$ (3.5)

Multiply the identities of (3.5) with $\hat{\phi}(p^{-1}\xi)w_m(\xi)$, $m = 0, 1, \ldots, p-1$, we obtain

$$\hat{\phi}(p^{-1}\xi)w_m(\xi) = \sum_{k\in\mathbb{Z}^+}\left(c_{m-pk}w_{pk}(\xi)h_0(p^{-1}\xi)\hat{\phi}(p^{-1}\xi) + \sum_{\ell=1}^{L}d_{m-pk}^{\ell}w_{pk}(\xi)h_\ell(p^{-1}\xi)\hat{\phi}(p^{-1}\xi)\right). \qquad (3.6)$$

Therefore, the system (3.5) can be written as

$$\begin{cases} \hat{\phi}(p^{-1}\xi)w_0(\xi) = \displaystyle\sum_{k\in\mathbb{Z}^+}\left(c_{-pk}\,w_{pk}(\xi)\hat{\phi}(\xi) + \sum_{\ell=1}^{L}d_{-pk}^{\ell}\,w_{pk}(\xi)\hat{\psi}^\ell(\xi)\right), \\[2mm] \hat{\phi}(p^{-1}\xi)w_1(\xi) = \displaystyle\sum_{k\in\mathbb{Z}^+}\left(c_{1-pk}\,w_{pk}(\xi)\hat{\phi}(\xi) + \sum_{\ell=1}^{L}d_{1-pk}^{\ell}\,w_{pk}(\xi)\hat{\psi}^\ell(\xi)\right), \\[2mm] \vdots \\[2mm] \hat{\phi}(p^{-1}\xi)w_{p-1}(\xi) = \displaystyle\sum_{k\in\mathbb{Z}^+}\left(c_{p-1-pk}\,w_{pk}(\xi)\hat{\phi}(\xi) + \sum_{\ell=1}^{L}d_{p-1-pk}^{\ell}\,w_{pk}(\xi)\hat{\psi}^\ell(\xi)\right). \end{cases}$$

This system of equations can be written in time domain as

$$\begin{cases} \phi(x) = \displaystyle\sum_{k\in\mathbb{Z}^+}\left(c_{-pk}\,\phi(x\ominus k/p) + \sum_{\ell=1}^{L}d_{-pk}^{\ell}\,\psi^\ell(x\ominus k/p)\right), \\[2mm] \phi(x\ominus 1/p) = \displaystyle\sum_{k\in\mathbb{Z}^+}\left(c_{1-pk}\,\phi(x\ominus k/p) + \sum_{\ell=1}^{L}d_{1-pk}^{\ell}\,\psi^\ell(x\ominus k/p)\right), \\[2mm] \vdots \\[2mm] \phi(x\ominus (p-1)/p) = \displaystyle\sum_{k\in\mathbb{Z}^+}\left(c_{p-1-pk}\,\phi(x\ominus k/p) + \sum_{\ell=1}^{L}d_{p-1-pk}^{\ell}\,\psi^\ell(x\ominus k/p)\right). \end{cases}$$

On the reformulation of above system, we obtain

$$\phi(x\ominus m/p) = \sum_{k\in\mathbb{Z}^+}\left(c_{m-pk}\,\phi(x\ominus k/p) + \sum_{\ell=1}^{L}d_{m-pk}^{\ell}\,\psi^\ell(x\ominus k/p)\right), \quad m\in\mathbb{Z}^+. \qquad (3.7)$$

Using (2.5) and its corresponding wavelet equation, we can rewrite formula (3.7) as

$$\sum_{m\in\mathbb{Z}^+}\alpha_{m,n}\phi(x\ominus m/p) = 0, \quad \forall\, n\in\mathbb{Z}^+. \qquad (3.8)$$

Thus, the UEP condition (3.1) is equivalent to (3.8). In conclusion, the proof of the theorem reduces to the proof of the equivalence of (3.2), (3.3), and (3.8).

It is obvious that (3.2) implies (3.8) which implies (3.3). In order to prove (3.3) $\Longrightarrow$ (3.2), we assume that f be a function of compact support, i.e. $f \in \mathcal{E}(\mathbb{R}^+)$. By using the properties that for every fixed m, $\alpha_{m,n} = 0$ except for finitely many n, the functional

$$\beta_n(f) = \sum_{m\in\mathbb{Z}^+}\alpha_{m,n}\big\langle f, \phi(\cdot\ominus m/p)\big\rangle, \quad n\in\mathbb{Z}^+,$$

just has finite nonzero's for $n \in \mathbb{Z}^+$. Since $\hat{\phi}(\xi)$ is nontrivial function, by taking the Fourier transform of (3.3), it follows that the polynomial $\sum_{n\in\mathbb{Z}^+}\beta_n(f)w_n(\xi)$ is identically zero. Obviously, $\beta_n(f) = 0$, $n \in \mathbb{Z}^+$. In other words, we say that

$$\Big\langle f, \sum_{m\in\mathbb{Z}^+}\alpha_{m,n}\,\phi(x\ominus m/p)\Big\rangle = 0, \quad n\in\mathbb{Z}^+.$$

Thus, the series in the above equation is a finite sum and hence represents a compactly supported function in $L^2(\mathbb{R}^+)$. By choosing f to be this function, it follows that

$$\sum_{m\in\mathbb{Z}^+}\alpha_{m,n}\,\phi\!\left(x\ominus m/p\right)=0,$$

which implies that the polynomial $\sum_{m\in\mathbb{Z}^+}\alpha_{m,n}w(\xi)$ is identically equal to 0 so that $\alpha_{m,n}=0$, $m,n\in\mathbb{Z}^+$. This completes the proof of the theorem. $\qquad\square$

Now we shall present a necessary condition for minimum-energy wavelet frames generated by the Walsh polynomials in terms of their wavelet masks.

THEOREM 3.2 *Let $\phi\in L^2(\mathbb{R}^+)$ be a compactly supported refinable function with refinement mask $h_0(\xi)$ such that $\hat{\phi}$ is continuous at 0 and $\hat{\phi}(0)=1$. If $\Psi=\{\psi_1,\psi_2,\ldots,\psi_L\}$ is the minimum-energy wavelet frame associated with $\phi(x)$, then*

$$\sum_{m=0}^{p-1}\left|h_0(\xi\oplus m/p)\right|^2\le 1,\quad\text{for all }\xi\in\mathbb{R}^+. \tag{3.9}$$

Proof Let $Q(\xi)$ be the first column of the modulation matrix $\mathcal{M}(\xi)$, as defined in (2.13). Then, $\mathcal{M}(\xi)=\big(Q(\xi),R(\xi)\big)$, where

$$R(\xi)=\begin{pmatrix} h_1(\xi) & h_1(\xi\oplus 1/p) & \cdots & h_1(\xi\oplus(p-1)/p) \\ h_2(\xi) & h_2(\xi\oplus 1/p) & \cdots & h_2(\xi\oplus(p-1)/p) \\ \vdots & \vdots & \ddots & \vdots \\ h_L(\xi) & h_L(\xi\oplus 1/p) & \cdots & h_L(\xi\oplus(p-1)/p) \end{pmatrix} \tag{3.10}$$

and

$$Q(\xi)=\Big[h_0(\xi)\ h_0(\xi\oplus 1/p)\ \ldots\ h_0(\xi\oplus(p-1)/p)\Big].$$

Therefore, the condition (3.1) can be reformulated as

$$Q(\xi)Q^*(\xi)+R(\xi)R^*(\xi)=I_p,$$

or equivalently,

$$I_p-Q(\xi)Q^*(\xi)=R(\xi)R^*(\xi).$$

Since $R(\xi)R^*(\xi)$ is a Hermitian matrix, the matrix $I_p-Q(\xi)Q^*(\xi)$ is positive semi-definite, so that

$$\det\left(I_p-Q(\xi)Q^*(\xi)\right)\ge 0,$$

and this gives

$$\sum_{m=0}^{p-1}\left|h_0(\xi\oplus m/p)\right|^2\le 1,\quad\text{for all }\xi\in\mathbb{R}^+.$$

In fact, we have

$$\begin{pmatrix} I_p & Q^*(\xi) \\ Q(\xi) & 1 \end{pmatrix}\begin{pmatrix} I_p & -Q^*(\xi) \\ -Q(\xi) & 1 \end{pmatrix}=\begin{pmatrix} I_p-Q(\xi)Q^*(\xi) & 0 \\ 0 & 1-Q(\xi)Q^*(\xi) \end{pmatrix},$$

$$\det\begin{pmatrix} I_p & Q^*(\xi) \\ Q(\xi) & 1 \end{pmatrix}=\det\begin{pmatrix} I_p & Q^*(\xi) \\ 0 & 1-Q(\xi)Q^*(\xi) \end{pmatrix},$$

$$\det\begin{pmatrix} I_p & -Q^*(\xi) \\ -Q(\xi) & 1 \end{pmatrix}=\det\begin{pmatrix} I_p & -Q^*(\xi) \\ 0 & 1-Q(\xi)Q^*(\xi) \end{pmatrix}.$$

Therefore

$$\det\left(I_p - Q(\xi)Q^*(\xi)\right)\left(1 - Q(\xi)Q^*(\xi)\right) = \left(1 - Q(\xi)Q^*(\xi)\right)^2,$$

and it gives $1 - Q(\xi)Q^*(\xi) \geq 0$. The proof of the Theorem 3.2 is completed. $\qquad\square$

According to the Theorem 3.2, there may not exist minimum-energy wavelet frame associated with a given compactly supported refinable function ϕ and in case if it exist, then the refinement mask must satisfy (3.9). In this context, we provide a sufficient condition for minimum-energy wavelet frames related to the Walsh polynomials based on the polyphase representation of the wavelet masks $h_\ell(\xi), \ell = 0, 1, \ldots, L$.

The *polyphase representation* of the refinement mask $h_0(\xi)$ can be derived by using the properties of Walsh polynomials as

$$h_0(\xi) = \sum_{k=0}^{p^n-1} c_k \overline{W_k(\xi)}$$

$$= \sum_{k=0}^{p^n-1} \sum_{m=0}^{p-1} c_{pk+m} \overline{W_{pk+m}(\xi)}$$

$$= \sum_{m=0}^{p-1} \overline{W_m(\xi)} \sum_{k=0}^{p^n-1} c_{pk+m} \overline{W_k(p\xi)}$$

$$= \frac{1}{\sqrt{p}} \sum_{m=0}^{p-1} \mu_{0,m}(p\xi)\overline{W_m(\xi)},$$

where

$$\mu_{0,m}(\xi) = \sqrt{p} \sum_{k=0}^{p^n-1} c_{pk+m} \overline{W_k(\xi)}, \quad m = 0, 1, \ldots, p-1. \tag{3.11}$$

Similarly, the wavelet masks $h_\ell(\xi), 1 \leq \ell \leq L$, as defined in (2.12) can be splitted into polyphase components as

$$h_\ell(\xi) = \frac{1}{\sqrt{p}} \sum_{m=0}^{p-1} \mu_{\ell,m}(p\xi) \overline{W_m(\xi)}, \tag{3.12}$$

where

$$\mu_{\ell,m}(\xi) = \sqrt{p} \sum_{k=0}^{p^{n-1}-1} d^\ell_{pk+m} \overline{W_k(\xi)}, \quad m = 0, 1, \ldots, p-1. \tag{3.13}$$

With the polyphase components given by (3.11) and (3.13), we formulate the *polyphase matrix* $\Gamma(\xi)$ as:

$$\Gamma(\xi) = \begin{pmatrix} \mu_{0,0}(\xi) & \mu_{1,0}(\xi) & \cdots & \mu_{L,0}(\xi) \\ \mu_{0,1}(\xi) & \mu_{1,1}(\xi) & \cdots & \mu_{L,1}(\xi) \\ \vdots & \vdots & \ddots & \vdots \\ \mu_{0,p-1}(\xi) & \mu_{1,p-1}(\xi) & \cdots & \mu_{L,p-1}(\xi) \end{pmatrix}.$$

Therefore, the modulation matrix $\mathcal{M}(\xi)$ can be expressed as

$$\mathcal{M}(\xi) = \Gamma(p\xi)\,\mathcal{W}^*(\xi), \tag{3.14}$$

where $\mathcal{W}(\xi)$ is the Walsh matrix given by

$$\mathcal{W}(\xi) = \frac{1}{\sqrt{p}} \begin{pmatrix} w_0(\xi) & w_1(\xi) & \cdots & w_{p-1}(\xi) \\ w_0(\xi \ominus 1/p) & w_1(\xi \ominus 1/p) & \cdots & w_{p-1}(\xi \ominus 1/p) \\ \vdots & \vdots & \ddots & \vdots \\ w_0(\xi \ominus (p-1)/p) & w_1(\xi \ominus (p-1)/p) & \cdots & w_{p-1}(\xi \ominus (p-1)/p) \end{pmatrix}.$$

Thus, we have

$$\mathcal{M}(\xi)\mathcal{M}^*(\xi) = \mathcal{W}(\xi)\Gamma^*(p\xi)\Gamma(p\xi)\mathcal{W}^*(\xi),$$

and, hence we conclude that

$$\mathcal{M}(\xi)\mathcal{M}^*(\xi) = pI_p \iff \Gamma^*(p\xi)\Gamma(p\xi) = pI_p,$$

or equivalently, we say that

$$\sum_{\ell=0}^{L} h_\ell(\xi \oplus m/p)\overline{h_\ell(\xi \oplus n/p)} = \delta_{m,n} \iff \sum_{\ell=0}^{L} \mu_{\ell,m}(p\xi)\overline{\mu_{\ell,n}(p\xi)} = \delta_{m,n}, \quad 0 \le m, n \le p-1.$$

Since the Walsh matrix $\mathcal{W}(\xi)$ is a unitary matrix, therefore, we have

$$\mathcal{M}(\xi)\mathcal{M}^*(\xi) = \Gamma^*(p\xi)\big(\mathcal{W}(\xi)\mathcal{W}^*(\xi)\big)\Gamma(p\xi) = \Gamma(p\xi)\Gamma^*(p\xi), \tag{3.15}$$

which implies that

$$\sum_{k=0}^{p-1} h_\ell(\xi \oplus k/p)\overline{h_{\ell'}(\xi \oplus k/p)} = \sum_{k=0}^{p-1} \mu_{\ell,k}(p\xi)\overline{\mu_{\ell',k}(p\xi)}, \quad 0 \le \ell, \ell' \le L. \tag{3.16}$$

Therefore, it follows from (3.9) and (3.16) that

$$\sum_{m=0}^{p-1} \left|\mu_{0,m}(\xi)\right|^2 \le 1, \quad \xi \in \mathbb{R}^+, \tag{3.17}$$

which further yields

$$\sum_{m=0}^{p-1} \left|b_{0,m}^{n,s}(\xi)\right|^2 \le 1, \quad s = 0, 1, \ldots, p^n - 1, \tag{3.18}$$

where $b_{0,m}^{n,s} = \mu_{0,m}(p^{1-n}\xi_{[s]})$. Since the polynomial $\mu_{\ell,m}(p\xi)$ is constant on the intervals $I_{n,s}$, $0 \le s \le p^n - 1$, so the polyphase components $\mu_{\ell,m}(\xi)$ can also be written as

$$\mu_{\ell,m}(p\xi) = \sum_{s=0}^{p^n-1} b_{\ell,m}^{n,s} 1_{I_{n,s}}(\xi), \quad m = 0, 1, \ldots, p-1 \tag{3.19}$$

where

$$\sum_{\ell=0}^{L} b_{\ell,m}^{n,s} \overline{b_{\ell,m'}^{n,s}} = \delta_{m,m'}, \quad 0 \le m, m' \le p-1, s = 0, 1, \ldots, p^n - 1. \tag{3.20}$$

Now, if there exists $\mu_{0,p}(\xi)$ such that

$$\sum_{m=0}^{p} \left|\mu_{0,m}(\xi)\right|^2 = 1. \tag{3.21}$$

then, we have the following theorem which provides a sufficient condition for minimum-energy wavelet frames generated by the Walsh polynomials in $L^2(\mathbb{R}^+)$.

THEOREM 3.3   Let $h_0(\xi)$ be the refinement mask of a compactly supported refinable function $\phi(x)$ and satisfy inequality (3.17). Furthermore, if there exist $\mu_{0,p}(\xi)$ of the form (3.21), then there exists a minimum-energy wavelet frame associated with $\phi(x)$.

Proof   Under the given assumptions, it is easy to verify that

$$\mathbf{f} = \left(\mu_{0,0}(\xi), \mu_{0,1}(\xi), \dots, \mu_{0,p-1}(\xi), \mu_{0,p}(\xi) \right)^T \tag{3.22}$$

is a unit vector, where T stands for the transpose of a given vector. By multiplying the diagonal matrix $D_0 = \text{diag}(\xi^{t_0}, \xi^{t_1}, \dots, \xi^{t_p})$ to the left side of (3.22), we obtain

$$\mathbf{f}_1 = \left(\xi^{t_0}\mu_{0,0}(\xi), \xi^{t_1}\mu_{0,1}(\xi), \dots, \xi^{t_{p-1}}\mu_{0,p-1}(\xi), \xi^{t_p}\mu_{0,p}(\xi) \right)^T = \sum_{j=0}^{J} \mathbf{u}_j\, \xi^j, \quad t_0, t_1, \dots, t_p \in \mathbb{Z}^+,$$

where $\mathbf{u}_j \in \mathbb{R}^+$, with $\mathbf{u}_0 \neq 0$ and $\mathbf{u}_J \neq 0$. It is also clear that $\mathbf{f}_1$ is a unit vector as

$$\mathbf{f}_1^* \mathbf{f}_1 = \left(\sum_{j=0}^{J} \mathbf{u}_j\, \xi^j \right)^* \left(\sum_{j=0}^{J} \mathbf{u}_j\, \xi^j \right) = 1, \quad \text{for all } \xi \in L^2[0,1]$$

and consequently, $\mathbf{u}_0^T \mathbf{u}_J = 0$.

Consider the $(p+1) \times (p+1)$ Householder matrix

$$\mathcal{H}_1 = I_{p+1} - \frac{2}{|\mathbf{v}|^2} \mathbf{v}\mathbf{v}^T, \tag{3.23}$$

where $\mathbf{v} = \mathbf{u}_J \pm \|\mathbf{u}_J\|\mathbf{e}_1$, with $\mathbf{e}_1 = (1,0,\dots,0)_{p+1}^T$, and the $+$ and $-$ signs are so chosen that $\mathbf{v} \neq 0$. Then

$$\mathcal{H}_1 \mathbf{u}_J = \pm\|\mathbf{u}_J\|\mathbf{e}_1.$$

By the orthogonal property of the Householder matrix, we have

$$(\mathcal{H}_1 \mathbf{u}_0)^T (\mathcal{H}_1 \mathbf{u}_J) = \mathbf{u}_0^T \mathcal{H}_1^T \mathcal{H}_1 \mathbf{u}_J = \mathbf{u}_0^T \mathbf{u}_J = 0.$$

Using previous equation, it follows that the first component of $\mathcal{H}_1 \mathbf{u}_0$ is 0. Since $\mathcal{H}_1\mathbf{f}_1 = \sum_{j=0}^{J}(\mathcal{H}_1\mathbf{u}_j)\xi^j$, therefore, we can construct a diagonal matrix $D_1 = \text{diag}(\xi^{t_{(1)}}, 1, \dots, 1), t_{(1)} \in \mathbb{Z}^+$ such that

$$\mathbf{f}_2 = D_1\mathbf{u}_1\mathbf{f}_1 = D_1 \sum_{j=0}^{J}(\mathcal{H}_1\mathbf{u}_j)\xi^j = \sum_{j=0}^{J} \mathbf{u}_j^{(1)}\xi^j$$

is also a unit vector and $J_1 < J$, $\mathbf{u}_0^{(1)} \neq 0$, $\mathbf{u}_{J_1}^{(1)} \neq 0$.

Similarly, we define the Householder matrix

$$\mathcal{H}_2 = I_{p+1} - \frac{2}{|\tilde{\mathbf{v}}|^2} \tilde{\mathbf{v}}\tilde{\mathbf{v}}^T, \tag{3.24}$$

where $\tilde{\mathbf{v}} = \mathbf{u}_{J_1} \pm \|\mathbf{u}_{J_1}\|\mathbf{e}_1 \neq 0$, and $D_2 = \text{diag}(\xi^{t_{(2)}}, 1, \dots, 1), t_{(2)} \in \mathbb{Z}^+$ such that

$$\mathbf{f}_3 = D_2\mathcal{H}_2\mathbf{f}_2 = D_2 \sum_{j=0}^{J_1}\left(\mathcal{H}_2\mathbf{u}_j^{(1)}\right)\xi^j = \sum_{j=0}^{J_1} \mathbf{u}_j^{(2)}\xi^j$$

is also a unit vector and $J_2 < J_1, \mathbf{u}_0^{(2)} \neq \mathbf{0}, \mathbf{u}_{J_1}^{(2)} \neq \mathbf{0}$. Since every component of $\mathbf{f}$ is a finite sum, we repeat this procedure finite times to get some unitary matrices $\mathcal{D}_N, \mathcal{H}_N, \mathcal{D}_{N-1}, \mathcal{H}_{N-1}, \ldots, \mathcal{H}_2, \mathcal{D}_1, \mathcal{H}_1$ such that

$$\mathcal{D}_N \mathcal{H}_N \mathcal{D}_{N-1} \mathcal{H}_{N-1} \ldots \mathcal{H}_2 \mathcal{D}_1 \mathcal{H}_1 \mathbf{f} = \mathbf{e}_1. \tag{3.25}$$

Therefore, it is clear that $\mathbf{f}$ is the first column of the unitary matrix

$$\mathcal{H} = \mathcal{D}_0^* \mathcal{H}_0^* \mathcal{D}_1^* \mathcal{H}_1^* \cdots \mathcal{D}_{N-1}^* \mathcal{H}_{N-1}^* \mathcal{D}_N^* \mathcal{H}_N^*.$$

By setting,

$$\mathcal{H} = \begin{pmatrix} \mu_{0,0}(\xi) & \mu_{0,1}(\xi) & \cdots & \mu_{0,p}(\xi) \\ \mu_{1,0}(\xi) & \mu_{1,1}(\xi) & \cdots & \mu_{1,p}(\xi) \\ \vdots & \vdots & \ddots & \vdots \\ \mu_{p-1,0}(\xi) & \mu_{p-1,1}(\xi) & \cdots & \mu_{p-1,p}(\xi) \end{pmatrix}.$$

It is immediate that $\mathcal{H}$ satisfies the equality $\Gamma^*(\xi)\Gamma(\xi) = I_p$. Further, if we choose polyphase representation of wavelet masks $h_\ell(\xi)$, $\ell = 1, 2, \ldots, L$ as defined by (3.13) or even (3.19) in Equation (2.13), then we can obtain the UEP condition (2.15). Therefore, Theorem 3.1 implies that Ψ generates a minimum-energy wavelet frame for $L^2(\mathbb{R}^+)$. This completes the proof of the Theorem 3.3. □

4. Decomposition and reconstruction algorithms

Suppose $\Psi = \{\psi_1, \psi_2, \ldots, \psi_L\}$ is the minimum-energy wavelet frame associated with the compactly supported refinable function $\phi(x)$. Then, for each $j \in \mathbb{Z}$, we consider

$$V_j = \overline{\text{span}}\left\{\phi_{j,k} : k \in \mathbb{Z}^+\right\} \quad \text{and} \quad W_j = \overline{\text{span}}\left\{\psi_{j,k}^\ell : k \in \mathbb{Z}^+, \ell = 1, 2, \ldots, L\right\}. \tag{4.1}$$

Thus,

$$V_{j+1} = V_j + W_j, \quad j \in \mathbb{Z}. \tag{4.2}$$

Note that decomposition (4.2) is not a direct sum decomposition since in general $V_j \cap W_j \neq \{0\}$. Thus, it follows from (4.1) and (4.2) that any $f \in V_{j+1}$ can be expressed as

$$f(x) = P_j f(x) + Q_j f(x), \tag{4.3}$$

where

$$P_j f(x) = \sum_{k \in \mathbb{Z}^+} \langle f, \phi_{j,k} \rangle \phi_{j,k}(x), \tag{4.4}$$

$$Q_j f(x) = P_{j+1} f(x) - P_j f(x) = \sum_{\ell=1}^{L} \sum_{k \in \mathbb{Z}^+} \langle f, \psi_{j,k}^\ell \rangle \psi_{j,k}^\ell(x), \tag{4.5}$$

are the *projection* and *detailed operators* defined on V_j and W_j, respectively. The importance of this frame expansion as compared to any other expansion

$$Q_j f = \sum_{\ell=1}^{L} \sum_{k \in \mathbb{Z}^+} a_{j,k} \psi_{j,k}^\ell. \tag{4.6}$$

of the same $Q_j f$ is that the energy in (4.5) is minimum in the sense that

$$\sum_{\ell=1}^{L} \sum_{k \in \mathbb{Z}^+} |\langle f, \psi_{j,k}^\ell \rangle|^2 \leq \sum_{\ell=1}^{L} \sum_{k \in \mathbb{Z}^+} |a_{j,k}|^2. \tag{4.7}$$

Therefore, by using (4.5) and (4.6), we have

$$\langle Q_j f, f \rangle = \sum_{\ell=1}^{L} \sum_{k \in \mathbb{Z}^+} |\langle f, \psi_{j,k}^{\ell} \rangle|^2 = \sum_{\ell=1}^{L} \sum_{k \in \mathbb{Z}^+} a_{j,k} \overline{\langle f, \psi_{j,k}^{\ell} \rangle}, \tag{4.8}$$

and this derives

$$0 \leq \sum_{\ell=1}^{L} \sum_{k \in \mathbb{Z}^+} |a_{j,k} - \langle f, \psi_{j,k}^{\ell} \rangle|^2$$

$$= \sum_{\ell=1}^{L} \sum_{k \in \mathbb{Z}^+} |a_{j,k}|^2 - 2 \sum_{\ell=1}^{L} \sum_{k \in \mathbb{Z}^+} a_{j,k} \overline{\langle f, \psi_{j,k}^{\ell} \rangle} + \sum_{\ell=1}^{L} \sum_{k \in \mathbb{Z}^+} |\langle f, \psi_{j,k}^{\ell} \rangle|^2$$

$$= \sum_{\ell=1}^{L} \sum_{k \in \mathbb{Z}^+} |a_{j,k}|^2 - \sum_{\ell=1}^{L} \sum_{k \in \mathbb{Z}^+} |\langle f, \psi_{j,k}^{\ell} \rangle|^2.$$

This inequality means that the coefficients of the error term $Q_j f$ in (4.5) have minimal l^2-norm among all sequences $\{ a_{j,k} \}$ which satisfy (4.6).

We now discuss the decomposition and reconstruction algorithms associated with minimum-energy wavelet frames on positive half-line. For any $f \in L^2(\mathbb{R}^+)$, we consider

$$a_{j,k} = \langle f, \phi_{j,k} \rangle; \quad b_{j,k}^{\ell} = \langle f, \psi_{j,k}^{\ell} \rangle, \quad \ell = 1, 2, \dots, L. \tag{4.9}$$

Then, by two scale relations (2.5) and the corresponding wavelet equation, we obtain

$$\phi_{j,i} = \sum_{k \in \mathbb{Z}^+} c_{k-pi} \phi_{j+1,k}, \quad \psi_{j,i}^{\ell} = \sum_{k \in \mathbb{Z}^+} d_{k-pi}^{\ell} \psi_{j+1,k}^{\ell}, \quad \ell = 1, 2, \dots, L, i \in \mathbb{Z}^+. \tag{4.10}$$

By taking the inner products with f on both sides of the two equations in (4.10), we have a *tight minimum-energy wavelet frame decomposition*:

$$a_{j,i} = \sum_{k \in \mathbb{Z}^+} c_{k-pi} a_{j+1,k}, \quad b_{j,i}^{\ell} = \frac{1}{\sqrt{p}} \sum_{k \in \mathbb{Z}^+} d_{k-pi}^{\ell} b_{j+1,k}^{\ell}, \quad \ell = 1, 2, \dots, L, j \in \mathbb{Z}^+. \tag{4.11}$$

Using the fact that $\phi_{j,k} \in V_j$ and relations (2.4) and wavelet equation, from (4.3) we also have

$$\phi_{j+1,i} = \sum_{k \in \mathbb{Z}^+} \left(c_{i-pk} \phi_{j,k} + \sum_{\ell=1}^{L} d_{i-pk}^{\ell} \psi_{j,k}^{\ell} \right), \quad i \in \mathbb{Z}^+. \tag{4.12}$$

By taking the inner products with f on both sides of (4.12), we have a *tight minimum-energy wavelet frame reconstruction*:

$$a_{j+1,i} = \sum_{k \in \mathbb{Z}^+} \left(c_{i-pk} a_{j,k} + \sum_{\ell=1}^{L} d_{i-pk}^{\ell} b_{j,k}^{\ell} \right), \quad i \in \mathbb{Z}^+. \tag{4.13}$$

Acknowledgements
The authors thank the referees for numerous suggestions which helped to improve the paper considerably.

Funding
The authors received no direct funding for this research.

Author details
Sunita Goyal[1]
E-mail: sunitagoel2011@gmail.com
ORCID ID: http://orcid.org/0000-0001-8461-869X
Firdous A. Shah[2]
E-mail: fashah79@gmail.com

[1] Department of Mathematics, JJT University, Jhunjhunu 333 001, Rajasthan, India.
[2] Department of Mathematics, University of Kashmir, South Campus, Anantnag 192 101, Jammu and Kashmir, India.

References
Christensen, O. (2003). *An introduction to frames and Riesz bases*. Boston, MA: Birkhäuser.

Chui, C. K., & He, W. (2000). Compactly supported tight frames associated with refinable functions. *Applied Computational and Harmonic Analysis, 8,* 293–319.

Daubechies, I. (1992). *Ten lectures on wavelets.* Philadelphia, PA: SIAM.

Daubechies, I., Grossmann, A., & Meyer, Y. (1986). Painless non-orthogonal expansions. *Journal of Mathematical Physics, 27,* 1271–1283.

Daubechies, I., Han, B., Ron, A., & Shen, Z. (2003). Framelets: MRA-based constructions of wavelet frames. *Applied Computational and Harmonic Analysis, 14,* 1–46.

Debnath, L., & Shah, F. A. (2015). *Wavelet transforms and their applications.* New York, NY: Birkhäuser.

Dong, B., Ji, H., Li, J., Shen, Z., & Xu, Y. (2012). Wavelet frame based blind image inpainting. *Applied Computational and Harmonic Analysis, 32,* 268–279.

Duffin, R. J., & Shaeffer, A. C. (1952). A class of nonharmonic Fourier series. *Transactions of American Mathematical Society, 72,* 341–366.

Farkov, Y. A. (2009). On wavelets related to Walsh series. *Journal of Approximation Theory, 161,* 259–279.

Farkov, Y. A., Maksimov, A. Y., & Stroganov, S. A. (2011). On biorthogonal wavelets related to the Walsh functions. *International Journal of Wavelets, Multiresolution and Information Processing, 9,* 485–499.

Farkov, Y. A., Lebedeva, E. A., & Skopina, M. A. (2015). Wavelet frames on Vilenkin groups and their approximation properties. *International Journal of Wavelets, Multiresolution and Information Processing, 13*(5), 19 p., 1550036.

Gao, X., & Cao, C. H. (2008). Minimum-energy wavelet frame on the interval. *Science in China F, 51,* 1547–1562.

Golubov, B. I., Efimov, A. V., & Skvortsov, V. A. (1991). *Walsh series and transforms: Theory and applications.* Dordrecht: Kluwer.

Han, B. (2012). Wavelets and framelets within the framework of non-homogeneous wavelet systems. In M. Neamtu & L. Schumaker (Eds.), *Approximation theory XIII* (pp. 121–161). Springer.

Huang, Y., & Cheng, Z. (2007). Minimum-energy frames associated with refinable function of arbitrary integer dilation factor. *Chaos, Solitons and Fractals, 32,* 503–515.

Huang, Y., Li, Q., & Li, M. (2012). Minimum-energy multiwavelet frames with arbitrary integer dilation factor. *Mathematical Problems in Engineering,* 37 p., 640789.

Kozyrev, S. V. (2002). Wavelet analysis as a p-adic spectral analysis. *Izvestiya: Mathematics, 66,* 149–158.

Lang, W. C. (1996). Orthogonal wavelets on the Cantor dyadic group. *SIAM Journal of Mathematical Analysis, 27,* 305–312.

Meenakshi, M. P., & Siddiqi, A. H. (2012). Wavelets associated with nonuniform multiresolution analysis on positive half-line. *International Journal of Wavelets, Multiresolution and Information Processing, 12*(2), 27 p., 1250018.

Ron, A., & Shen, Z. (1997). Affine systems in $L^2(\mathbb{R}^+)$: The analysis of the analysis operator. *Journal of Functional Analysis, 148,* 408–447.

Schipp, F., Wade, W. R., & Simon, P. (1990). *Walsh series: An introduction to Dyadic harmonic analysis.* Bristol: Adam Hilger.

Shah, F. A. (2009). Construction of wavelet packets on p-adic field. *International Journal of Wavelets, Multiresolution and Information Processing, 7,* 553–565.

Shah, F. A. (2012a). Non-orthogonal p-wavelet packets on a half-line. *Analysis in Theory and Applications, 28,* 385–396.

Shah, F. A. (2012b). Biorthogonal wavelet packets related to the Walsh polynomials. *Journal of Classical Analysis, 1,* 135–146.

Shah, F. A. (2013). Tight wavelet frames generated by the Walsh polynomials. *International Journal of Wavelets, Multiresolution and Information Processing, 11*(6), 15 p., 1350042.

Shah, F. A. (2015). p-Frame multiresolution analysis related to the Walsh functions. *International Journal of Analysis and Applications, 7,* 1–15.

Shah, F. A., & Debnath, L. (2011a). Dyadic wavelet frames on a half-line using the Walsh–Fourier transform. *Integral Transforms and Special Functions, 22,* 477–486.

Shah, F. A., & Debnath, L. (2011b). p-Wavelet frame packets on a half-line using the Walsh–Fourier transform. *Integral Transforms and Special Functions, 22,* 907–917.

Sharma, V., & Manchanda, P. (2013). Wavelet packets associated with nonuniform multiresolution analysis on positive half-line. *Asian-European Journal of Mathematics, 6*(1), 16 p., 1350007.

Zhu, F., Li, Q., & Huang, Y. (2013). Minimum-energy bivariate wavelet frame with arbitrary dilation matrix. *Journal of Applied Mathematics,* 10 p., 896050.

Permissions

The contributors of this book come from diverse backgrounds, making this book a truly international effort. This book will bring forth new frontiers with its revolutionizing research information and detailed analysis of the nascent developments around the world.

We would like to thank all the contributing authors for lending their expertise to make the book truly unique. They have played a crucial role in the development of this book. Without their invaluable contributions this book wouldn't have been possible. They have made vital efforts to compile up to date information on the varied aspects of this subject to make this book a valuable addition to the collection of many professionals and students.

This book was conceptualized with the vision of imparting up-to-date information and advanced data in this field. To ensure the same, a matchless editorial board was set up. Every individual on the board went through rigorous rounds of assessment to prove their worth. After which they invested a large part of their time researching and compiling the most relevant data for our readers.

The editorial board has been involved in producing this book since its inception. They have spent rigorous hours researching and exploring the diverse topics which have resulted in the successful publishing of this book. They have passed on their knowledge of decades through this book. To expedite this challenging task, the publisher supported the team at every step. A small team of assistant editors was also appointed to further simplify the editing procedure and attain best results for the readers.

Apart from the editorial board, the designing team has also invested a significant amount of their time in understanding the subject and creating the most relevant covers. They scrutinized every image to scout for the most suitable representation of the subject and create an appropriate cover for the book.

The publishing team has been an ardent support to the editorial, designing and production team. Their endless efforts to recruit the best for this project, has resulted in the accomplishment of this book. They are a veteran in the field of academics and their pool of knowledge is as vast as their experience in printing. Their expertise and guidance has proved useful at every step. Their uncompromising quality standards have made this book an exceptional effort. Their encouragement from time to time has been an inspiration for everyone.

The publisher and the editorial board hope that this book will prove to be a valuable piece of knowledge for researchers, students, practitioners and scholars across the globe.

List of Contributors

Prashantkumar Patel
Department of Applied Mathematics & Humanities, S. V. National Institute of Technology, Surat, 395 007, Gujarat, India
Department of Mathematics, St. Xavier's College, Ahmedabad, 380 009, Gujarat, India

Vishnu Narayan Mishra
Department of Applied Mathematics & Humanities, S. V. National Institute of Technology, Surat, 395 007, Gujarat, India
L. 1627 Awadh Puri Colony Beniganj, Phase-III, Opposite - Industrial Training Institute (ITI), Ayodhya Main Road, Faizabad, Uttar Pradesh, 224 001, India

Mediha Örkcü
Faculty of Sciences, Department of Mathematics, Gazi University, 06500, Teknikokullar, Ankara, Turkey

V.K. Bhat and Meeru Abrol
School of Mathematics, SMVD University, P/o SMVD University, Katra 182320, Jammu and Kashmir, India

Johan Kok
Tshwane Metropolitan Police Department, City of Tshwane, Republic of South Africa

N.K. Sudev
Department of Mathematics, Vidya Academy of Science & Technology, Thalakkottukara, Thrissur, 680501, India

K.P. Chithra
Naduvath Mana, Nandikkara, Thrissur, 680301, India

Vakeel A. Khan, Mohd Shafiq and Rami Kamel Ahmad Rababah
Department of Mathematics, Aligarh Muslim University, Aligarh 202002, India

M.H. Rashid
Faculty of Science, Department of Mathematics, University of Rajshahi, Rajshahi-6205, Bangladesh

Ouyang Difei
College of Mathematics, Changsha University of Science and Technology, Changsha 410077, Hunan, P.R. China

Hacer Bozkurt
Department of Mathematics, Batman University, 72100 Batman, Turkey

Yılmaz Yılmaz
Department of Mathematics, İnönü University, 44280 Malatya, Turkey

Dan Kucerovsky
Department of Mathematics, University of New Brunswick, Fredericton, Canada, NB, E3B 5A3

Kaveh Mousavand
Département de mathématiques, L'Université du Québec à Montréal, Montréal, Canada, H3C 3P8

Aydin Sarraf
Department of Computer Science, University of New Brunswick, Fredericton, Canada, NB, E3B 5A3

Lily Li Liu and Xiaoli Li
School of Mathematical Science, Qufu Normal University, Qufu 273165, P.R. China

Ismail Ibedou
Faculty of Science, Department of Mathematics, Benha University, 13518 Benha, Egypt
Faculty of Science, Department of Mathematics, Jazan University, KSA

R. Uthayakumar and A. Gowrisankar
Department of Mathematics, The Gandhigram Rural Institute – Deemed University, Gandhigram – 624302, Dindigul, Tamil Nadu, India

Muhammad Gulistan, Shah Nawaz and Syed Zaheer Abbas
Department of Mathematics, Hazara University, Mansehra, Pakistan

Arvind Kumar Sinha and Anand Prakash
Department of Mathematics, National Institute of Technology Raipur, Raipur 492010, India

Ambuj K. Mishra
Department of Mathematics, Institute of Applied Sciences and Humanities, GLA University, Mathura, Uttar Pradesh, India

Jugal K. Prajapat and Sudhananda Maharana
Department of Mathematics, Central University of Rajasthan, NH-8, Bandarsindri, Kishangarh, 305817 Ajmer, Rajasthan, India

Mika Koskenoja
Department of Mathematics and Statistics, University of Helsinki, P.O. Box 68 (Gustaf Hällströmin katu 2b), FI-00014 Helsinki, Finland

Michael Milgram
Geometrics Unlimited Ltd., Box 1484, Deep River, Ontario, Canada, K0J 1P0

Wei-Dong Jiang
Department of Information Engineering, Weihai Vocational College, Weihai City, Shandong Province, 264210, China

Feng Qi
College of Mathematics, Inner Mongolia University for Nationalities, Tongliao City, Inner Mongolia Autonomous Region, 028043, China
Department of Mathematics, College of Science, Tianjin Polytechnic University, Tianjin City, 300387, China
Institute of Mathematics, Henan Polytechnic University, Jiaozuo City, Henan Province, 454010, China

Sunita Goyal
Department of Mathematics, JJT University, Jhunjhunu 333001, Rajasthan, India

Firdous A. Shah
Department of Mathematics, University of Kashmir, South Campus, Anantnag 192101, Jammu and Kashmir, India

Index

A

Advanced Mathematics, 1, 11, 18, 29, 41, 60, 69, 79, 91, 103, 116, 128, 138, 145, 153, 168, 199, 205

Algebra, 11, 17, 32, 38, 79, 87-90, 102, 128-130, 136-139, 143-144, 169

Analytic and Univalent Functions, 145

Anomalous Zeros, 168-169, 188-190, 196

Armendariz Rings, 11, 13-15, 17

Asymptotics, 169

Attractor, 116-119, 121-122, 124, 126

B

B+-chromatic Sum, 18, 26

B-chromatic Number, 18-19, 28

B-chromatic Sum, 18-20, 26, 28

Bmo Space, 60

C

Catalan Numbers, 91-93, 102

Catalan-like Numbers, 91-95, 102

Chromatic Number, 18-19, 26-28

Circulant Matrices, 79-80, 88

Classical Solution, 153

Combinatorics, 18, 91-92, 102, 199

Compact Operator, 29-30, 33

Continuation, 103, 153, 166

Contraharmonic Mean, 199, 204

Critical Line, 168-170, 173, 176, 180-181, 187, 189-190, 193-194, 196-197

D

Direct Product, 125, 128-129, 133, 136

Divided Difference, 41-42, 44

Dynamical Systems, 116

E

Endomorphism, 11, 15-17

Extended Newton-type Method, 41-43, 45-46, 49, 53, 58

Extension Principles, 205, 207

F

Filter, 29, 32, 103-105, 107, 113, 115, 206

Frame, 1, 205-207, 211, 213-214, 216, 219

Fuzzy Contraction Mapping, 116-117

Fuzzy Filters, 103-107, 113

Fuzzy Iterated Function System, 117-118

Fuzzy Metric Space, 116-119, 121, 125, 127

Fuzzy Neighbourhood Filters, 103, 105, 107

Fuzzy Self-similar Group, 117, 121, 125-126

Fuzzy Topological Spaces, 103, 107-108, 115

H

Hankel Determinant, 145, 148, 152

Hilbert Quasilinear Space, 69

Homogeneous Equation, 153

I

I-bounded Sequence, 29

I-convergent Sequence, 29, 40

I-null Sequence, 29

Ideal, 29, 31-33, 130, 133, 136

Injective Module, 138-140, 143

Inner Product Quasilinear Space, 69-70, 74-75

Intuitionistic Fuzzy Subalgebra, 133

L

Lemma, 3, 13, 15, 26, 33, 38, 44-46, 49, 51, 55, 57, 62-63, 71-72, 76, 106-107, 142, 148, 151, 156, 166, 202

Lemmas, 33, 63

Lipschitz-like Mappings, 41, 59

M

Minimum-energy Frame, 205

Minimum-energy Wavelet, 205, 207, 211, 213-214, 216, 219

Modulus Function, 29-31, 37-38, 40

Motzkin Numbers, 91-93, 102

Mra based Wavelet, 206-207

Multiplier Operator, 60-63, 68

N

Neighbourhood Systems, 103, 106

Neuman-sándor Mean, 199, 204

Noetherian Ring, 11, 14

O

Operator Theory, 79, 89

Orlicz Space, 60-61

Orthogonality, 68-69, 74

Orthonormality, 69, 205

P
P-multiresoltion Analysis, 205
Polynomial Weight, 1
Power, 28, 31, 80-81, 152, 199, 204

Q
Quasilinear Space, 69-72, 74-75

R
Rate of Convergence, 1
Real Functions, 170, 199
Real Monge-ampère Equation, 153-154
Reflection Principle, 153-154, 156, 166-167
Riemann Hypothesis, 168-169, 181, 197
Riemann's Functional Equation, 168-169
Riemann's Zeta Function, 168-169, 196-197
Rough Module, 138, 140
Rough Set Theory, 138, 144

S
Scaling Function, 205
Schröder Numbers, 91-93, 102
Self-similar Group, 116-117, 121-127
Semilocal Convergence, 41-43, 59
Separation Axioms, 103-104
Sharp Bound, 199
Solid and Monotone Space, 29

Starlike and Convex Functions With Respect to Symmetric Points, 145, 147
Subtraction Algebras, 128-129, 136
Symmetric Space, 29
Szász-mirakyan Operators, 1

T
Taylor Expansion, 91-93, 95
Thue Chromatic Number, 18, 26, 28
Toeplitz Determinant, 145, 147
Toeplitz Matrices, 79-81, 85-87
Toeplitz Type Operator, 60, 62-63

V
Variational Inclusions, 41-42, 59
Volchkov Criterion, 168-169, 181
Volchkov Equivalence, 181, 196

W
Walsh Polynomial, 205, 208
Walsh-fourier Transform, 205, 207-208
Wavelet Frame, 205-207, 211, 213-214, 216, 219
Weighted Approximation, 1

Z
Zeros, 168-169, 176, 180-181, 185-190, 193-197

Lightning Source UK Ltd.
Milton Keynes UK
UKHW05n1526230518
323014UK00003B/61/P